Programming for Unified Communications with Microsoft® Office Communications Server 2007 R2

Rui Maximo, Kurt De Ding, Vishwa Ranjan, Chris Mayo, Oscar Newkerk, Albert Kooiman, Mark Parker, and the Microsoft Office Communications Server Team

PUBLISHED BY
Microsoft Press
A Division of Microsoft Corporation
One Microsoft Way
Redmond, Washington 98052-6399

Copyright © 2009 by Microsoft Corporation (All)

All rights reserved. No part of the contents of this book may be reproduced or transmitted in any form or by any means without the written permission of the publisher.

Library of Congress Control Number: 2009927403

Printed and bound in the United States of America.

1 2 3 4 5 6 7 8 9 QWT 4 3 2 1 0 9

Distributed in Canada by H.B. Fenn and Company Ltd.

A CIP catalogue record for this book is available from the British Library.

Microsoft Press books are available through booksellers and distributors worldwide. For further information about international editions, contact your local Microsoft Corporation office or contact Microsoft Press International directly at fax (425) 936-7329. Visit our Web site at www.microsoft.com/mspress. Send comments to mspinput@microsoft.com.

Microsoft, Microsoft Press, Access, Active Directory, ActiveX, Expression, Forefront, IntelliSense, Microsoft Dynamics, MS, MSDN, Outlook, RoundTable, SharePoint, SQL Server, Visual Basic, Visual C#, Visual C++, Visual Studio, Win32, Windows, Windows Mobile, Windows NT, Windows PowerShell, Windows Server, Windows Vista and WinFX are either registered trademarks or trademarks of the Microsoft group of companies. Other product and company names mentioned herein may be the trademarks of their respective owners.

The example companies, organizations, products, domain names, e-mail addresses, logos, people, places, and events depicted herein are fictitious. No association with any real company, organization, product, domain name, e-mail address, logo, person, place, or event is intended or should be inferred.

This book expresses the author's views and opinions. The information contained in this book is provided without any express, statutory, or implied warranties. Neither the authors, Microsoft Corporation, nor its resellers, or distributors will be held liable for any damages caused or alleged to be caused either directly or indirectly by this book.

Acquisitions Editor: Ben Ryan
Developmental Editor: Devon Musgrave
Project Editor: Victoria Thulman
Editorial Production: Custom Editorial Productions, Inc.
Technical Reviewer: Mitch Duncan; Technical Review services provided by Content Master, a member of CM Group, Ltd.
Cover: Tom Draper Design

Body Part No. X15-53123

Contents at a Glance

Part I Understanding Unified Communications
1. Microsoft Unified Communications 3
2. Microsoft Unified Communications APIs Foundation 13

Part II Office Communicator Automation API
3. Programming a Microsoft Office Communicator Automation API Application 49
4. Embedding Contextual Collaboration 87

Part III Unified Communications Managed API Workflow
5. Unified Communications Managed API (UCMA) Workflow . . . 113
6. Business Process Communication 149

Part IV Unified Communications Managed API
7. Structure of a UCMA Application 183
8. Publishing Custom Presence with UCMA 205

Part V Debugging, Tuning, and Deploying Unified Communications Applications
9. Preparing the UC Development Environment 219
10. Debugging a Unified Communications Application 293

Table of Contents

Foreword	xi
Acknowledgments	xiii
Introduction	xvii
Why We Wrote This Book	xvii
What This Book Is About	xvii
Who This Book Is For	xviii
Companion Content	xviii
Hardware and Software Requirements	xix
Servers	xix
Client Computers	xix
Database Requirements	xx
Office Communications Server 2007 R2	xx
Administrative Tools	xx
Development Tools	xx
Sample Test Topology	xxi
Find Additional Content Online	xxi
Support for This Book	xxi
Questions and Comments	xxi

Part I Understanding Unified Communications

1 Microsoft Unified Communications 3

Unified Communications: Challenges and Opportunities	3
Challenges in Unified Communications	4
Opportunities in Unified Communications	5
The Unified Communications Platform	7
Unified Communications APIs	8
Summary	12
Additional Resources	12

What do you think of this book? We want to hear from you!

Microsoft is interested in hearing your feedback so we can continually improve our books and learning resources for you. To participate in a brief online survey, please visit:

www.microsoft.com/learning/booksurvey

2 Microsoft Unified Communications APIs Foundation 13

Unified Communications Managed API 2.0 13
- Scenarios 14
- Considerations 14
- API Architecture 16
- Object Model 17

UCMA 2.0 Workflow API 18
- Scenarios 19
- Windows Workflow Foundation 20
- Considerations 21
- Workflow Architecture 22
- Object Model 22

Office Communicator Automation API 23
- Scenarios 23
- Considerations 24
- Application Architecture 24
- Object Model 25

Unified Communications Client API 27
- Scenarios 27
- Considerations 28
- Application Architecture 28
- The UCC API Object Model 29

Unified Communications AJAX API 32
- Scenarios 32
- Considerations 32
- Application Architecture 33
- XML Model 34

Office Communications Server 2007 Speech Server Developer Edition 37
- Scenarios 38
- Considerations 40
- Application Architecture 40

Summary 46
Additional Resources 46

Part II Office Communicator Automation API

3 Programming a Microsoft Office Communicator Automation API Application 49

Signing In to and Out of Office Communicator 49
    Using the *Messenger* Class 49
    Determining Whether Office Communicator Is Running 50
    Checking Local User Status 51
    Signing In to Office Communicator 52
    Signing Out of Office Communicator............................... 54
    Putting It All Together.. 55
Working with Contact Information and Contact Presence................. 58
    Displaying Local User Information 58
    Retrieving Contact Information................................... 63
    Publishing and Subscribing to Contact Presence................... 68
    Putting It All Together.. 72
Working with the Office Communicator Contact List 77
    Putting It All Together.. 79
Starting Conversations 81
    Using the *IMessengerAdvanced* Interface 81
    Putting It All Together.. 83
Summary 85
Additional Resources... 85

4 Embedding Contextual Collaboration 87

Introduction to Contextual Collaboration 87
Scenario 90
Business Value... 90
Choice of Technology 91
Test Environment 91
Overall Code Structure 92
    Displaying Application-Specific Contact Lists...................... 92
    Starting Application-Specific Conversations........................ 97
    Accepting Application-Specific Conversations..................... 104
Summary 110
Additional Resources.. 110

Part III Unified Communications Managed API Workflow

5 Unified Communications Managed API (UCMA) Workflow 113
UCMA Workflow. 113
    Using Project Templates 114
    Selecting a Workflow Language 115
    Using Workflow Designer. 115
    Workflow Runtime Services 116
    General Activities. 120
    Call Control Activities 126
    Dialog Activities. 128
    Command Activities 137
    Call Control Communications Event Activities 141
    Dialog Communications Event Activities. 144
    Presence-Related Activity. 147
Summary 148
Additional Resources 148

6 Business Process Communication. 149
Scenario 149
Business Value. 149
Choice of Technology 150
Overall Code Structure 150
Test Environment 150
Building the Application. 151
    Task 1: Create a New Communication Workflow Project 151
    Task 2: Configure the Application to Connect to Office
    Communications Server 151
    Task 3: Allow User Input to the Workflow Instance 152
    Task 4: Get the Approver's Presence Information. 154
    Task 5: Implement Branching Logic Based on
    the Approver's Presence 155
    Task 6: Update *cantBeContactedBranch* 157
    Task 7: Update *canBeContactedBranch* 158
Summary 179
Additional Resources 179

Part IV Unified Communications Managed API

7 Structure of a UCMA Application............................183

Creating a UCMA Application 183
    CollaborationPlatform..184
    Endpoints 187
    Conversation, Call, and Call Flow 190
    Creating Calls 191
    Conferences 195
    Publish and Subscribe to Presence 197
Summary 203
Additional Resources...204

8 Publishing Custom Presence with UCMA 205

Creating Custom Presence Categories.............................205
Common Custom Presence Application Scenario 206
Choice of Technology 206
Overall Code Structure 207
Test Environment 207
Detailed Code..208
Summary 216
Additional Resources...216

Part V Debugging, Tuning, and Deploying Unified Communications Applications

9 Preparing the UC Development Environment 219

UC Application Development Environment Components 219
    AD DS for Managing a Network 221
    Office Communications Server Roles............................223
    UC APIs 224
Deploying Office Communications Server Standard Edition 226
    Building an AD DS Forest.......................................226
    Preparing AD DS for UC 237
    Configuring DNS for Automatic Sign-In 245
    Setting Up the Office Communications Server Host Computer 248
    Installing and Configuring Office Communications Server Standard Edition..251
    Configuring UC User Accounts 272
    Validating Server Functionality 278

Configuring Application Development Components 279
        Configuring the Office Communicator Automation API. 280
        Configuring UCMA Core. 281
        Configuring UCMA Workflow 291
    Summary 291
    Additional Resources 292

10 Debugging a Unified Communications Application 293
    Debugging in the UC Platform 293
        Sources of Errors and Failures 293
        Error Codes and Exception Classes. 294
        Session Initiation Protocol Error Codes 295
        Tracing 296
        Debugging Tools for UC Applications 297
        Best Practice for Debugging or Troubleshooting a UC Application 301
    Debugging Office Communicator Automation API Applications 302
        Enabling Tracing. 302
        Handling Exceptions Using *HRESULT* Error Codes 305
        Troubleshooting Office Communicator Automation
        API Applications 307
    Debugging UCMA Core Applications. 314
        Enabling Tracing. 314
        Handling Exceptions Using the UCMA Core Exception Model 316
        Debugging UCMA Core Applications. 321
    Debugging UCMA Workflow Applications 330
        Enabling Tracing. 330
        Handling Exceptions Using the Fault Handler Activity 332
        Debugging UCMA Workflow Applications 333
    Summary 338
    Additional Resources 339

Glossary 341

Index 371

What do you think of this book? We want to hear from you!

Microsoft is interested in hearing your feedback so we can continually improve our books and learning resources for you. To participate in a brief online survey, please visit:

www.microsoft.com/learning/booksurvey

Foreword

The Microsoft Unified Communications (UC) platform, which includes Microsoft Exchange Server and Office Communications Server, provides a unified infrastructure for communication and offers a single-user experience for e-mail, instant messaging (IM), Voice over Internet Protocol (VoIP), application sharing, audio/video (A/V), and Web conferencing. This user experience is centered on a single identity (provided by Active Directory Domain Services) combined with an inbox (managed by Exchange Server) and enhanced presence (provided by Office Communications Server). This integrated solution realizes the unified view in UC as opposed to a collection of disparate solutions. As a software-based platform, UC provides a rich and open API set for developers to extend the UC platform. In fact, our own products are built using these same APIs. These APIs make presence, e-mail, IM, and VoIP a programmable platform that developers can use to create innovative solutions easily.

With rich and open APIs, we're infusing UC into new business applications, workflow technologies, and content management. This effort wouldn't be possible without the rapidly growing number of Microsoft Most Valued Professionals (MVPs), developer partners who are building these enterprise applications for customers. It is, therefore, important to help the developer community by providing the necessary information, training, and support to make them successful. Already, many customers have enhanced Office Communications Server with custom applications. This book launches an important series that UC developers will want to add to their library of programming books. Written by experts from the Unified Communications Group at Microsoft, you'll find informative and valuable technical coverage, complete with examples to illustrate best practices.

I want to recognize the efforts of members of the Unified Communications Group who demonstrated perseverance and dedication in producing this book. Publishing this title required vision and commitment from Tim Toyoshima, Ben Ryan, Rui Maximo, Susan Bradley, Kurt De Ding, Vishwa Ranjan, Oscar Newkerk, Chris Mayo, Albert Kooiman, Victoria Thulman, Mitch Duncan, 15 technical reviewers, and a large support crew.

It is my pleasure to introduce the first book on programming for Office Communications Server.

Gurdeep Singh Pall
Corporate Vice President
Unified Communications Group

Acknowledgments

The authors of *Programming for Unified Communications with Microsoft Office Communications Server 2007 R2* want to thank the numerous members of the Office Communications Server product team, Microsoft Learning team, and numerous other key contributors who helped make this first edition of the book as thorough, comprehensible, and accurate as possible. These individuals have contributed their time and effort to this project in several important ways, which include:

- Reviewing each chapter for technical accuracy and completeness
- Providing project management, technical management, lab management, art management, and editorial support
- Providing vision, leadership, advice, encouragement, resource support, and funding support

We acknowledge and thank the following product experts for their extensive and invaluable technical reviews:

- Ajay Soni
- Chloe Brussard
- Dongkyun Nam
- Ellen Zehr
- George Durzi
- John Austin
- K. Ganesan
- Marshall Harrison
- Niko Schuessler
- Sara Morgan
- Srivatsa Srinivasan
- Srividya Mohan
- Vincent Bannister
- Weimin Li

If we have forgotten anyone, please forgive us!

Acknowledgments

We acknowledge and thank the following key contributors, without whom this book would be a dream rather than a reality.

Rui Maximo, senior technical writer, for his role as technical lead for the first-ever programming book about Office Communications Server. Rui worked with dedication, perseverance, and passion, reviewing each chapter draft multiple times and working closely with authors to ensure technical accuracy, logical presentation, consistency, cohesion, and coherence across 10 chapters, 7 writers, and 20 reviewers. Rui, we could not have delivered this book without you.

Chris Mayo, UC technical evangelist, for his role in creating and managing the lab environment used by authors to take their screenshots.

Diane Forsyth, senior technical editor, for her role as glossary and forward link editor. Thanks to Diane's earlier efforts, we have a glossary with an extensive list of key terms and definitions as well as links that point to the latest resources.

Janet Lowen, lead technical editor, for her thorough editorial reviews and oversight, as well as her able assistance with the glossary, forward links, template, and style guidelines.

Kate Gleeson, technical editor, for her editorial assistance during the end game when chapters were flying so fast that we could barely keep up with them.

Victoria Thulman, Microsoft Learning project editor, for managing our delayed deliveries, endless rounds of questions, and hundreds of e-mails a day with good humor and focused professionalism. Vicky, this would have been a far more difficult journey without your expertise and support.

John Clarkson, senior programming writer, for his role in managing the art across all chapters and ensuring that all art met quality standards and guidelines.

Mitch Duncan, Microsoft Learning technical reviewer, who tested all procedures and identified errors so that this book is an accurate and comprehensive resource for our customers.

Susan S. Bradley, senior content project manager and "Master of the Universe," who managed this project from inception to delivery. Her infectious laughter and positive energy kept everyone motivated to work through the details of producing the book and tracking to the finish line. The number of moving parts to track across a large cast is mind-numbing. Without her leadership and perseverance, this book wouldn't exist.

Ben Ryan, Microsoft Learning lead product manager, who championed and supported this first-edition book and new authoring team. Without Ben's vision and support, *Programming for Unified Communications with Microsoft Office Communications Server 2007 R2* would not exist.

Juan-Carlos Rivas, Office Communications Group senior group manager, for believing in the value of this book and for allocating writers from his team to assist with writing, chapter reviews, and art review and management.

Tim Toyoshima, Office Communications Group user assistance principal group manager, for envisioning the need for this book, evangelizing that need, and sponsoring this cross-team project. Without Tim's vision, leadership, and continuing support, which included both funding and resources for this book, *Programming for Unified Communications with Microsoft Office Communications Server 2007 R2* would not exist.

We also thank our outstanding and dedicated editorial team at Microsoft Learning, including **Devon Musgrave**, our development editor. Thanks also to **Custom Editorial Productions**, who handled the production aspects of this book, and to **Susan McClung**, our copy editor.

—The Author Team

Personal Acknowledgments

Rui Maximo I want to thank my coauthors, who endured multiple rigorous reviews and incorporated my feedback into their chapters. To Susan Bradley, thank you for managing this project as well as previous projects. It's a pleasure to be your sidekick as we work together on ever more ambitious projects. I've really enjoyed working with you since we started working together on the first edition of the *Microsoft Office Communications Server 2007 Resource Kit*. To Mitch Duncan, you've become a good friend and respected colleague. To Vicky and Janet, whom we've been fortunate to work with over the past two books, thank you! To Tim Toyoshima, thanks for your continual support and for trusting in me to deliver. To my friends in the product group, thank you for reviewing and sharing your expertise. These books are dedicated to the great work you do on Office Communications Server. To my wife and kids, you are my inspiration.

Kurt De Ding First, I want to express my gratitude to all current and former members of the Unified Communications Group who helped me understand and appreciate the technologies of the Office Communications Server products family. In particular, I want to thank Sam Bedekar, Subbu Chandrasekaran, K. Ganesan, Adrian Potra, and Stephane Taine for their insights, their helpfulness, and their patient explanations. My gratitude also goes to Sara Morgan for her insightful comments and expert suggestions on the earlier drafts of Chapters 9 and 10. I want to thank Juan-Carlos Rivas for giving me the opportunity to contribute to this book and creating a supportive work environment. I also want to thank Tim Toyoshima for letting me be part of this wonderful team working on interesting UC technologies. Finally, for my colleagues on the SDK documentation team, thank you for all your support.

Vishwa Ranjan I have to start by thanking my wife, Rajpreet, who was very patient with me while I worked on this book during the last month of her pregnancy. I would also like to thank my brother, Peeyush, without whose efforts I would not be in the computer industry. I would like to thank all my peers on the product team, especially Dongkyun Nam, for making sure that the right technical content went into the book. Finally, I would like to extend my gratitude to all the team members working on this project who helped me through the rigors of writing a book. A special thanks to Rui Maximo for turning the pile of information that I put on paper into something that is comprehensible and to Susan S. Bradley for driving us in the right direction.

Chris Mayo In 1981, Naperville School District 203 purchased computers for my elementary school. My third-grade science teacher, Mike Manolakes, taught me how to write software after school. That investment in technology in the classroom and the extra effort by a generous teacher changed my life forever. My parents, Calvin and Judy, were way ahead of their time when they bought me my first computer in 1983, a time when a computer in the home was truly a luxury. That luxury turned an interest into a career I truly love. Marg and Bill Brannen listen with genuine interest when I blather on for hours about computers, software, and new technologies and deal with constant "upgrades" that often break something that worked previously. I appreciate their patience and hope I fix more things than I break. With regard to this book, Susan Bradley, Rui Maximo, Kurt De Ding, Diane Forsyth, Janet Lowen, Kate Gleeson, John Clarkson, and George Durzi did the impossible by taking something I wrote and turning it into something worth reading. That was no small feat. Kerry Brannen supported me through the long hours I spent locked in my office working on this book. I couldn't have done it without her. She makes anything possible and everything more fun.

Oscar Newkerk Thanks to all the people in the Unified Communications Group for the opportunity to work with such talented people on such a fun product. I want to especially thank Kyle Marsh for his help, knowledge, and all the brainstorming sessions.

Introduction

Why We Wrote This Book

Microsoft Office Communications Server is a relatively new product. Although it has its origins in the Enterprise Instant Messaging products, such as Exchange Instant Messaging, Live Communications Server 2003, and Live Communications Server 2005, it has evolved to become a comprehensive platform for all real-time communications. The current release, Office Communications Server 2007 R2, not only supports enterprise instant messaging (IM) and rich presence, but also offers a powerful Voice over Internet Protocol (VOIP)–based telephony system, multiparty audio conferencing, Web conferencing, and application sharing. This server offers tangible benefits in direct cost savings (for example, by eliminating costly audio conferencing services provided by telecom carriers) and improves productivity by providing more efficient ways for people to contact each other. Office Communications Server is one of the fastest-growing server products in Microsoft history, with tens of millions of licenses sold.

Office Communications Server is a software-based solution that runs on standard computing and networking hardware. This server offers a rich, open API platform, making it an open and extensible part of the Microsoft Unified Communications (UC) platform. There are many opportunities for developers to build new applications on this platform. We know that developers are looking for resources to help them develop applications using the UC APIs, and this book is the only one on the market today that addresses this need. Written by experts from the product group, *Programming for Unified Communications with Microsoft Office Communications Server 2007 R2* offers an easy-to-read exploration of the APIs. We hope it serves you well.

What This Book Is About

This book is organized into five parts.

Part I, "Understanding Unified Communications," introduces the UC platform and provides an overview of the APIs.

Part II, "Office Communicator Automation API," explains the Office Communicator Automation API in depth and provides a detailed walkthrough of an example.

Part III, "Unified Communications Managed API Workflow," explains the UCMA Workflow API in detail and walks through an example of a business process communication.

Part IV, "Unified Communications Managed API," covers the Unified Communications Managed API architecture and shows how to extend the Office Communications Server Enhanced Presence model by using this API.

Part V, "Debugging, Tuning, and Deploying Unified Communications Applications," explains how to debug, tune, and deploy UC applications.

Who This Book Is For

This book is intended for developers who want to create enterprise applications that include communications functionality built on the UC platform. Familiarity and experience with Microsoft Windows COM, Microsoft .NET Framework, and Windows Workflow Foundation development is recommended. This book is written on the assumption that the reader has this knowledge. Code examples in this book are written in C# unless otherwise noted. For clarity and to better illustrate how to use the APIs, the code samples are not written with defensive coding practices in mind. Please apply defensive code practices when reusing the samples in your own production applications.

For an in-depth resource on the internals of Office Communications Server 2007 R2, see the *Microsoft Office Communications Server 2007 R2 Resource Kit* (Microsoft Press, 2009), which you can purchase in a bookstore or order from *http://www.microsoft.com/learning/en/us/Books/13113.aspx*. That book also covers the Office Communications Server Software Development Kit (SDK), which is intended for administering and controlling compliance of the conversations (for example, ethical walls and custom disclaimers) with Office Communications Server, and therefore is outside the scope of this book.

We sincerely hope that you find the technical information within this book useful and lucrative to your work.

Companion Content

This book features a companion Web site that makes available to you all of the code used in the book. This code is organized by chapter, and you can download it from the companion site at *http://code.msdn.microsoft.com/programmingocs*.

Hardware and Software Requirements

You need the following hardware and software to work with the companion content that is included with this book:

Servers

Hardware Use only a 64-bit computer that is running a 64-bit edition of Windows Server (see more about the operating system below). Other technical specifications include the following:

- **CPU** Dual-core 2.0-gigahertz (GHz) 4-way processor
- **RAM** 2 gigabytes (GB) of memory
- **Hard drive** 100-GB hard drive
- **Network adapter** 100 megabit-per-second (Mbps) network adapter

Operating System Use only the 64-bit edition of Windows Server 2003 SP2, Windows Server 2003 R2 SP2, or Windows Server 2008. Supported editions include Standard, Enterprise, and Data Center versions of Windows Server 2003 and Windows Server 2008.

Client Computers

Hardware Use any 32-bit or 64-bit computer that is running Windows Vista SP1. Other technical specifications include the following:

- **CPU** A minimum 1.6-GHz Pentium III+ processor
- **RAM** 1 GB of memory
- **Hard drive** 50-GB hard drive
- **Network adapter** 100-Mbps network adapter
- **Video** A video monitor with 800 × 600 or higher resolution and at least 256 colors
- A CD-ROM or DVD-ROM drive
- A Microsoft mouse or compatible pointing device

Operating System The 32-bit or 64-bit edition of Windows Vista SP1 or later. Windows Vista Home Premium Edition, Windows Vista Business Edition, or Windows Vista Ultimate Edition.

Database Requirements

Use the 32-bit version of Microsoft SQL Server 2005 Express Edition SP2, which is included with Office Communications Server 2007 R2.

Office Communications Server 2007 R2

Deploy Office Communications Server 2007 R2 Standard Edition on a private network.

> **More Info** For more information about deployment for the UC test environment, see the "Deploying Office Communications Server Standard Edition" section in Chapter 9, "Preparing the UC Development Environment."

Administrative Tools

Install the Office Communications Server Administrative Tools. The administrative tools can be installed independent of the Office Communications Server deployment on a computer that is running the 32-bit or 64-bit edition of Windows Server 2003 SP2, Windows Server 2003 R2 SP2, Windows Server 2008, Windows Vista Business, or Windows Vista Enterprise with SP1.

Development Tools

The software development environment and tools required to build UC applications include the following:

- Microsoft Visual Studio 2008 SP1
- Microsoft .NET Framework 3.5 SP1
- Microsoft Visual C++ 2008 Redistributable Package
- Office Communicator Automation API SDK
- Unified Communications Managed API (UCMA) Core 2.0 SDK

Visual Studio 2008 SP1–supported software includes Visual Studio 2008 Standard Edition, Visual Studio 2008 Enterprise Edition, Visual C# 2008 Express Edition, and Microsoft Visual Web Developer 2008 Express Edition.

> **More Info** For more information about configuring the UC software development environment, see the "Configuring Application Development Components" section in Chapter 9.

Sample Test Topology

To build and test the samples included in this book, a typical test topology includes the following clients and servers:

- A Windows Server 2008 domain controller, including the Domain Name System (DNS) and Certificate Authority (CA) roles
- Office Communications Server 2007 R2 Standard Edition deployment
- A Windows Server 2008 member server serving as the application server to run the sample applications
- Two Windows Vista clients running Office Communicator

Find Additional Content Online

As new or updated material becomes available that complements your book, it will be posted online on the Microsoft Press Online Developer Tools Web site. The type of material you might find includes updates to book content, articles, links to companion content, errata, sample chapters, and more. This material is available at *www.microsoft.com/learning/books/online/developer*, and the site is updated periodically.

Support for This Book

Every effort has been made to ensure the accuracy of this book and the contents of the Web site. As corrections or changes are collected, they will be added to a Microsoft Knowledge Base article.

Microsoft Press provides support for books and companion content at the following Web site:

http://www.microsoft.com/learning/support/books/.

Questions and Comments

If you have comments, questions, or ideas regarding the book or the companion content, or questions that are not answered by visiting the sites mentioned previously, please send them to Microsoft Press via e-mail to

mspinput@microsoft.com

or via postal mail to

Microsoft Press
Attn: *Programming for Unified Communications with Microsoft® Communications Server® 2007 R2* Editor
One Microsoft Way
Redmond, WA 98052-6399.

Please note that Microsoft software product support is not offered through these addresses.

Part I
Understanding Unified Communications

In this part, Chapter 1, "Microsoft Unified Communications," introduces the Microsoft Unified Communications (UC) platform. This chapter highlights the challenges of UC and the opportunities that these challenges offer developers. UC opens up a new category of applications that were previously reserved for specialized systems with limited public interfaces. UC makes it easier to build instant messaging (IM) and voice or video solutions into existing and new applications. Chapter 1 also provides insight into the future direction of the UC platform.

Chapter 2, "Microsoft Unified Communications APIs Foundation," provides an overview of the APIs that are available. If you are unsure which API to use to build your application, this chapter helps you select the most appropriate one by providing examples of the types of applications each API can be used to build, the constraints of each API, and a high-level view of each API's architecture and object model.

If you are already familiar with the UC platform and know which API you want to use, you can skip directly to the chapter that covers that API in Part II, "Office Communicator Automation API."

Chapter 1
Microsoft Unified Communications

This chapter will help you to:

- Understand the challenges and opportunities for application developers in unified communications.
- Understand the features and values of the Microsoft Unified Communications (UC) platform.

Unified Communications: Challenges and Opportunities

What is unified communications? The term *unified communications* refers to technologies that integrate various and often disparate communication systems and capabilities in order to provide a more seamless and enhanced communication experience. The goal of unified communications can be summarized with the following statement:

Connect to the right individuals at the right time with the right information using the right capability.

For example, UC technology enables a user to place a call to another user, to receive a call from another user, and to transfer a call to another user using Microsoft Office Communicator. This call can be an instant messaging (IM) conversation, a computer-to-computer audio/video (A/V) call, or a voice call conducted between a computer and a Public Switched Telephone Network (PSTN) or Private Branch eXchange (PBX) phone. The call can be between two users or between three or more conference participants. With Office Communicator, users can make and take calls anywhere with an Internet connection as if they were using the phone in their office.

Before each call, the user can utilize Office Communicator to check the presence of the called party and determine whether that user is available and willing to take the call. The presence information can also include contact information, such as the primary and alternate phone numbers that can be used to direct an outgoing call. Without knowing a contact's presence, callers may have to make several attempts before reaching their contact.

Before getting into the specifics of building UC applications, let's explore the challenges and opportunities that application developers can seize in the UC space. Applications that help people and businesses communicate more efficiently can provide a compelling package.

Challenges in Unified Communications

Over the last decade, the number of available communication modalities has multiplied rapidly. A business card today barely has enough space to list them all. Professionals can be reached on their PBX work phone, work and personal cell phone by voice or Short Message Service (SMS), work and personal e-mail address, corporate IM address, personal IM provider, Facebook account, LinkedIn account, blog URL, and many other options. With so many ways of contacting people, how do you know which method is the most effective to reach them or what is the best time to reach them? Is the contact in the same time zone (or even in the same country) as you? Maybe the person is in a meeting and will not respond to a phone call but will respond to an IM or SMS message. Depending on the modality, the person might be responsive depending on what he or she is doing at that moment.

In today's world of high-tech communication, UC is beginning to address three main challenges:

- Reducing communications overload
- Connecting disparate and disconnected communications systems
- Using communications to improve collaboration

The first challenge is exacerbated by the explosion of media and devices that can be used to communicate. Therefore, to reach your contact, it's important to understand what methods of communication are most effective at any given time. The converse of this challenge is also a problem. Given the various ways you can be reached, how do you make it easy for others to contact you? It's no wonder that address book management software offers so many fields that can be filled for each contact. Imagine the frustration your family, friends, coworkers, and professional contacts experience trying to reach you with all of the various methods you are providing them. Although the available modalities that can be used to contact a person are intended to make individuals more accessible, the proliferation of communication modalities can affect communication negatively.

The second challenge illustrates the state of today's communications systems. Because of the explosion of communication solutions, they still work in silos, and it is hard to connect the communication experience across applications. Communications systems are disconnected. Until recently, enterprises bought a PBX and phones from one vendor, bought an additional voice mail system from another vendor or as part of the PBX bundle, and then obtained audio-conferencing solutions from another vendor. The contact center side of the enterprise bought an Automatic Call Distributor (ACD) and Interactive Voice Response (IVR) system from yet a different set of vendors. Meanwhile, the IT side of the enterprise bought an e-mail system and a corporate IM system, as well as line-of-business applications from yet another vendor or vendors. Most of these systems offer APIs with limited access to the entire functionality of their platform or APIs that are not designed to exchange data between the silos to enable them to work together. Consequently, users are forced to learn and work across different devices or must buy expensive applications to tie together the data from various communications system for their needs. The interaction between various silo solutions becomes taxing and reduces efficiency.

The third challenge highlights the difficulty of communicating with remote parties. Whether communicating by e-mail, phone, or IM, communicating effectively requires sufficient context. Often, context is best conveyed in face-to-face interaction; however, this requires teams to be collocated or to incur costly travel expenses. As costs, regulations, and personal preferences make it more and more difficult to relocate workers to the same region, companies must find ways for employees to work together effectively and productively even if they are in different places. Remote communication and collaboration can be improved substantially by making it simpler to share context automatically and easily. Contextual information can include visually sharing information (application sharing), providing a link to an e-mail of interest for discussion, or synchronizing the same application across all parties in the call.

Certainly, there are more challenges to effective and efficient communication; however, these three are the major sources of the frustration that users experience when communicating and collaborating in remote scenarios.

Opportunities in Unified Communications

Microsoft Office Communications Server and Microsoft Exchange Server are foundational pieces of UC because they manage two important communication repositories: Enhanced Presence and the unified inbox.

Office Communications Server uses presence to address the premise that individuals want actual contact with other users rather than reaching a mechanized inbox, voice mail, or IM response. Users have no interest in dialing a number only to get a voice mail greeting. Users do not want to send an e-mail if the likelihood of a timely response is low. To avoid such scenarios, it's necessary to know the user's availability. Office Communications Server makes it possible to expose user presence information so that the contact can be reached more quickly and easily.

Alternatively, users do not want to be reachable at every moment because they want to focus on a project, meet with a client, or perform other necessary business functions without being interrupted. With Office Communications Server, users can control the level of availability they want to have at any particular time. In the same way that users can control how their time is allocated through their calendar, users can manage their presence.

Presence information is a set of metadata about a person's availability or, more precisely, their willingness to be available to others. Office Communications Server manages presence as an Extensible Markup Language (XML) schema that can be extended to include additional information, such as location or skills, which can be relevant to the context of communications.

Now that a user can view your presence (at least the level of presence you're willing to advertise), how does that improve communication? With any of the Office Communications Server clients, such as Office Communicator, communicating—whether by e-mail, IM, phone, or application sharing—is only a click away. The user no longer needs to track all of the devices on which you've published and are accessible. This addresses the first challenge discussed in the previous section. From the presence icon, users can launch additional modalities of communication (such as escalating an IM conversation to a voice call or adding application sharing) without having to start multiple independent applications. Because the process of communicating becomes simpler, Office Communications Server reduces human latency—the time it takes to establish a call with another user.

While tens of millions of information workers use Enterprise IM, the majority of enterprise applications that exist today do not offer embedded presence and the real-time communications functionality that comes with it. There's a wide opportunity to embed Office Communications Server presence and real-time communications modalities into existing and new applications. As the popularity increases and enterprises recognize the value that UC offers, demand for UC-enabled applications will continue to rise.

Microsoft builds UC as a pure software solution running on standard general-purpose computing and networking hardware. Building UC as a software solution overcomes the second challenge of integrating disparate and disconnected communications systems. By using the same programming paradigms (such as Component Object Model [COM], Microsoft .NET Framework, and Web Services) and the same tools (such as Microsoft Visual Studio) familiar to Microsoft Windows developers, programmers can create and integrate communications into new and existing applications as easily as developing Windows applications without having to deal with the deeper underpinnings of communications, such as Session Initiation Protocol (SIP) protocols, media stacks, and speech technology.

While the key elements of the UC platform are Office Communications Server and Exchange Server, this software platform integrates and includes many other Microsoft products such as Microsoft Windows, Microsoft SharePoint, Microsoft Dynamics, Microsoft Office, and Visual Studio. By extension, this software-based communications experience can be embedded in any line-of-business application that developers are creating. Because all of UC is built as a software solution, new innovative scenarios are possible.

When communicating over the phone, IM, or e-mail, explaining the context of what you're trying to communicate can be laborious at best. Users sometimes have to go into a long explanation to convey the scenario or problem before they can get into their question. Conveying context across a modality, whether it is IM, e-mail, or voice, is challenging at best. Meeting face to face is universally the preferred mode of communication, and for good reason. However, this is not always possible, and optimizing remote communications is sometimes necessary. UC tackles this third challenge by providing contextual collaboration. Contextual collaboration is about sending metadata with the actual communication. This metadata can be as simple as a link to the e-mail that the caller is interested in discussing. The caller can click the link to open the specific e-mail in her or his inbox. A caller can set the subject line so that the recipient of the call knows what the caller wants to discuss before answering the call. Metadata can be more elaborate, such as starting an application-sharing or A/V session or synchronizing the view of a design within an application used by both parties. All of these ways of providing context (and more) are possible with Office Communications Server. Because Office Communications Server uses the data or Internet Protocol (IP) network to send IM, audio, and video, it can be used to send metadata (documents, Web pages, application-specific protocols, and so on) easily with the communication. Building these contextual collaboration experiences is a major opportunity for developers to enhance existing applications. Contextual collaboration can significantly improve the collaboration experience and help streamline communications between people.

The UC platform also offers key opportunities to use communications in business processes. For example, business processes often require human intervention. Business processes can use communications to cut down the latency period when a human decision is required. For example, to notify users, you can build alerts and notifications that use the IM or voice channels supported by Office Communications Server or the e-mail channel offered by Exchange Server. Such systems provide a push model wherein users no longer need to check whether a document is ready for signing but can be notified when it's ready for the user to sign.

Communications also can be automated in reverse: a caller can contact IVR applications via the telephone or automated agents (query/response bots). This allows organizations to reduce costs by having automated systems answer customers' frequently asked questions (FAQs) instead of using human representatives. The ability to create automated agents easily is another key capability that comes as an integral part of Office Communications Server.

There are many more types of applications that developers will undoubtedly imagine and create using UC APIs. These APIs are discussed in the rest of this chapter.

The Unified Communications Platform

UC is a software-based platform. Integrated with Active Directory Domain Services, this platform provides a unified view of the user's identity, presence, and inbox. All means of communications with the user evolve around this single identity and inbox. Users are not interested in remembering a long list of phone numbers (work, cell, and so on), IM handles (one for each public IM provider), or e-mail addresses to reach a contact. With Exchange Server and Office Communications Server, enterprises can create a standardized e-mail, IM address, and phone number for every user. This is illustrated in Figure 1-1.

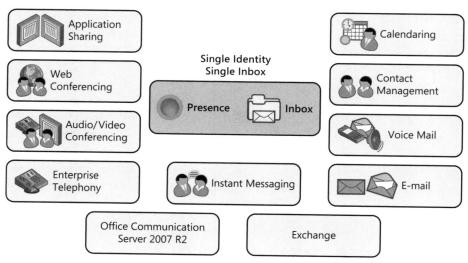

FIGURE 1-1 The UC platform.

Unified Communications APIs

Just as Windows makes it easy to develop graphical user interface (GUI) applications, Office Communications Server extends this user interface to voice and IM applications. Because of the software-based nature of the platform and its rich public API set offering, UC APIs can undoubtedly be used to create many different applications that build on Office Communications Server and Exchange Server. There are at least three types of applications that developers can build with UC APIs. These are as follows:

- **Contextual collaboration**   Integrating modalities and context into existing and new enterprise applications to facilitate communications between humans.

- **Business process communications**   Applications that contact users based on triggers defined by the business, such as alerts. For example, a server failure in the datacenter can alert the set of administrators who are currently online and available.

- **Anywhere information access**   Employees using various forms of communication to obtain information within the enterprise, such as bots and IVR systems. For example, users can use the phone or IM to request information.

The UC platform offers client-side and server-side APIs to fit the type of application that you want to build. These APIs are either .NET Framework–based or COM-based. The COM-based APIs integrate into the .NET Framework using the COM-Interoperability wrappers, which means that you can use any of these APIs in a .NET Framework application. Where appropriate, these APIs are well integrated with Visual Studio.

Programming Office Communications Client Applications

The Office Communications Server client API is UC Client API (UCC API) 1.0. It is a COM-based API, although it is accessible to .NET Framework applications through an interop assembly. The UCC API offers all of the functionality used by Office Communicator, including presence, IM, voice, video, and conferencing. This API provides signaling, media, and the capability to traverse firewalls using the Interactive Connectivity Establishment (ICE) standard to connect media channels. In the Office Communications Server 2007 R2 release, the API has been updated with only QFEs (Quick Fix Engineering) to make the UCC API 1.0 compatible with the new features of R2.

The Office Communicator Automation API is a client API that uses the business logic built into Office Communicator. It is an automation API, and its functionality is less comprehensive than the UCC API. With the Office Communicator Automation API, you can program an Office Communicator instance from a third-party application, as well as extend and customize user experiences with Office Communicator. The Office Communicator Automation API is used by the sample presence controls and a sample ActiveX control that Microsoft published on MSDN, available at *http://msdn.microsoft.com/en-us/office/cc718982.aspx*, as well as the Name Control that ships and installs with Office.

In the "Microsoft Office Communicator 2007 R2 Deployment Guide" located at *http://go.microsoft.com/fwlink/?LinkID=133744*, you can find information on how to customize the Office Communicator look and feel to some extent. Custom Tabs is one of these customizable capabilities. A Custom Tab uses a browser window in the lower third of the Office Communicator main window with the contact list. In Office Communicator 2007 R2, context-sensitive information related to a person can be linked to a tab button in the Office Communicator contact card. Note that the functionality is more streamlined in the R2 release and behaves different from OCS 2007.

The UC Asynchronous JavaScript and XML (AJAX) API is an API that talks to the Communicator Web Access server role in Office Communications Server 2007. Application developers can use it to create Office Communicator–compatible clients that allow users to manage and share presence information, to manage contacts and groups, to send and receive instant messages, and to search for users within an enterprise. Such clients can be a browser-based Web application (for example, an ASP or ASP.NET application) or a standalone network application (for example, a .NET Framework executable). The client applications can be written in a wide range of programming languages, including JavaScript, C#, Microsoft Visual Basic .NET, Perl, and C/C++.

The UC AJAX API supports only presence, IM, and call deflection. The API does not support voice, video, or conferencing.

In Office Communications Server 2007 R2, the UC AJAX API is deemphasized. It will not be available in the next release of Office Communications Server. Because the Unified

Communications Managed API (UCMA) scales better than the UC AJAX API, it is preferable to use UCMA. Therefore, it's recommended to build your own Web Service using UCMA server-side to extend the reach to non-Windows platforms.

Figure 1-2 illustrates the client APIs for building client-based applications.

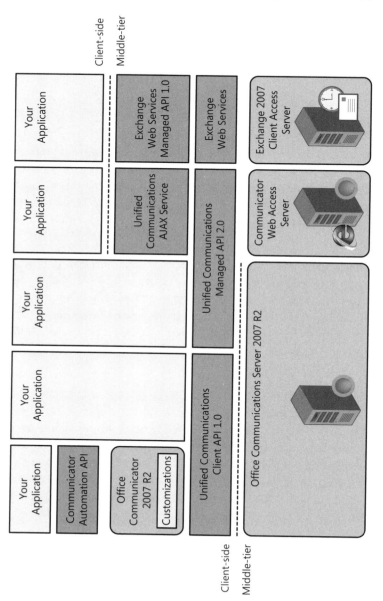

FIGURE 1-2 Client APIs.

Programming Office Communications Server Applications

On the server side, the UCMA offers a SIP stack, a media stack, and powerful speech engines for both Automatic Speech Recognition (ASR) and speech synthesis (Text-to-Speech [TTS]). UCMA 2.0 can be used to build outbound applications, such as alerts, notifications, and surveys, and inbound applications, such as conferencing services, IVR solutions, automated agents (query response bots), and ACDs, which perform skill-based routing. The API provides access to the presence information available in Office Communications Server 2007 R2 and can be used to build role agents that use this presence information to streamline communications between people.

The UCMA 2.0 Workflow API is a higher API abstraction layer of UCMA, which provides Windows Workflow Activities for Office Communications Server. UCMA 2.0 Workflow builds on top of the .NET Framework 3.5 SP1 Windows Workflow Foundation. This VUCMA Workflow API workflow defines new Workflow Activities for querying presence and creating IM and voice dialogs in workflow-based applications. Simple drag-and-drop operation development is now possible to build sophisticated applications. UCMA Workflow simplifies development of voice and IM-based communications into applications that facilitate business processes. Ideally, developers start developing a project on the UCMA 2.0 Workflow and then fall back to the lower-level UCMA in case the UC Workflow Activities do not offer the capabilities they required.

The Microsoft Office Communications Server 2008 Speech Server APIs target the development of Interactive Voice Response applications using the UC platform. Speech Server includes speech recognition and speech synthesis engines. Speech Server (2007) provides the following APIs:

- Speech Server Workflow Activities
- Speech Server Managed API
- VoiceXML (VXML) 2.1
- Speech Application Language Tags (SALT) 1.0

Speech Server (2007) is certified for VXML 2.0/2.1 by the VoiceXML Forum. And Speech Server (2007) Developer Edition (*http://go.microsoft.com/fwlink/?LinkID=70208*) comes with a comprehensive toolset for building speech applications and is tightly integrated with Visual Studio 2005.

Speech Server (2007) is the last release of a standalone voice recognition platform. Applications developed using VXML 2.1 are supported on UCMA. The tools for developing speech grammars, lexicons, tuning tools, and other elements is integrated with UCMA.

Figure 1-3 illustrates the server APIs for building server-based applications.

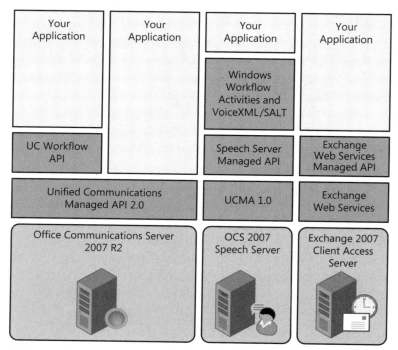

FIGURE 1-3 Server APIs.

Summary

With Office Communications Server 2007 R2, Microsoft innovates by using the .NET Framework and tools familiar to developers, such as the Visual Studio Integrated Development Environment (IDE), to provide a development platform familiar to Windows developers. The UC platform makes it easy for Windows developers to extend their expertise into building IM-based and voice-based applications previously reserved to developers with a very specialized skill set.

Additional Resources

- "Microsoft Office Communicator 2007 R2 Deployment Guide" (*http://go.microsoft.com/fwlink/?LinkID=133744*)
- Office Communications Developer Portal (*http://go.microsoft.com/fwlink/?LinkID=133627*)

Chapter 2
Microsoft Unified Communications APIs Foundation

This chapter will help you to:

- Understand the scenarios that you can build using the API.
- Understand the considerations to help you determine whether this API meets your requirements.
- Understand the API's architecture.
- Understand the API's design model.

This chapter provides a technical overview of the Microsoft Unified Communications (UC) APIs so that you can understand them, how they relate to each other, and what purpose they serve.

The intent of this chapter is to provide you with sufficient information to help you decide which UC APIs best fulfill your needs. If you already know which APIs meet your needs, you can look only at the sections that cover the APIs of interest to you, or skip this chapter altogether if you wish.

Unified Communications Managed API 2.0

The Unified Communications Managed API (UCMA) is a code platform managed by Microsoft .NET Framework, which provides access to presence, instant messaging (IM), telephony, and audio/video (A/V). UCMA is a Session Initiation Protocol (SIP)–based platform. SIP is a signaling protocol that is used for setting up and tearing down multimedia communication sessions. This API abstracts the details of the communication protocols used by Microsoft Office Communications Server.

Scenarios

UCMA is used to build scalable middle-tier applications that work with Office Communications Server 2007 R2, provide large-scale message throughput, and represent multiple endpoints. You can use this API to build the following types of applications:

- Highly scalable notification and alert systems that perform the following actions:
    - Send outbound alert messages.
    - Use the Enhanced Presence feature of the Office Communications Server 2007 R2 platform to determine the appropriate channel and media to deliver alerts, such as instant messages, e-mail, or voice calls.
- Interactive automated agents (query/response bots) that perform the following actions:
    - Respond to user requests for information by means of IM or voice sessions.
    - Create custom call routing and interactive voice response systems.

More advanced applications include the following:

- Contact center or help desk applications that do the following:
    - Route incoming communications sessions to available agents.
    - Use the Enhanced Presence capabilities of the Office Communications Server 2007 R2 platform to route to agents based on specific skill sets.
    - Provide "music on hold" functionality for voice sessions.
    - Create back-to-back user agents (B2BUAs) for help desk scenarios; for example, so that the specific identity of the help desk agent is not exposed to the customer.
- Conferencing portal applications that do the following:
    - Create custom conference bridging.
    - Record the contents of conference calls.
    - Schedule and manage instances of conferences.

Considerations

UCMA 2.0 is considered a middle-tier API written completely in C#. Therefore, it runs only in environments where the .NET Framework is supported. It provides the following characteristics:

- **Scalability**   UCMA 2.0 is able to support thousands of endpoints and concurrent communications and collaboration sessions. It is designed for building server applications (the recommended operating system is Microsoft Windows Server 2008 on 64-bit hardware). UCMA 2.0 is multithreaded, and operations are performed asynchronously to maximize throughput.

- **Availability**   The deployment model supports running multiple instances of the UCMA 2.0 application for load balancing and failover across multiple computers with the use of hardware load balancers.

- **Extensibility**   New modalities can be added in the conversation framework. Extension headers and Uniform Resource Identifier (URI) parameters can be supplied and consumed through the APIs to support the creation of custom sessions.

UCMA 2.0 supports two types of SIP endpoints that are designed for distinct application scenarios: the *ApplicationEndpoint* class and the *UserEndpoint* class. You use the *ApplicationEndpoint* class in applications that represent automated applications, such as bots that interact with users. You use the *UserEndpoint* class in applications that connect to Office Communications Server on behalf of users and perform operations on behalf of those users.

To configure Office Communications Server to trust connections from UCMA applications that use the *ApplicationEndpoint* and *UserEndpoint* classes, the application provisioning process must define a Globally Routable User Agent URI (GRUU). For more information about GRUUs, see Chapter 9, "Preparing the UC Development Environment." This is all that is necessary if the application creates only *UserEndpoints*. After the application has authenticated the user, it does not need to supply those credentials to Office Communications Server for authentication. For applications that create *ApplicationEndpoints*, the provisioning process also must create a *Contact* object that defines the application's SIP URI. You also have the option to create a display name and a TEL URI. The *ApplicationEndpoint* uses this *Contact* object to register with Office Communications Server.

Examples of applications that use the *ApplicationEndpoint* class are Automatic Call Distributor (ACD), interactive IM or voice bots, and conference bridges. For more information about these applications, see the "UCMA 2.0 Workflow API" section later in this chapter. These applications use a *Contact* object to identify the application in Active Directory Domain Services. The *Contact* object specifies the application's SIP URI and phone number. Examples of applications that use the *UserEndpoint* class are those that publish additional presence information. Examples of additional presence information include showing a Global Positioning System (GPS) location on behalf of a user or acting as a proxy when the user is not available and routing incoming IM messages through a Short Message Service (SMS) gateway.

API Architecture

UCMA is composed of the following two interfaces:

- UCMA 2.0 Core API
- UCMA 2.0 Speech API

This architecture can be represented as shown in Figure 2-1.

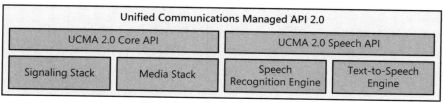

FIGURE 2-1 UCMA version 2.0 architecture.

UCMA 2.0 Core API

The UCMA 2.0 Core API hides the complexity of most of the Office Communications Server SIP/SIMPLE (Session Initial Protocol for Instant Messaging and Presence Leveraging Extensions)-Based protocols by offering an API that exposes almost all of the features of the protocol but is simpler to understand and use.

The UCMA Core API provides access to the signaling and media stack as follows:

- The SIP stack in UCMA 2.0 offers a managed code SIP endpoint API.
- The Media stack in UCMA 2.0 provides a protocol abstraction over the multiple real-time media protocols. This protocol abstraction is exposed in the *Microsoft.Rtc.Collaboration* namespace. This interface provides access to the following functionality:
    - Publication of and subscription to Enhanced Presence
    - Creation and management of multiparty conference sessions
    - Creation, modification, and deletion of Contacts and Groups
    - Call Control for routing audio sessions
    - Management of audio media in sessions

UCMA 2.0 Speech API

The UCMA 2.0 Speech API is a server-grade speech API that allows developers to build multichannel speech recognition– and speech synthesis–enabled applications using Microsoft state-of-the-art speech technology. The UCMA 2.0 Speech API supports 12 different languages, including English (North America, United Kingdom), French (France, Canada), German, American Spanish, Portuguese (Brazil), Italian, Japanese, Korean, and Chinese (Simplified and Traditional).

The Speech API provides access to the Automatic Speech Recognition (ASR) engine and Text-to-Speech (TTS) engine that ship as part of UCMA 2.0:

- **ASR Engine**   This state-of-the art speech recognizer supports triphone phonetics, as well as whole-word modeled speech recognition, for optimal speech recognition, not only for natural language recognition but also for command and control and number recognition.
- **TTS Engine**   This very accurate speech synthesizer uses Hidden Markov Model (HMM)–based Speech Synthesis (HTS) for maximum intelligibility.

Object Model

In UCMA 2.0 Core API, the entry point class is *CollaborationPlatform*. An application can create multiple *CollaborationPlatform* instances. Each instance can host multiple endpoints. An endpoint is the basis for communication and collaboration functionality with Office Communications Server. The properties and methods of these classes provide access to the functionality of the collaboration sessions.

The *CollaborationPlatform* class can be used to:

- Initiate and manage a conversation (*Conversation* class).
- Schedule and manage conferences (*ConferenceServices* class).
- Subscribe to the presence of remote users (*RemotePresence* class).
- Publish presence for the endpoint owner (*LocalOwnerPresence* class).
- Manage contacts and groups (*ContactGroupServices* class).

Figure 2-2 shows the relationships among the principal classes of the architecture, as well as the personas involved in each type of object. The numbers shown between two objects indicate the mapping between classes. For example, one local *endpoint* object can be associated with zero or more *Presence Subscription* objects.

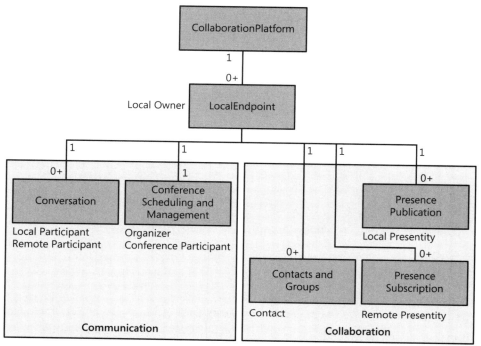

FIGURE 2-2 UCMA 2.0 Object Model.

In UCMA 2.0 Speech API, *Microsoft.Speech* is the main namespace. It is modeled very closely after the *System.Speech* client namespace in the .NET Framework, yet it provides the ability to run many recognizers and synthesizers in parallel.

The two key classes in this namespace are:

- *Microsoft.Speech.Recognition*   This class controls the Recognizer.
- *Microsoft.Speech.Synthesis*   This class controls the *Synthesizer*.

Microsoft.Speech.Recognition.SrgsGrammar provides a class to optimize performance of handling speech recognition grammars.

UCMA 2.0 Workflow API

The UCMA Workflow API and Workflow Activities extend the Windows Workflow Foundation with additional activities that provide access to UCMA functionality. This API consists of a set of custom workflow activities and supporting classes on top of the Windows Workflow Foundation of .NET Framework 3.5 SP1. It consists of the following items:

- A set of custom activities for unified communications (for example, *AcceptCall*)

- Two Workflow Runtime services (*CommunicationsWorkflowRuntimeService* and *TrackingDataWorkflowRuntimeService*) that enable the custom activities to run

Scenarios

The UCMA Workflow API and Workflow Activities are a higher-level API built on top of UCMA 2.0. Using the .NET Framework 3.5 SP1 Windows Workflow Foundation, it offers developers an abstraction layer of UCMA that is easier to use.

UCMA Workflow Activities can add value at communication points in business processes. You can use the UCMA Workflow API to build the following types of applications:

- Presence queries for individuals
- Alerts and notification applications that use the IM or voice channel
- IM-based automated agents (query/response bots)
- Speech- or dual-tone multifrequency (DTMF)–enabled, voice-based automated agents (simple Interactive Voice Response [IVR] applications)

When writing a presence-aware application, the business logic in the workflow can make presence-based, intelligent modality decisions. For example, it is better to send an instant message to a person who is in the middle of a phone call than to call the person.

Alerts and notifications can make more intelligent decisions based on the user's presence state, such as whether to start a phone call or IM session with a user when triggered by an event.

IM-based automated agents understand text-based input using grammar-modeled speech technology. IM-based automated agents can be used to navigate through menu-based systems as well as provide information services, such as querying a database or knowledge base system. For example, an automated IM bot can spell out acronyms, such as converting "MSFT" to "Microsoft."

Incoming voice calls can be understood using the same grammars used for IM calls, powered by ASR using UCMA 2.0 Speech API and DTMF-based touch-tone input. UCMA Workflow voice applications respond by prerecorded speech, or speech synthesis.

The UCMA Workflow API creates sophisticated voice- and text-based dialogs easily. It provides a great dialog experience to the user by understanding commands (for example, "Help" or "Repeat") or by providing guidance in its intelligent responses for invalid inputs (for example, "I'm sorry I didn't understand you").

The UCMA Workflow API provides a set of Workflow Activities for call control, such as accepting incoming IM invitations, transferring a call to another party, and disconnecting the call.

Windows Workflow Foundation

To better understand UCMA Workflow, it is important to understand the Windows Workflow Foundation. The Windows Workflow Foundation is a framework, available as part of .NET Framework 3.0, .NET Framework 3.5, and .NET Framework 3.5 SP1, that enables developers to create applications that can be modeled as a workflow. These workflows can be automated (for example, a workflow that alerts an administrator about specific events) or can require human interaction (for example, an expense report–processing workflow that requires human approval). The Windows Workflow Foundation API supports the Microsoft Visual Basic .NET and C# languages. It consists of the following high-level components:

- **A workflow runtime**   Provides capacity to the host application for the executing workflows.

- **A workflow compiler**   Used to compile workflows, developed using C#, Visual Basic .NET, or Microsoft Extensible Application Markup Language (XAML), into an assembly.

- **A graphical Workflow Designer**   A graphical user interface that you can use to design the workflow. You can drag the activities to this designer canvas to define the flow of the application logic. The Workflow Designer can also be rehosted in any application outside Microsoft Visual Studio. For more information about rehosting the Workflow Designer, see *http://msdn.microsoft.com/en-us/library/cc835242.aspx*.

- **A workflow debugger**   A debug engine that enables you to debug a workflow application. You can also use this to set breakpoints on the activities on the designer canvas.

- **A rules engine**   Conditions in workflow (for example, as used in *if-else* or *while* constructs) can be specified either in code or as a declarative rule condition. You use the rules engine to evaluate these conditions at run time.

- **A set of Windows Workflow Activities**   An out-of-the-box set of activities that range from simple activities, such as the *Code*, *IfElse*, and *While* activities, to a more complex set of activities like Conditional Group Activity. An activity is the building block of a workflow. A custom activity, such as the ones defined in UCMA Workflow Activities, is a class that is derived from the *Activity* class defined in the Windows Workflow Foundation. You can create the workflows either by using code or markup language. The framework also provides an extensible model to build custom activities that you can reuse across projects. For details about Windows Workflow Foundation, see the Windows Workflow Foundation tutorials at *http://msdn.microsoft.com/en-us/library/ms735927.aspx*.

Note Even though Windows Workflow Foundation is available in .NET Framework 3.0 and .NET Framework 3.5, UCMA Workflow Activities is supported only on .NET Framework 3.5 SP1.

For more information about these components, see *http://msdn.microsoft.com/en-us/netframework/aa663328.aspx*.

Considerations

The Windows Workflow Runtime must satisfy two requirements for the UCMA Workflow Activities to execute. These requirements are as follows:

- The Windows Workflow Runtime must allow custom Workflow Runtime services, such as the UCMA Workflow Activities, to be added.

    The two Workflow Runtime services (*CommunicationsWorkflowRuntimeService* and *TrackingDataWorkflowRuntimeService*) provide infrastructure for the UCMA Workflow Activities to execute. Therefore, you should configure the workflow run time to allow these services to be added for UCMA Workflow Activities to execute properly.

- Disable persistence during the execution of UCMA Workflow Activities.

    Some workflow run times allow persistence by using custom workflow services that are derived from the *WorkflowPersistenceService* class. You use these services to store workflow state information on disk when idle and then recreate it when needed. UCMA Workflow Activities do not support persistence. Therefore, persistence should be disabled when running these activities.

From a high level, UCMA Workflow Activities offers the following functionalities:

- Call control functionalities for both phone and IM calls (for example, accepting the call, disconnecting the call, or creating an outbound call).
- Specific call control functionalities for phone calls (for example, blind transfers, hold events, and retrieve events).
- Play messages to the user over the phone. These messages can be recorded prompts or can be synthesized by using the TTS engine.
- Recognize user input over the phone, both by means of speech and DTMF.
- Send, receive, and recognize messages over an IM channel.
- Query for presence information from the workflow.
- Enable moving the execution of the workflow from one activity to another using a *GotoActivity* object.
- Support for multiple calls (that is, phone or IM) in one workflow.

You should use UCMA Workflow Activities to create or enhance applications that are state engine workflows. Examples of such applications are applications that interact with the user over the phone or IM channel.

Workflow Architecture

Figure 2-3 illustrates a high-level architecture of UCMA Workflow and its dependencies.

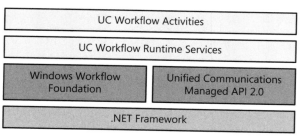

FIGURE 2-3 UCMA Workflow architecture.

- **Windows Workflow Foundation** The Windows Workflow Foundation is a framework that is part of .NET Framework 3.0 and .NET Framework 3.5.
- **UCMA** This API consists of the following interfaces:
  - **UCMA Core** The UCMA 2.0 collaboration platform. It provides functionalities such as calling, presence, and conferencing.
  - **UCMA Speech** The UCMA 2.0 speech platform. It provides functionalities such as TTS, speech recognition, and DTMF recognition.
- **UCMA Workflow Runtime Services** These are custom Workflow Runtime services built on top of the Windows Workflow Foundation that enable UCMA Workflow Activities to execute properly. For details about these run-time services, see Chapter 4, "Embedding Contextual Collaboration."
- **UCMA Workflow Activities** These are custom activities that encapsulate UCMA Core (for example, call and IM) and UCMA Speech (for example, recognizing user input and TTS synthesis) functionality into an activity format. Examples of such functionalities are having a question-and-answer-based dialog with the user over the phone call or IM session. However, all functionalities of UCMA Core and UCMA Speech are not provided in the form of workflow activities as part of UCMA Workflow Activities. Developers can create their own custom workflow activities to capture the functionalities not available from UCMA Workflow Activities.

Object Model

The UCMA Workflow provides managed classes, events, and enumerations to allow you to build communication-enabled workflow applications.

When you install the UCMA Software Development Kit (SDK), the following two libraries are included:

- **Microsoft.Rtc.Workflow.dll**   This library contains all of the workflow activities and custom Workflow Runtime services that are used in the development of the communications-enabled workflow application.

- **Microsoft.Rtc.Workflow.Toolbox.dll**   This library contains the package that is installed in Visual Studio so that the UCMA Workflow Activities show up in the Visual Studio toolbox.

Office Communicator Automation API

The Office Communicator Automation API provides programmatic access to Microsoft Office Communicator 2007 R2 so that you can automate this software running on the client. It is a quick and easy way to integrate Office Communicator functionality into your applications. This API is used in Microsoft Office 2007 and Microsoft SharePoint 2007 to integrate presence information and real-time communications from Office Communicator into these products.

Scenarios

The Office Communicator Automation API provides access to most of the functionalities in Office Communicator 2007 R2 programmatically. Your code can sign in the user to Office Communicator and perform actions on behalf of this user (for example, starting an IM session or calling a contact), change user preferences (for example, tagging a contact), and manage contacts (for example, retrieving the contact list, adding and removing contacts from the list, and working with contact groups). The API can also raise events from Office Communicator to alert you to things like changes in contact presence or an incoming call.

With these capabilities in mind, the following examples are the types of features that you can add to your applications using the API:

- **Embedding presence with application-specific contact lists**   You can use the API to add Office Communicator contact presence to your applications by building a custom contact list. This application-specific contact list can show the presence of contacts even if they are not in the user's Office Communicator contact list.

- **Enhancing communications**   You can enhance your applications by using the API to allow users to communicate by IM, voice, and video directly from your application.

- **Creating application context-specific communications**   You can use the API to integrate data from your application into the conversations your application creates to provide application-specific context for the conversation.

Considerations

The Office Communicator Automation API was created for integrating presence and communication features into client applications. The API is not appropriate for server-side solutions because it requires Office Communicator to be running on the local machine and signed in to Office Communications Server 2007 R2 with a valid user account.

Because the Office Communicator Automation API automates Office Communicator, very little code is required to provide sophisticated communication features. While this increases your productivity as a developer by allowing you to create communication features quickly, keep in mind that the API requires Office Communicator to be installed on the client machine for communication features built with the API to work. Also note that when using this API, your code shows elements of the Office Communicator user interface (UI) (for example, when you add a contact or start a voice conversation) rather than allowing you to provide your own UI for these actions.

Consider using the Office Communicator Automation API if you want to integrate presence and collaboration functionality into your application quickly without having to write your own client or to understand SIP and real-time communications protocols.

Application Architecture

Only a single instance of Office Communicator 2007 R2 can be installed and run on the client. Office Communicator (that is, Communicator.exe) is a Component Object Model (COM) server that runs out of process with your application and provides a programmatic interface to Office Communicator by using the libraries supplied by the API. The Office Communicator Automation API libraries (that is, CommunicatorAPI.dll and CommunicatorPrivate.dll) run in your application process. Figure 2-4 illustrates the run-time architecture of your application when using the API.

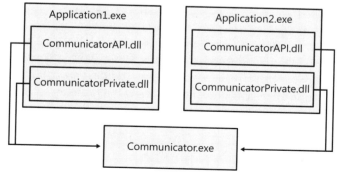

FIGURE 2-4 Office Communicator Automation API application architecture.

For more information about COM and interoperating with COM from managed code, see the "Additional Resources" section at the end of this chapter.

> **Note** The Office Communicator Automation API automates either Office Communicator 2007 R2 or Office Communicator 2007 depending on which client is installed on the computer.

Object Model

The Office Communicator Automation API provides two classes, *Messenger* and *MessengerPriv*. Each of these classes exposes interfaces for you to build your communication features.

Figure 2-5 shows the relationship between the classes supported by the API and the interfaces each class implements.

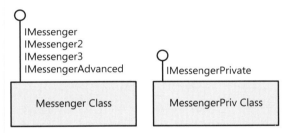

FIGURE 2-5 *Messenger* class and *MessengerPriv* class interface implementation.

The *Messenger* Class

The *Messenger* class represents the instance of Office Communicator 2007 R2 running on the local computer and is the entry point to most of the information and functionalities you can access in the API by using the interfaces it supports. *Messenger* also supports a number of key events for keeping your application in sync with Office Communicator. When using the *Messenger* class, some methods display Office Communicator dialog windows (for example, calling the method to add a contact shows the Add Contact dialog box).

Messenger **class interfaces**   The *Messenger* class implements the following interfaces:

- *IMessenger* **interface**   Supports the most basic features of the API, such as working with the local user's information (that is, displayed name and phone number) and contact list.

- *IMessenger2* **interface**   Inherits from the *IMessenger* interface (and thus supports all of its properties and methods) and adds methods and properties for working with Office Communicator contact groups.

- **_IMessenger3_ interface** Inherits from the _IMessenger2_ interface (and thus supports all of its properties and methods) and adds two unsupported properties. This interface is not used.

- **_IMessengerAdvanced_ interface** Inherits from the _IMessenger3_ interface (and therefore inherits all of the properties and methods of _IMessenger3_, _IMessenger2_, and _IMessenger_) and adds a method for starting conversations programmatically.

> **Note** Because _IMessengerAdvanced_ includes all of the properties and methods available in the _Messenger_ class, it is the interface used most often with UCMA.

Specialized _Messenger_ class interfaces The following _Messenger_ class interfaces provide specialized interfaces to access contacts, contact groups, and active IM conversations:

- **_IMessengerContactAdvanced_ interface** Represents a contact in Office Communicator and allows you to work with contact information (that is, displayed name and phone number) and presence information for that contact programmatically.

- **_IMessengerContacts_ interface** Provides access to the contacts list in Office Communicator 2007 R2 by letting you iterate through the contact list as well as remove contacts from the contact list. Contacts are added to the contact list by using the _IMessengerAdvanced_ interface.

- **_IMessengerGroup_ interface** Provides programmatic access to an individual contact group defined in Office Communicator 2007 R2, including enabling you to change the name of the group and to manage contacts within that group programmatically.

- **_IMessengerGroups_ interface** Provides access to the collection of contact groups in Office Communicator as a collection of _IMessengerGroup_ by enabling you to access the collection of groups, retrieve an individual group (that is, as an instance of _IMessengerGroup_), and remove a group from the collection. Groups are added to Office Communicator by using the _IMessengerAdvanced_ interface.

- **_IMessengerConversationWndAdvanced_ interface** Provides methods and properties for working with an active conversation in Office Communicator, including the ability to send IM text and read the IM conversation history.

> **Note** The conversations are multimodal, so _IMessengerConversationWndAdvanced_ applies to all conversation modalities (that is, IM, voice, and video).

The _MessengerPriv_ Class

Like the _Messenger_ class, the _MessengerPriv_ class represents the instance of Office Communicator 2007 R2 running on the local computer. However, when using the

MessengerPriv class, your method calls suppress any Office Communicator UI and performs each action silently. The *MessengerPriv* class supports a single interface, *IMessengerPrivate*, which provides the ability to add contacts to the signed-in user's contact list without showing the Office Communicator Add Contact dialog box.

Unified Communications Client API

The Unified Communications Client (UCC) API is a comprehensive client-side API that provides connectivity and access to the functionality of Office Communications Server. It's the API that is used to build Office Communicator, and it is a SIP-based application framework for building and deploying real-time communications client applications against Office Communications Server 2007. In addition to the standard features (such as IM, voice calling, video chatting, contact managing, and presence tracking), this framework allows the application to provide users with telephony integration, conferencing, encrypted A/V calls, and the publication and subscription of custom presence information and other application-specific data. The flexible publication and subscription framework makes the application framework appealing for custom applications that do not need to expose the full functionality of Office Communicator 2007 or use the SIP protocol in innovative ways to provide new features.

Scenarios

This API enables you to create the following types of applications:

- **A custom Office Communications Server client** Similar to Office Communicator, but customized for your specific application. For example:
    - Call center client
    - Conference-only client
- **Integrated real-time communications into line-of-business applications** For example, UC client functionality can be deeply integrated into CAD/CAM applications, financial trading applications, and other line-of-business applications. You can create custom presence categories and clients that use these custom categories in innovative ways. For example:
    - A collaboration client that synchronizes three-dimensional views between remote medical imaging applications.
    - Enhance user attributes with GPS information.
- **A standalone Office Communications Server–enabled application** Unlike the Office Communicator Automation API, the UCC API enables you to redistribute the UCC dynamic link libraries (DLLs) with your application. This enables you to ship your application as a standalone Office Communications Server–enabled application.

Considerations

Before embarking on building your solution using the UCC API, it is advisable to consider the intended purpose of this API. The API is intended to be used for client-side, single endpoint Windows applications. Office Communicator 2007 was built using the UCC API, and consequently the UCC API provides access to all of the functionality available in Office Communicator 2007 and more. The UCC API is a low-level client-side API. It does not include any of the business logic that is embedded in Office Communicator. Developers must build their own business logic and user experience. To embed presence and real-time communications into existing applications, therefore, it is advisable to use the Office Communicator Automation API instead of the UCC API. The Office Communicator Automation API requires less code to embed a UC experience into an existing application. For more information, see the "Office Communicator Automation API" section earlier in this chapter.

Application Architecture

The UCC API encapsulates two major functional features in real-time communications: signaling and media handling. Signaling is responsible for setting up and tearing down multimedia communication sessions, such as voice and video calls over an Internet Protocol (IP) network in this context. Media refers to the various real-time modes of communication: audio, video, IM, or e-mail. The encapsulation of signaling provides a higher-level object abstraction of the following protocols:

- SIP
- Centralized Conferencing Control Protocol (CCCP)
- Computer Supported Telephony Applications (CSTA)

The UCC API also provides a set of signaling interfaces for developers interested in working at a lower SIP abstraction level. The encapsulation of media handling provides an object representation of the following protocols:

- Real-Time Transport Protocol (RTP)
- Real-Time Transport Control Protocol (RTCP)
- Secure Real-Time Transport Protocol (SRTP)
- Interactive Connectivity Establishment (ICE)

The UCC API design separates the signaling and media handling between an application and the underlying SIP stack and media management over the RTP stack. Figure 2-6 illustrates the UCC API application architecture stack.

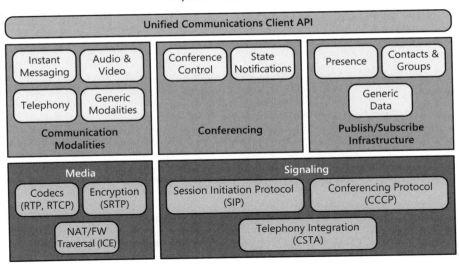

FIGURE 2-6 UCC API architecture.

The SIP stack handles signaling following the standard SIP. It is responsible for carrying out all low-level SIP operations, such as sending a session request, dispatching and receiving provisional responses, and accepting, forwarding, or rejecting an invitation. These operations are necessary for establishing communications and conference sessions in which participants can communicate and collaborate with each other.

The media stack (that is, Media Manager) is responsible for the low-level media management functions, including establishing communication channels to transmit audio, video, or other application data between endpoints.

The UCC API exposes a set of COM-based APIs encapsulating the low-level functionalities and provides applications with SIP and media functionalities in object-oriented programming patterns. With the UCC API, developers can create unified multimodal communication applications, including IM, voice calling (computer-to-computer, computer-to-phone, and phone-to-phone), video chatting, application sharing, and conferencing. It also works with Office Communications Server 2007 and other SIP registrar or proxy servers to manage presence information, and it facilitates communications among communication parties.

The UCC API Object Model

UCC API objects can be grouped logically into the following feature-based categories:

- *Platform* objects   The entry point to all other UCC API functionalities.
- *Endpoint* objects   The object representation of a user in real-time communications and collaborations.

- *Session* objects   Encapsulation of signaling and collaboration sessions, including IM, A/V, application-sharing, and conference.

- *Publication* and *Subscription* objects   Encapsulation of the general framework for publishing and subscribing to data or information.

- *Device Management* objects   Encapsulation of the management functionalities for local devices to render media.

- *Media Connectivity* objects   Encapsulation of the management functionality for enabling media transmission across firewalls.

Figure 2-7 illustrates a diagram of the UCC API object model.

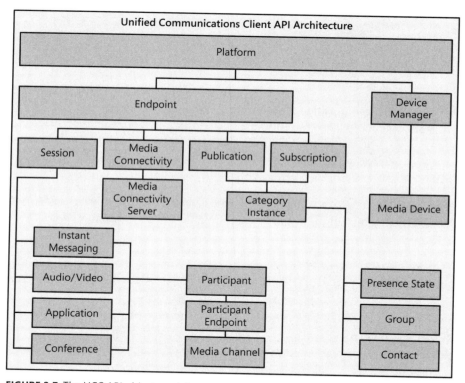

FIGURE 2-7 The UCC API object model.

The *Platform* object corresponds to the UCC API application framework. It is the starting point for such an application to access API functionalities, including the following:

- **Enabling an endpoint**   When you enable an endpoint, you are registering a user with Office Communications Server. An enabled endpoint means a user who is logged on to the UC network successfully.

- **Creating a session to invite other participants and communicate with each other using text messaging, A/V calls, and other communication means** No session can be created unless an endpoint is enabled.

- **Publishing, subscribing, or querying category instances** A user cannot publish, subscribe, or query category instances unless its endpoint is enabled.

- **Maintaining media connectivity for establishing media channels across firewalls** This also requires that the endpoint be enabled.

- **Managing local devices for rendering or capturing media**

Figure 2-8 shows the UCC API interfaces.

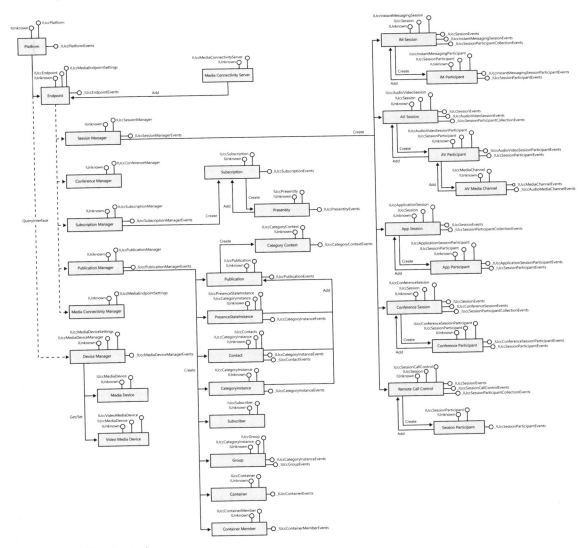

FIGURE 2-8 UCC API interfaces.

Unified Communications AJAX API

The UC Asynchronous JavaScript and XML (AJAX) API is an API to the Office Communicator Web Access server. Although it is not Simple Object Access Protocol (SOAP)–based, it is a Web service API for building applications against the Office Communicator Web Access server, which in turn communicates with Office Communications Server 2007 and Office Communications Server 2007 R2. These APIs allow the application to provide users with IM, presence tracking, publication and subscription of custom presence information, and call deflection. These APIs are based on JavaScript, Extensible Markup Language (XML), and the browser's support for the XMLHTTPRequest API or for the .NET C# System.Net.HttpWebRequest API. The client sends requests to the server in the form of XML payloads, and the server returns XML responses.

The UC AJAX API supports only presence, IM, and call control capabilities. It does not offer A/V, Web conferencing, or telephony support.

Scenarios

The UC AJAX API enables you to create your own non-Windows and Web-based clients, similar to the Office Communicator Web Access client.

The following examples are the types of applications that you can create:

- Mobile/phone clients
- Web applications that combine information from different Web services into a single user interface (known as a *mashup*)
- IM- and presence-enabled Web portals
- High-touch customer service by means of a personalized Web portal

Considerations

Before deciding to build your solution using the UC AJAX API, it is advisable to consider the intended purpose of this API. This API is targeted at building client-side, single-endpoint Web or non-Windows applications. The UC AJAX API makes it possible to build Office Communicator–like applications for users running operating systems that are not Windows and access a subset of the functionalities available in Office Communicator.

The UC AJAX API is supported only on Office Communications Server 2007. For Office Communications Server 2007 R2 deployments, customers are required to deploy an Office Communications Server 2007, Office Communicator Web Access Server to use your UC AJAX API–based application. This lack of forward compatibility occurs because the UC AJAX API is being deemphasized in the Office Communications Server 2007 R2 release and will not be supported in future releases.

Due to the performance aspects, these APIs are not intended for multiendpoint applications that need to scale out. A better choice is the UCMA.

Application Architecture

The UC AJAX API architecture consists of XML messages sent over two Hypertext Transfer Protocol Secure (HTTPS) channels to the Office Communicator Web Access servers that are persistent for the duration of the session while the user is logged on. The user logs on to Office Communicator Web Access, which authenticates the user by using integrated Windows authentication (NT LAN Manager [NTLM]), forms-based authentication, or a custom single sign-on method, and then an authentication (Auth) ticket is issued to the user. This Auth ticket must be provided in every request to the server. The server can refresh this Auth ticket anytime during the lifetime of the user session. Therefore, it is important that the application always uses the latest Auth ticket sent by the server. Note that the user is logged on to only Office Communicator Web Access, not Office Communications Server. Office Communicator Web Access connects to the user's home computer running Office Communications Server once a session is initiated and retrieves the user's contact list, in-band provisioning settings, and so on, which it then sends to the client application.

After the user is authenticated, the application must establish two secure channels (HTTPS) to Office Communicator Web Access. The XMLHTTPRequest browser API is used to establish these communication channels with the server. One channel is used to issue requests to the server. This is referred to as the *command channel*. The client application uses the command channel to send commands to the server. It is recommended to batch requests before issuing the command to the server. When a command is sent to Office Communicator Web Access, the server immediately responds to acknowledge whether the request was received successfully or whether an error occurred. The servicing of the client request is performed by the server asynchronously. This response is sent by the server by using another communication channel. Figure 2-9 illustrates the UC AJAX API communications architecture.

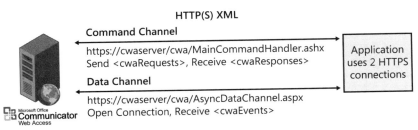

FIGURE 2-9 The UC AJAX API communications architecture.

For the server to be able to push events to the client application (for example, responses to requests sent over the command channel), the client must create a long-held HTTPS connection with the server named the *data channel*. Anytime Office Communicator Web

Access has an event for the user, it sends it immediately on the data channel without the client application having to request it explicitly. The server holds this connection open for up to 35 seconds. If there are no events for the client application, the server dynamically requests the client to delay the next open *GET* call on the data channel. This mechanism (referred to as Comet) makes it possible for the server to push events to a Web client.

All of the APIs are defined as an XML schema. When the client application logs on the user, the user is signed in to Office Communicator Web Access only. After the user is logged on, the user can create multiple sessions. Each session represents an endpoint registered with Office Communications Server. At this point, user presence is made known to Office Communications Server by means of the server running Office Communicator Web Access. Because Office Communicator Web Access does not keep user state, it transforms SIP traffic from Office Communications Server into XML payloads that are sent to the Web clients. The client must cache any information that Office Communicator Web Access returns (for example, contact list, in-band provisioning settings, and roaming user settings). All caching must be performed on the client side.

XML Model

When communicating with Office Communicator Web Access, the client application uses the following URLs where *<server>* refers to the base URL of the server running Internet Information Services (IIS). The Logon channel is used to sign in the user. Depending on the type of authentication used, this URL is slightly different. After the user is signed, this channel is no longer needed. The Command and Data channels persist for the duration of the session. Each session begins with the XML request, *cwaRequests*, that contains the *initiateSession* element and ends with another XML request, *cwaRequests*, that contains the *terminateSession* element. Every request and response occurs within a session.

Logon channel:

- Forms-based authentication: *https://<server>/forms/logon.html*
- Integrated Windows authentication: *https://<server>/iwa/logon.html*
- Custom single sign-on authentication: *https://<server>/sso/logon.html*

Command channel:

- *https://<server>/cwa/MainCommandHandler.ashx*

Data channel:

- *https://<server>/cwa/AsyncDataChannel.ashx?Sid=<sid>&AckId=<ackId>&UA=<ua>*

UC AJAX APIs are defined as XML elements. These XML elements are organized into the following categories:

- **cwaRequests**   This element defines the type of requests a Web client application can submit to the Office Communicator Web Access server on the Command channel.
- **cwaResponses**   This element defines the response that Office Communicator Web Access returns immediately after receiving a request. The server returns only whether the request was accepted or rejected and why it was rejected.
- **cwaEvents**   This element defines the type of events the application can expect to receive from Office Communicator Web Access. These events are the results from the request sent to the server. Because the result is returned asynchronously to the application, there needs to be a way to match the event to the request. This is done by matching the event ID (*eid*) with the request ID (*rid*) of the corresponding *cwaRequests* element.

cwaRequests

Every request to the server is a *cwaRequests* element of type *CwaRequestsType*. This complex type consists of a *sid* attribute of type *CwaSessionId* (unsigned long) and one or more of the subelements listed in Table 2-1.

TABLE 2-1 Subelements for *CwaRequestType*

Element	Type
Logon	CwaLogonRequestType
initiateSession	CwaInitiateSessionRequestType
terminateSession	CwaRequestBaseType
addGroup	CwaAddGroupRequestType
updateGroup	CwaUpdateGroupRequestType
deleteGroup	CwaDeleteGroupRequestType
addContact	CwaContactRequestType
updateContact	CwaContactRequestType
deleteContact	CwaUriRequestType
acknowledgeSubscriber	CwaUriRequestType
conference	CwaRequestBaseType
updateContainer	CwaUpdateContainerRequestType
publishSelfPresence	CwaPublishSelfPresenceRequestType
subscribePresence	CwaUriListRequestType
unsubscribePresence	CwaUriListRequestType
queryPresence	CwaUriListRequestType
publishRawCategories	CwaPublishRawCategoriesRequestType
Search	CwaSearchRequestType

Each of the types listed in Table 2-1 inherits from the base type *CwaRequestBaseType*. This complex type, *CwaRequestBaseType*, consists of a single attribute, *rid*, which is a request ID that is used to uniquely identify every request. The XML schema (*xsd*) defines this attribute to be of type *CwaRequestId* (string).

cwaResponses

Office Communicator Web Access immediately responds to *cwaRequests* to indicate the status of the request. These statuses are defined as elements of the *cwaResponses* element, which is of type *CwaResponsesType*. The element *CwaResponses* is composed of the attribute *requestProcessed* and zero or more of the subelements listed in Table 2-2.

TABLE 2-2 Subelements for *CwaResponses*

Element	Type
requestSucceeded	CwaRequestSucceededType
requestAccepted	CwaRequestAcceptedType
requestFailed	CwaRequestFailedResponseType
requestRejected	CwaRequestFailedResponseType
Error	CwaFailureType

Each of the types listed in Table 2-2 inherits from the base type *CwaResponseBaseType*, with the exception of the complex type *CwaFailureType*. This complex type, *CwaResponseBaseType*, consists of a single attribute, *rid*, which is the ID of the matching request. The XML schema (*xsd*) defines new complex types for these subelements.

cwaEvents

Office Communicator Web Access returns the results of each *cwaRequests* item as a *cwaEvents* element. A *cwaEvents* element, which is of type *CwaEventsType*, is transmitted by Office Communicator Web Access on the data channel. The element *CwaEvents* is composed of the attributes *sid, ackId, pollWaitTime, sessionTimeout,* and zero or more of the subelements listed in Table 2-3.

TABLE 2-3 Subelements for *CwaEvents*

Element	Type
pollFailed	CwaFailureType
requestSucceeded	CwaRequestSucceededEventType
requestFailed	CwaRequestFailedEventType
requestCancelled	CwaRequestCancelledEventType
contactGroup	CwaContactGroupEventType
Subscribers	CwaSubscribersEventType

TABLE 2-3 Subelements for *CwaEvents*

Element	Type
Conference	CwaConferenceEventType
Containers	CwaContainerListEventType
selfPresence	CwaSelfPresenceEventType
userPresence	CwaUserPresenceEventType
selfRawCategories	CwaSelfRawCategoriesEventType
userRawCategories	CwaUserRawCategoriesEventType
presenceSubscriptionState	CwaPresenceSubscriptionStateEventType
Configuration	CwaConfigurationsEventType
searchResult	CwaSearchResultsEventType
locationProfiles	CwaLocationProfilesEventType
contactGroup	CwaContactGroupEventType

Each of the types listed in Table 2-3 inherits from the base type *CwaEventBaseType*. This complex type, *CwaEventBaseType*, consists of a single attribute, *erid*, which is a unique ID of the event. The XML schema (*xsd*) defines new complex types for these subelements.

Office Communications Server 2007 Speech Server Developer Edition

Using Speech Server (2007) APIs, developers can build applications by using managed code, .NET Framework 3.0–based Windows Workflow Foundation Activities, and Web-based standards, such as the World Wide Web Consortium (W3C) specifications for VoiceXML 2.1 or Speech Application Language Tags (SALT) 1.0, as used in Speech Server 2004 and Speech Server 2004 R2.

Dialog flows can be implemented by using any of the following techniques:

- .NET Framework 3.0–based Windows Workflow Foundation Speech Dialog Workflow Activities
- The Speech Server Managed API
- VoiceXML 2.1
- SALT 1.0

The Speech Dialog Workflow Activities available with Speech Server (2007) are very elaborate. They can be used to perform a wide variety of activities related to telephone call management, dialog flow, and application logic and structure.

In case the Workflow Activities do not suffice, developers can use the lower-level Speech Server Managed API. The Speech Server Managed API covers the following core areas, which are needed to produce speech-enabled telephony applications:

- **Speech synthesis**   When the application prompts the caller for information or simply provides the caller with information, the prompt is handled through the *Synthesizer* property. This property is a reference to a *SpeechSynthesizer* instance, which provides capabilities for converting text into speech.

- **Speech recognition**   When a caller responds to a prompt, the recognition is handled through the *SpeechRecognizer* property. This property is a reference to a *SpeechRecognizer* instance, which parses speech by using a set of grammars and by extracting meaningful information.

- **Call handing and control**   The *TelephonySession* property provides methods for call control, such as answering or transferring a call.

- **DTMF processing**   When a caller presses a keypad button (DTMF input), the application processes it by using the *DtmfRecognizer* property.

- **Application hosting**   With the *IApplicationHost* and *IHostedSpeechApplication* interfaces, Speech Server manages the lifetime of an application instance.

Speech Server (2007) is a VoiceXML Forum–certified platform that supports the W3C VoiceXML 2.1 standard. For more information about the VoiceXML standard, see the W3C site at *http://www.w3.org/TR/voicexml20/* and MSDN at *http://msdn.microsoft.com/en-us/library/bb857664.aspx*.

Speech Server 2004 and Speech Server 2004 R2 support the ASP.NET-based SALT 1.0 standard. Speech Server (2007) still supports running these SALT applications for backward compatibility.

Scenarios

Office Communications Server 2007 Speech Server (Speech Server 2007) is the IVR platform that is part of Office Communications Server 2007 and Office Communications Server 2007 R2. The Developer Edition is available as a free download and contains the Speech Server APIs, speech technology tools that are fully integrated into Visual Studio 2005, and a data warehousing solution for processing call log files. The product is licensed as Office Communications Server when it is deployed.

Speech Server (2007) supports sophisticated speech technology, such as Conversational Understanding. This technology is suitable for building sophisticated speech-enabled IVR applications that support human interaction with callers. It supports mixed-initiative form filling that lets users control dialog flow by providing all needed information in a single utterance or as a sequence of utterances. By using the mixed-initiative style to fill a form (a *FormFillingDialog* instance), a user can answer multiple questions at once. The application can accept an answer in response to a specific question, but it can also accept and fill a form with extra answers that apply to questions the application has not yet asked. This style enables a nonsequential dialog. Each question-and-answer cycle includes one question and one or more answers. Mixed-initiative dialogs are typically more difficult to design than system-initiative dialogs (in which the system asks a question and expects a single answer), but they provide users with greater flexibility when answering questions. Mixed-initiative dialogs simulate human interaction more closely than system-initiative dialogs and can recognize either DTMF keypad presses or speech input. Speech Server (2007) is the most appropriate API for building complex IVR applications, such as Voice Portal applications that include multislot speech recognition where callers enter multiple items all in one utterance. Of course, Speech Server (2007) also supports simple call routing applications and DTMF menu-driven applications.

Speech Server (2007) differs from the UCMA 2.0 Workflow speech capabilities not only in the conversational understanding technology used, but also in its support of different types of IVR and speech technology–specific tools. Speech Server (2007) provides a Conversational Grammar Builder, a Prompt Recording and Editing tool, lexicon tools, and tuning tools, such as tools that detect words in utterances that are outside the grammar.

Speech Server supports speech recognition and speech synthesis in five languages:

- English (North America, United Kingdom)
- American Spanish
- Canadian French
- German

Speech Server supports DTMF applications and speech synthesis in nine additional languages:

- Australian English
- French
- Castilian Spanish
- Portuguese (Brazil)
- Italian
- Japanese
- Korean
- Chinese (Simplified and Traditional)

Considerations

Before you create an interactive voice response telephony application, consider the following points:

- **Complexity of the speech technology**   For simple speech applications, often the UCMA Workflow API suffices. For more sophisticated speech technology, such as conversational understanding, use Speech Server (2007).

- **Standards support**   Currently, VoiceXML support is not available on UCMA, so if VoiceXML support is mandatory, Speech Server (2007) provides the functionality.

- **Language support**   Speech Server supports ASR and TTS in 5 languages, and UCMA supports 12 languages. Depending on your language needs, one or both might be a better fit.

Application Architecture

Figure 2-10 illustrates the Speech Server components and the relationships between them as applied to the design stage of application development. This diagram assumes that you have a single computer that is running Visual Studio 2005, Speech Server, and Web server software.

The SIP peer represents all possible client endpoints, including IP Private Branch eXchange (IP/PBX) telephony clients, Voice over Internet Protocol (VoIP) gateways, SIP phones and softphones, and Telephony Interface Manager Connector (TIMC). The SIP peer communicates with the application by using SIP for signaling data and RTP for audio data.

The developer uses Visual Studio 2005 to create a speech application, choosing one of the following application types:

- **Voice Response Workflow Application**   To create a managed code interactive voice response application, choose this application type from the New Project dialog box.

- **Voice Response Web Application**   To create a Web-based SALT interactive voice response application, choose this application type from the New Web Site dialog box.

- **VoiceXML Speech Application**   To create a Web-based VoiceXML application, choose this application type from the New Web Site dialog box.

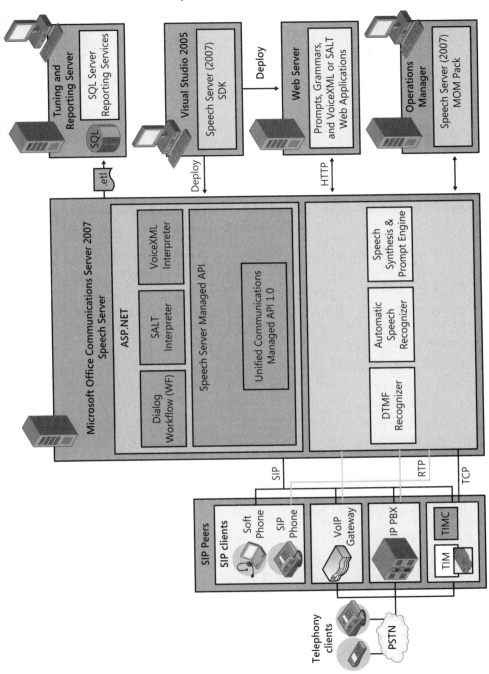

FIGURE 2-10 Speech Server (2007) components.

Application Components

Following the Web model for which they are designed, SALT voice response applications and VoiceXML applications are deployed to the Web server. However, because the code that drives a managed code assembly from one state to the next is located on the computer running Speech Server, the assembly must be deployed to the computer running Speech Server itself.

After deploying the application, the developer tests it with a connected SIP peer or simulator.

The principal application components depicted in Figure 2-10 are described in the next two sections.

ASP.NET Components   The ASP.NET run time hosts IVR applications in any of four types:

- **Speech Server Managed API**   Runs the assembly that contains the code that is used by the managed code IVR application. The Speech Server Managed API is the lowest-level API. It includes the signaling stack (UCMA 1.0) and provides access to speech and media resources.

- **Dialog Workflow Run Time**   For a Dialog Workflow–based application, the Dialog Workflow run time executes dialog workflow activities. The Dialog Workflow activities are based on the .NET Framework 3.0 Windows Workflow Foundation.

- **VoiceXML Interpreter**   For a VoiceXML speech application, the VoiceXML interpreter is responsible for loading, parsing, and running VoiceXML code, which is stored on the Web server.

- **SALT Interpreter**   For a Voice Response Web application, the SALT interpreter manages dialog flow with the caller and controls telephone calls as it interprets SALT code, which is stored on the Web server.

Web Server Components   A Web server running IIS 6.0 or IIS 7.0 is an integral part of a complete Speech Server deployment. IIS is included with Windows Server 2003, Microsoft Windows XP Professional SP2, and Windows Vista. However, you must install IIS explicitly the first time it is used with Speech Server.

- **Grammars**   A grammar file contains a structured list of words and phrases that the Speech Server API parses for the Speech Engine Services (SES) speech recognition engine. Grammars are specific to applications developed for Speech Server.

- **Prompts**   A prompt database is an application-specific repository of prerecorded sound files used by the SES speech output engines. To improve prompt quality and therefore the quality of the speech output, consider hiring a professional to record the prompts for the database.

- **VoiceXML Code**   Developer-written VoiceXML code for the Web-based voice response application.
- **SALT + JScript + HTML Code**   Developer-written SALT, JScript, and Hypertext Markup Language (HTML) code for the Web-based voice response application.

Object Model

A simplified Speech Server API object model is shown in Figure 2-11. At the top of the hierarchy is a *SpeechSequentialWorkflowActivity* object, which has an *ApplicationHost* property (as well as others not shown in the illustration), which in turn has a *TelephonySession* property. Three of the properties on the *TelephonySession* object are *DtmfRecognizer*, *SpeechRecognizer*, and *Synthesizer*. These properties are references to instances of the *DtmfRecognizer*, *SpeechRecognizer*, and *Synthesizer* classes, respectively. An application can use these *TelephonySession* properties to access the members of these classes. Each of the two *Recognizer* objects has a *Grammars* property, which is a collection of *Grammar* objects that can be used by the *Recognizer*. Each *Grammar* object contains one or more rules that a speech recognizer or DTMF recognizer can use to extract semantic meaning from user input.

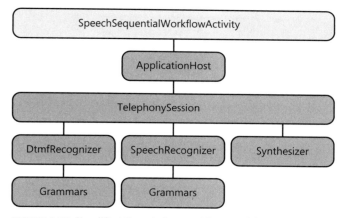

FIGURE 2-11 Simplified Speech Server object model.

Speech Server API

The Speech Server API, shown in Figure 2-12, consists of five namespaces that developers can use to create managed-code voice response applications.

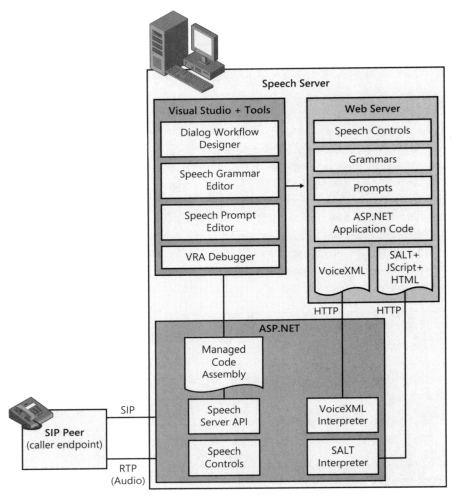

FIGURE 2-12 Speech Server API.

- *Microsoft.SpeechServer* **namespace**   Provides low-level control of core speech services, such as creating hosted application containers, creating telephony and conference sessions, manipulating caller information, controlling logging, and controlling recording.

- *Microsoft.SpeechServer.Dialog* **namespace**   Provides a number of classes derived from Windows Workflow Foundation activities. Applications can use the classes provided in the *Microsoft.SpeechServer.Dialog* namespace to do any of the following:

    ❑ Perform tasks related to managing phone calls, such as answering a call, disconnecting a call, transferring a call to a third party, and recording a call. These classes are *AnswerCallActivity*, *MakeCallActivity*, *BlindTransferActivity*,

DeclineCallActivity, *DisconnectCallActivity*, *RecordAudioActivity*, *RecordMessageActivity*, and *DetectAnsweringMachineActivity*.

- Perform tasks related to dialog flow, such as issuing simple prompts, responding to commands or requests for help, asking a question and receiving an answer, specifying how silences or nonrecognitions are handled, playing a menu of choices from which the user can select one, playing a list of choices through which the user can navigate, validating user input, and responding to system events. These classes are *StatementActivity*, *CommandActivity*, *HelpCommandActivity*, *RepeatCommandActivity*, *QuestionAnswerActivity*, *GetAndConfirmActivity*, *ConsecutiveNoInputsSpeechEventActivity*, *ConsecutiveNoRecognitionsSpeechEventActivity*, *ConsecutiveSilencesSpeechEventActivity*, *MenuActivity*, *FormFillingDialogActivity*, *NavigableListActivity*, *ValidatorActivity*, *SpeechEventActivity*, and *SpeechEventsActivity*.

- Support complex application logic with activities for branching, grouping related activities into tasks, and partitioning applications into modules. These classes are *GoToActivity*, *SetTaskStatusActivity*, *InvokeWorkflowActivity*, *SpeechSequenceActivity*, *SpeechCompositeActivity*, *SpeechSequentialWorkflowActivity*, *SaltInterpreterActivity*, and *VoiceXMLInterpreterActivity*.

- **Microsoft.SpeechServer.Recognition namespace**   Provides a number of classes that can be used in conjunction with grammars to parse user speech and extract semantic information from it. The classes in this namespace can be used to do any of the following:

    - Create instances of speech or DTMF recognizers
    - Construct and load grammar objects into a speech or DTMF recognizer
    - Create objects that contain recognized words and phrases as well as recognition results

- **Microsoft.SpeechServer.Recognition.SrgsGrammar namespace**   The classes in this namespace provide the ability to create and compile grammars that adhere to the Speech Recognition Grammar Specification (SRGS). Most of the classes in this namespace map directly to SRGS elements, such as *item*, *one-of*, *rule*, and *ruleref*.

    Grammars can be created manually using a text editor, programmatically in an application, or dynamically at run time and can be compiled to binary context-free grammar (CFG) files for optimization purposes.

- **Microsoft.SpeechServer.Synthesis namespace**   The classes in this namespace can be used to create TTS output. The text can take the form of character strings and can include Speech Synthesis Markup Language (SSML) markup, bookmarks, "say as" information, pronunciation cues, and audio output. For speech output, there are a variety of options, such as voice gender, age, speaking rate, culture, and others.

Summary

The Microsoft UC platform offers several APIs that you can choose among depending on your development needs, whether they are client-side or server-side, Windows-based or non–Windows-based, and so on. This chapter provides a condensed technical overview of each of these APIs to help you decide which API or APIs best suit your application needs. From this starting point, you can refer to the individual API SDK documentation for more in-depth details of every method, event, and property that the API exposes.

Additional Resources

- "Microsoft Office Communications Server 2007 R2 Developer References" (*http://technet.microsoft.com/en-us/library/dd425166(Office.13).aspx*)
- Microsoft Unified Communications AJAX SDK download (*http://go.microsoft.com/fwlink/?LinkId=142478*)
- Microsoft Unified Communications Client API SDK download (*http://go.microsoft.com/fwlink/?LinkID=141197*)
- "Office Communicator 2007 and Office Communicator 2007 R2 Automation API Documentation" (*http://msdn.microsoft.com/en-us/library/bb758719.aspx*)
- Microsoft Unified Communications Managed API 2.0 SDK (32 bit) download (*http://go.microsoft.com/fwlink/?LinkID=140790*)
- Microsoft Unified Communications Managed API 2.0 SDK (64 bit) download (*http://go.microsoft.com/fwlink/?LinkID=139195*)
- Microsoft Office Communications Server 2007 Speech Server Developer Edition download (*http://go.microsoft.com/fwlink/?LinkID=70208*)
- "Office Communications Developer Portal" (*http://msdn.microsoft.com/ocdev*)
- "W3C VoiceXML 2.0 Introduction" (*http://msdn.microsoft.com/en-us/library/bb857664.aspx*)
- "W3C VoiceXML Version 2.0" (*http://www.w3.org/TR/voicexml20/*)
- "WF Scenarios Guidance: Workflow Designer Re-Hosting" (*http://msdn.microsoft.com/en-us/library/cc835242.aspx*)
- "Getting Started with Workflow Foundation" (*http://msdn.microsoft.com/en-us/netframework/aa663328.aspx*)
- "Using Unified Communications Client API" (*http://msdn2.microsoft.com/en-us/library/bb878217.aspx*)
- Integrating Web Chat Functionality - Microsoft Unified Communications AJAX API Sample (*http://www.microsoft.com/downloads/details.aspx?FamilyId=C8C3F762-7BE4-4541-9B18-82499DB61293&displaylang=en*)

Part II
Office Communicator Automation API

After reviewing the APIs that are available in the Microsoft Unified Communications (UC) platform in Part I, "Understanding Unified Communications," Part II covers the Microsoft Office Communicator Automation API, as shown in the following figure.

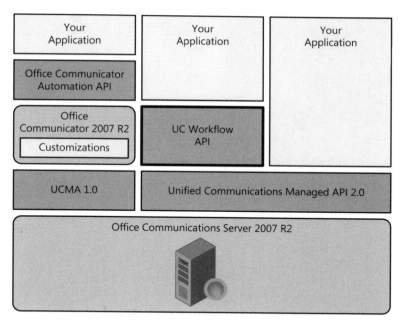

Chapter 3, "Programming a Microsoft Office Communicator Automation API Application," explains how to use this API. Chapter 4, "Embedding Contextual Collaboration," covers an example of how to embed contextual collaboration into an application and includes a complete walkthrough of the source code.

Chapter 3
Programming a Microsoft Office Communicator Automation API Application

This chapter will help you to:

- Sign in and sign out of Microsoft Office Communicator 2007 R2 programmatically and react to sign-in status events.
- Work with local user and contact information programmatically.
- Subscribe to presence events for the local user and contacts to display up-to-date presence in applications.
- Manage the contact list programmatically.
- Start conversations using instant messaging (IM), audio, and video.

Signing In to and Out of Office Communicator

Chapter 2, "Microsoft Unified Communications APIs Foundation," introduces the Office Communicator Automation API and the classes, interfaces, and events that you can use to automate Office Communicator. In this chapter, you learn how to use this API.

When using the Office Communicator Automation API, you can provide Office Communicator features in your application. For your code to automate Office Communicator successfully, you must meet the following conditions:

- Office Communicator is running on the local computer.
- The local user is signed in to Office Communicator.

Using the *Messenger* Class

The *Messenger* class encapsulates Office Communicator 2007 R2 running on the local computer and is the main entry point to the API. The *Messenger* class supports a number of properties, methods, and events for keeping track of the sign-in state of the local user.

After installing the Office Communicator 2007 Software Development Kit (SDK) and referencing the Office Communicator Automation API, the first thing you need to do is instantiate the *Messenger* class, as follows.

```
Messenger _messenger = new Messenger();
```

The *Messenger* class, like all classes in the Office Communicator Automation API, is an unmanaged Component Object Model (COM) class that you access from the Microsoft .NET Framework's managed code by using COM Interop. This means that you must explicitly release every reference of the *Messenger* class and other classes instantiated from this API. The reference can be released by calling the *System.Runtime.InteropServices.Marshal.ReleaseComObject()* method and then setting the reference to *NULL*, as shown in the following code example.

```
Messenger _messenger = new Messenger();

System.Runtime.InteropServices.Marshal
    .ReleaseComObject(_messenger);

_messenger = null;
```

If you fail to call *Marshal.ReleaseComObject()* and set the reference to *NULL*, this results in memory leaks in your application.

Note You must release references in this manner for all references that you create using the Office Communicator Automation API.

Determining Whether Office Communicator Is Running

To determine whether Office Communicator is running on the local computer, you can use the HKEY_CURRENT_USER\Software\IM Providers\Communicator registry key. When Office Communicator is running, the *UpAndRunning* value of that key is set to 2.

Use the following code to determine whether Office Communicator is running on the local computer.

```
// OC Automation API class can't be instantiated unless
//  Office Communicator is running.
if (Convert.ToInt32(Microsoft.Win32.Registry.CurrentUser
        .OpenSubKey("Software").OpenSubKey("IM Providers")
        .OpenSubKey("Communicator")
        .GetValue("UpAndRunning", 1)) == 2)
{
    Console.WriteLine("Office Communicator is running.");
}
else
{
    Console.WriteLine("Office Communicator is *not*
        running.");
}
```

Checking Local User Status

The *Messenger* class provides the *MyStatus* property to determine the sign-in status of the local user. Office Communicator displays the status of the user to the right of their name, as shown in Figure 3-1.

FIGURE 3-1 The status of the user in Office Communicator.

MyStatus returns a value from the *MISTATUS* enumeration. The values defined in *MISTATUS* are shown in Table 3-1.

TABLE 3-1 *MISTATUS* **Enumeration Values**

Element	Value
MISTATUS_UNKNOWN	0x0000
MISTATUS_OFFLINE	0x0001
MISTATUS_ONLINE	0x0002
MISTATUS_INVISIBLE	0x0006
MISTATUS_BUSY	0x000A
MISTATUS_BE_RIGHT_BACK	0x000E
MISTATUS_IDLE	0x0012
MISTATUS_AWAY	0x0022
MISTATUS_ON_THE_PHONE	0x0032
MISTATUS_OUT_TO_LUNCH	0x0042
MISTATUS_IN_A_MEETING	0x0052
MISTATUS_OUT_OF_OFFICE	0x0062
MISTATUS_DO_NOT_DISTURB	0x0072
MISTATUS_IN_A_CONFERENCE	0x0082
MISTATUS_ALLOW_URGENT_INTERRUPTIONS	0x0092

TABLE 3-1 *MISTATUS* Enumeration Values

Element	Value
MISTATUS_MAY_BE_AVAILABLE	0x00A2
MISTATUS_CUSTOM	0x00B2
MISTATUS_LOCAL_FINDING_SERVER	0x0100
MISTATUS_LOCAL_CONNECTING_TO_SERVER	0x0200
MISTATUS_LOCAL_SYNCHRONIZING_WITH_SERVER	0x0300
MISTATUS_LOCAL_DISCONNECTING_FROM_SERVER	0x0400

By using the following code, you can use the *MyStatus* property to check whether the local user is in any of the online states such as *Available, In a Call,* or *In a Meeting.*

```
Messenger _messenger = new Messenger();

// Check if the user is already signed in to
//   Office Communicator.
if ((_messenger.MyStatus & MISTATUS.MISTATUS_ONLINE) ==
    MISTATUS.MISTATUS_ONLINE)
{
    Console.WriteLine("Local user is signed in to
    Office Communicator");
}
else
{
    Console.WriteLine("Local user is *not* signed in to
    Office Communicator");
}
```

This code performs a logical bitwise AND of *Messenger.MyStatus* and the *MISTATUS. MISTATUS_ONLINE* value. If *Messenger.MyStatus* returns any of the status values that represent an online state (that is, any value from Table 3-1 other than *MISTATUS_UNKNOWN* or *MISTATUS_OFFLINE*), this operation is equal to *MISTATUS.MISTATUS_ONLINE.*

Signing In to Office Communicator

The *Messenger* class provides methods and events to sign the local user in to and out of Office Communicator.

Messenger.Signin() Method

The *Messenger.Signin()* method signs the local user in to Office Communicator using a given Session Initiation Protocol (SIP) Uniform Resource Identifier (URI) and password. For example, the following code signs in a user with the SIP URI chrism@uc.contoso.com and "password" as the password.

```
Messenger _messenger = new Messenger();

_messenger.Signin(0, "sip:chrism@uc.contoso.com",
    "password");
```

Messenger.AutoSignin() Method

The *Messenger.AutoSignin()* method signs in the local user based on the credentials cached in Office Communicator. For example, the following code signs in the local user automatically using the last used credentials.

```
Messenger _messenger = new Messenger();

_messenger.AutoSignin();
```

Messenger.OnSignin Event

Signing in to Office Communicator is performed asynchronously. When you call *Messenger.Signin()* or *Messenger.AutoSignin()*, you are making a request to Microsoft Office Communications Server 2007 R2 to sign in the local user to Office Communicator. When the local user is signed in successfully, the *Messenger* class raises the *OnSignin* event. The following code shows you how to subscribe to the *OnSignin* event using the *DMessengerEvents_OnSigninEventHandler* event handler.

```
Messenger _messenger = new Messenger();

_messenger.OnSignin += new
    DMessengerEvents_OnSigninEventHandler(
    _messenger_OnSignin);

...

void _messenger_OnSignin(int hr)
{
    if (hr == 0)
    {
        Console.WriteLine("OnSignin()");
    }
}
```

The *Messenger.OnSignin* event passes a single parameter, an integer to represent the success of the sign-in, to the corresponding event handler. A value equal to 0 represents a successful sign-in.

 Note The *Messenger.OnSignin* event fires every time the local user signs in either through Office Communicator or the Office Communicator Automation API.

Signing Out of Office Communicator

The following sections describe the methods and events associated with signing out of Office Communicator.

Messenger.Signout() Method

The *Messenger.Signout()* method signs out the local user from Office Communicator. Use the following code to sign out the local user.

```
Messenger _messenger = new Messenger();

_messenger.Signout();
```

Messenger.OnSignout Event

Similar to signing in, signing out of Office Communicator is done asynchronously. When you call *Messenger.Signout()*, you are making a request to Office Communications Server 2007 R2 to sign the local user out of Office Communicator. When the local user is actually signed out, the *Messenger* class raises the *OnSignout* event. The following code shows you how to subscribe to the *OnSignout* event using the *DMessengerEvents_OnSignoutEventHandler* event handler.

```
Messenger _messenger = new Messenger();

_messenger.OnSignout += new
    DMessengerEvents_OnSignoutEventHandler(
    _messenger_OnSignout);

...

void _messenger_OnSignout()
{
    Console.WriteLine("OnSignout()");
}
```

Note The *Messenger.OnSignout* event fires every time the local user signs out either through Office Communicator or the Office Communicator Automation API.

Messenger.AppShutdown Event

If the local user shuts down Office Communicator, the Office Communicator Automation API raises the *Messenger.OnSignout* event, followed by the *Messenger.AppShutdown* event. The following code shows you how to subscribe to the *AppShutdown* event using the *DMessengerEvents_AppShutdownEventHandler* event handler.

Chapter 3 Programming a Microsoft Office Communicator Automation API Application

```
Messenger _messenger = new Messenger();

_messenger.OnAppShutdown += new
    DMessengerEvents_OnAppShutdownEventHandler(
    _messenger_OnAppShutdown);
```

...

```
void _messenger_OnAppShutdown()
{
    Console.WriteLine("OnAppShutdown()");
}
```

Putting It All Together

Using the concepts from the preceding sections, you can quickly put together code to determine whether Office Communicator is running, as well as the sign-in status of the local user. In this section, you create a console application and use the *Messenger* class by performing the following steps:

1. Download and install the Office Communicator 2007 SDK. For details, see the "Additional Resources" section later in this chapter.

2. Start Microsoft Visual Studio 2008 and create a new Microsoft Visual C# Windows console application named SignInSignOut.

3. In Solution Explorer, right-click the References node, click Add Reference, and then, from the COM tab, add references to Microsoft Office Communicator 2007 API Type Library.

4. Open Program.cs and add the following *using* statements.

   ```
   using CommunicatorAPI;
   using System.Runtime.InteropServices;
   ```

   This code brings the *Office Communicator Automation API* and *InteropServices* namespaces into scope.

5. Add the following static declarations inside the *Program* class.

   ```
   private static Messenger _messenger;
   private static bool _signedIn = false;
   ```

   The *_messenger* member variable provides an instance of the *Messenger* class. The *_signedIn* variable represents the signed-in state of the local user in Office Communicator.

6. Add the following method to the *Program* class.

```
static bool IsCommunicatorRunning()
{
    return Convert.ToInt32(
        Microsoft.Win32.Registry.CurrentUser
            .OpenSubKey("Software").OpenSubKey("IM Providers")
            .OpenSubKey("Communicator")
            .GetValue("UpAndRunning", 1)) == 2;
}
```

If Office Communicator is running, the *IsCommunicatorRunning()* method returns *True*.

7. Add the following code to the *Main()* method in the *Program* class.

```
static void Main(string[] args)
{
    // OC Automation API classes can't be instantiated unless
    // Office Communicator is running.
    if (IsCommunicatorRunning())
    {
        Console.WriteLine("Office Communicator is running.");

        _messenger = new Messenger();

        _messenger.OnAppShutdown += new
            DMessengerEvents_OnAppShutdownEventHandler(
            _messenger_OnAppShutdown);
        _messenger.OnSignin += new
            DMessengerEvents_OnSigninEventHandler(
            _messenger_OnSignin);
        _messenger.OnSignout += new
            DMessengerEvents_OnSignoutEventHandler(
            _messenger_OnSignout);

        // Check if the user is already signed in to
        // Office Communicator and sign in if they are not.
        if ((_messenger.MyStatus & MISTATUS.MISTATUS_ONLINE) ==
            MISTATUS.MISTATUS_ONLINE)
        {
            _signedIn = true;

            Console.WriteLine("Local user is signed in to
            Office Communicator");
        }
        else
        {
            _signedIn = false;
            Console.WriteLine("Local user is signed in to
            Office Communicator");

            Console.WriteLine("Signing in local user to Office
            Communicator via AutoSignin().");

            _messenger.AutoSignin();
        }
```

Chapter 3 Programming a Microsoft Office Communicator Automation API Application

```
    // Sign out of Office Communications Server
    _messenger.Signout();

    Console.WriteLine("\nPress Enter key to exit the
    application.\n");
    Console.ReadLine();

    _messenger.OnAppShutdown -= new
        DMessengerEvents_OnAppShutdownEventHandler(
        _messenger_OnAppShutdown);
    _messenger.OnSignin -= new
        DMessengerEvents_OnSigninEventHandler
        _messenger_OnSignin);
    _messenger.OnSignout -= new
        DMessengerEvents_OnSignoutEventHandler(
        _messenger_OnSignout);

    Marshal.ReleaseComObject(_messenger);
    _messenger = null;
}
else
{
    Console.WriteLine("Office Communicator is *not*
    running.");
    Console.WriteLine("Please start Office Communicator and
    run the application again.");

    Console.WriteLine("\nPress Enter key to exit the
    application.");
    Console.ReadLine();
}
}
```

In the preceding code, the *IsCommunciatorRunning()* method checks to see whether Office Communicator is running. If *IsCommunicatorRunning()* returns *True*, an instance of the *Messenger* class is created and event subscriptions are established for the *OnSignin*, *OnSignout*, and *AppShowdown* events. Next, *Messenger.MyStatus* checks the status of the local user and sets the *_signedIn* variable. If the local user is not signed in, *Messenger.AutoSignin()* signs the local user in to Office Communicator. *Messenger.Signout()* is called to sign out the local user. When the user exits the application, references to the *Messenger* class are released using *Marshal.ReleaseComObject()*.

8. Add the following code to the *Program* class to define the event delegates defined in *Main()*.

```
static void _messenger_OnSignout()
{
    _signedIn = false;
    Console.WriteLine("OnSignout()");
}
```

```
static void _messenger_OnSignin(int hr)
{
    if (hr == 0)
    {
        _signedIn = true;
        Console.WriteLine("OnSignin()");
    }
}

static void _messenger_OnAppShutdown()
{
    Console.WriteLine("OnAppShutdown()");
}
```

In the preceding code, the *_signedIn* variable is set to *False* when the local user is signed out. The variable is set to *True* when the user signs in. Note that the *_signedIn* variable does not need to be set when the *OnAppShutdown* event fires because that event is always preceded by the *OnSignout* event.

9. Run the application by clicking Debug on the Visual Studio menu, and then click Start Debugging.

10. Sign in to and sign out of Office Communicator, noting the effect of these actions in the console application.

11. Close the console application and save your work in Visual Studio.

This console application provides the logic to create Office Communicator API classes only when Office Communicator is running and manages a variable to keep track of the current signed-in state of the local user.

Working with Contact Information and Contact Presence

With the Office Communicator Automation API, you can display information from Office Communicator for the local user and her or his contacts. This information includes information such as name, phone number, and presence.

Displaying Local User Information

The *Messenger* class has properties and methods to provide programmatic access to information about the local user signed in to Office Communicator. Use these properties to display information about the local user in your applications. The *Messenger* class also provides events to notify your application of changes to this information so you can keep information up to date in your application.

Messenger.MySigninName Property

You can get the sign-in name (or SIP URI) for the local user by using the *Messenger.MySigninName* property. For example, the following code writes the local user's sign-in name to the console.

```
Messenger _messenger = new Messenger();

Console.WriteLine("MySigninName: {0}",
    _messenger.MySigninName);
```

Messenger.MyFriendlyName Property

You can get the friendly name (that is, a name readable by humans) for the local user using the *Messenger.MyFriendlyName* property. For example, the following code writes the local user's friendly name to the console.

```
Messenger _messenger = new Messenger();

Console.WriteLine("MyFriendlyName: {0}",
    _messenger.MyFriendlyName);
```

Messenger.OnMyFriendlyNameChange Event

If the friendly name of the local user changes, the *Messenger* class raises the OnMyFriendlyNameChange event. The following code shows how to subscribe to the OnMyFriendlyNameChange event by using the *DMessengerEvents_OnMyFriendlyNameChange-EventHandler* event handler.

```
Messenger _messenger = new Messenger();

_messenger.OnMyFriendlyNameChange += new DMessengerEvents_
OnMyFriendlyNameChangeEventHandler(_messenger_OnMyFriendlyNameChange);

...

        static void _messenger_OnMyFriendlyNameChange(int hr,
            string bstrPrevFriendlyName)
    {
        if (hr == 0)
        {
            Console.WriteLine(
                "OnMyFriendlyNameChange: PrevFriendlyName: {0}
                MyFriendlyName: {1}",
                bstrPrevFriendlyName,
                _messenger.MyFriendlyName);
        }
    }
```

When the *OnMyFriendlyNameChange* event is raised, an integer is passed that represents the success of the change (that is, a value equal to 0 represents success) and the previous friendly name of the local user is passed as a string.

Messenger.get_MyPhoneNumber Property

You can get the phone numbers for the local user using the *Messenger.get_MyPhoneNumber* property. An enumeration named *MPHONE_TYPE* is used to specify which phone number is returned by the method. For example, the following code writes the local user's phone numbers to the console.

```
Messenger _messenger = new Messenger();

Console.WriteLine("\tMyPhoneNumber (Work): {0}",
    _messenger.get_MyPhoneNumber(
    MPHONE_TYPE.MPHONE_TYPE_WORK));
Console.WriteLine("\tMyPhoneNumber (Mobile): {0}",
    _messenger.get_MyPhoneNumber(
    MPHONE_TYPE.MPHONE_TYPE_MOBILE));
Console.WriteLine("\tMyPhoneNumber (Home): {0}",
    _messenger.get_MyPhoneNumber(
    MPHONE_TYPE.MPHONE_TYPE_HOME));
Console.WriteLine("\tMyPhoneNumber (Other): {0}",
    _messenger.get_MyPhoneNumber(
    MPHONE_TYPE.MPHONE_TYPE_CUSTOM));
```

Messenger.OnMyPhoneChange Event

If the phone number of the local user changes, the *Messenger* class raises the *OnMyPhoneChange* event. The following code shows how to subscribe to the *OnMyPhoneChange* event by using the *DMessengerEvents_ OnMyPhoneChangeEventHandler* event handler.

```
_messenger = new Messenger();

_messenger.OnMyPhoneChange += new
    DMessengerEvents_OnMyPhoneChangeEventHandler(
    _messenger_OnMyPhoneChange);
```

...

```
static void _messenger_OnMyPhoneChange(MPHONE_TYPE PhoneType,
    string bstrNumber)
{
    Console.WriteLine("OnMyPhoneChange: PhoneType: {0} Number:
    {1}", PhoneType.ToString(), bstrNumber);
}
```

When the *OnMyPhoneChange* event is raised, a *MPHONE_TYPE* value is passed to specify which phone number changed (that is, work, mobile, home, or other) along with the new phone number as a string.

Putting It All Together

Using the concepts in this section, you can display local user information in your application and keep that information up to date easily by using the events raised by the *Messenger* class. In this section, you create a console application and use the *Messenger* class to display local user information and react to changes in that information.

1. Start Visual Studio (if it's not already running) and create a new Visual C# Windows console application named LocalUserInfo.

2. In Solution Explorer, right-click the References node, click Add References, and then, from the COM tab, add a reference to Microsoft Office Communicator 2007 API Type Library.

3. Open Program.cs and add the following *using* statements.

   ```
   using CommunicatorAPI;
   using System.Runtime.InteropServices;
   ```

   The preceding code brings the *Office Communicator Automation API* and *InteropServices* namespaces into scope.

4. Add the following declaration inside the *Program* class.

   ```
   private static Messenger _messenger;
   ```

5. Add the following code to the *Main()* method in the *Program* class.

   ```
   static void Main(string[] args)
   {
       _messenger = new Messenger();
       _messenger.OnMyFriendlyNameChange += new
           DMessengerEvents_OnMyFriendlyNameChangeEventHandler(
           _messenger_OnMyFriendlyNameChange);
       _messenger.OnMyPhoneChange += new
           DMessengerEvents_OnMyPhoneChangeEventHandler(
           _messenger_OnMyPhoneChange);

       Console.WriteLine("Local User Info for {0}:",
           _messenger.MyFriendlyName);
       Console.WriteLine("\tMySigninName: {0}",
           _messenger.MySigninName);
       Console.WriteLine("\tMyPhoneNumber (Work): {0}",
           _messenger.get_MyPhoneNumber(
           MPHONE_TYPE.MPHONE_TYPE_WORK));
       Console.WriteLine("\tMyPhoneNumber (Mobile): {0}",
           _messenger.get_MyPhoneNumber(
           MPHONE_TYPE.MPHONE_TYPE_MOBILE));
       Console.WriteLine("\tMyPhoneNumber (Home): {0}",
           _messenger.get_MyPhoneNumber(
           MPHONE_TYPE.MPHONE_TYPE_HOME));
   ```

```
            Console.WriteLine("\tMyPhoneNumber (Other): {0}",
                _messenger.get_MyPhoneNumber(
                MPHONE_TYPE.MPHONE_TYPE_CUSTOM));

            Console.WriteLine("\nPress Enter key to exit the
            application.\n");
            Console.ReadLine();

            _messenger.OnMyFriendlyNameChange -= new
                DMessengerEvents_OnMyFriendlyNameChangeEventHandler(
                _messenger_OnMyFriendlyNameChange);
            _messenger.OnMyPhoneChange -= new
                DMessengerEvents_OnMyPhoneChangeEventHandler(
                _messenger_OnMyPhoneChange);

            Marshal.ReleaseComObject(_messenger);
            _messenger = null;
        }
```

The preceding code uses the *Messenger* class to write the friendly name, sign-in name, and phone numbers for the local user.

6. Add the following code to Program.cs to define the event delegates for the local user information events.

```
        static void _messenger_OnMyPhoneChange(MPHONE_TYPE PhoneType,
            string bstrNumber)
        {
            Console.WriteLine("OnMyPhoneChange: PhoneType: {0}
            Number: {1}", PhoneType.ToString(), bstrNumber);
        }

        static void _messenger_OnMyFriendlyNameChange(int hr,
            string bstrPrevFriendlyName)
        {
            if (hr == 0)
            {
                Console.WriteLine("OnMyFriendlyNameChange:
                PrevFriendlyName: {0} MyFriendlyName: {1}",
                bstrPrevFriendlyName, _messenger.MyFriendlyName);
            }
        }
```

7. Run the application by clicking Debug on the Visual Studio menu, and then click Start Debugging.

8. Change the phone numbers listed in Office Communicator to see the result of the event being raised in the console application.

9. Close the console application and save your work in Visual Studio.

This console application provides the logic to display local user information and present updated information in the event of changes.

Retrieving Contact Information

The *IMessengerContactAdvanced* class represents a contact in Office Communicator and enables you to display contact information, such as the contact's name and phone numbers.

Messenger.GetContact() Method

To obtain an instance of *IMessengerContactAdvanced*, you pass the SIP URI of the contact to the *Messenger.GetContact()* method. For example, the following code gets an instance of *IMessengerContactAdvanced* for the contact with the SIP URI adamb@uc.contoso.com.

```
Messenger _messenger = new Messenger();

IMessengerContactAdvanced contact =
    (IMessengerContactAdvanced)
    _messenger.GetContact("sip:adamb@uc.contoso.com",
    _messenger.MyServiceId);
```

In the preceding code, note the use of the *Messenger.MyServiceID* property. This property provides a globally unique identifier of the Office Communications Server 2007 R2 instance that the local user has signed in to.

Getting Contact Information

The *IMessengerContactAdvanced* class supports properties to display the friendly name, sign-in name, and phone numbers of a contact in a way that resembles the *Messenger* class support of similar properties for the local user. For example, the following code displays contact information for the contact adamb@uc.contoso.com.

```
Messenger _messenger = new Messenger();

IMessengerContactAdvanced contact =
    (IMessengerContactAdvanced)
    _messenger.GetContact("sip:adamb@uc.contoso.com",
    _messenger.MyServiceId);

if (contact != null)
{
    Console.WriteLine("Contact Info for {0}:",
        contact.FriendlyName);
    Console.WriteLine("\tSigninName: {0}",
        contact.SigninName);

    try
    {
        Console.WriteLine("\tPhoneNumber (Work): {0}",
            contact.get_PhoneNumber(
            MPHONE_TYPE.MPHONE_TYPE_WORK));
        Console.WriteLine("\tPhoneNumber (Mobile): {0}",
            contact.get_PhoneNumber(
            MPHONE_TYPE.MPHONE_TYPE_MOBILE));
```

```
                Console.WriteLine("\tPhoneNumber (Home): {0}",
                    contact.get_PhoneNumber(
                    MPHONE_TYPE.MPHONE_TYPE_HOME));
                Console.WriteLine("\tPhoneNumber (Other): {0}",
                    contact.get_PhoneNumber(
                    MPHONE_TYPE.MPHONE_TYPE_CUSTOM));
            }
            catch
            {
                // Exception logic goes here.
            }
        }
```

> **Note** Office Communications Server 2007 R2 and Office Communicator 2007 R2 support the implementation of a new presence model named Enhanced Presence. With Enhanced Presence, the local user has access to contact information based on the level of access granted by the contact. For example, if the local user has been granted team-level access by a contact, the local user has access to the work and mobile phone numbers, but not the home phone number of the contact. When calling *IMessengerContactAdvanced.get_PhoneNumber* on a contact, the method throws an exception when trying to access the home phone number. Therefore, it is important to wrap such calls in a *try/catch* block. For more information about the different levels of access defined by Office Communicator, see Office Communicator Help at *http://go.microsoft.com/fwlink/?linkid=143210*.

Messenger.OnContactFriendlyNameChange Event

If the friendly name of a contact in the local user's contact list changes, the *Messenger* class raises the *Messenger.OnContactFriendlyNameChange* event. The following code shows how to subscribe to this event by using the *DMessengerEvents_OnContactFriendlyNameChangeEventHandler* event handler.

```
        Messenger _messenger = new Messenger();

        _messenger.OnContactFriendlyNameChange += new
DMessengerEvents_OnContactFriendlyNameChangeEventHandler(
_messenger_OnContactFriendlyNameChange);

    ...

        static void _messenger_OnContactFriendlyNameChange(int hr,
            object pMContact, string bstrPrevFriendlyName)
        {
            IMessengerContactAdvanced contact =
                (IMessengerContactAdvanced)pMContact;
```

```
            if (hr == 0)
            {
                Console.WriteLine("OnMyFriendlyNameChange for {0}:
                    \n\tPrevFriendlyName: {1} FriendlyName: {2}",
                        contact.FriendlyName, bstrPrevFriendlyName,
                        contact.FriendlyName);
            }
        }
```

When the *OnContactFriendlyNameChange* event is raised, an instance of *IMessengerContactAdvanced* is passed as an object in the *pMContact* parameter. By casting this parameter to be of type *IMessengerContactAdvanced*, you provide access to a reference of the contact.

> **Note** The *Messenger.OnContactFriendlyNameChange* event fires only for contacts in the local user's contact list in Office Communicator.

Messenger.OnContactPhoneChange Event

If the phone number for a contact in the local user's contact list changes, the *Messenger* class raises the *Messenger.OnContactPhoneChange* event. The following code shows how to subscribe to this event using the *DMessengerEvents_OnContactPhoneChangeEventHandler* event handler.

```
            Messenger _messenger = new Messenger();

            _messenger.OnContactPhoneChange += new
DMessengerEvents_OnContactPhoneChangeEventHandler(
_messenger_OnContactPhoneChange);
```

...

```
        static void _messenger_OnContactPhoneChange(int hr,
            object pContact, MPHONE_TYPE PhoneType, string bstrNumber)
        {
            IMessengerContactAdvanced contact =
                (IMessengerContactAdvanced)pContact;

            if (hr == 0)
            {
                Console.WriteLine("OnContactPhoneChange for {0}:
                    \n\tPhoneType: {1} Number: {2}", contact.FriendlyName,
                    PhoneType.ToString(), bstrNumber);
            }
        }
```

When the *OnContactPhoneChange* event is raised, an instance of *IMessengerContactAdvanced* is passed as an object in the *pContact* parameter. By casting this parameter to be of type *IMessengerContactAdvanced*, you provide access to a reference of the contact.

> **Note** The *Messenger.OnContactPhoneChange* event fires only for contacts in the local users contact list in Office Communicator. Furthermore, the *Messenger.OnContactPhoneChange* event fires only for phone numbers that the local user has access to by using Enhanced Presence. For example, the *Messenger.OnContactPhoneChange* event fires for changes to the work and mobile numbers if the local user has been granted team-level access by the contact and does not fire when the contact changes her or his home phone number.

Putting It All Together

Using the concepts in this section, you can add contact information to your application and keep that information up to date easily by using the events raised by the *Messenger* class. In this section, you create a console application and use the *Messenger* class and its *IMessengerContactAdvanced* interface to display contact information by performing the following steps:

1. Start Visual Studio (if it's not already running) and create a new Visual C# Windows console application named ContactInfo.

2. In Solution Explorer, right-click the References node, click Add References, and then, from the COM tab, add a reference to Microsoft Office Communicator 2007 API Type Library.

3. Open Program.cs and add the following *using* statements.

   ```
   using CommunicatorAPI;
   using System.Runtime.InteropServices;
   ```

4. Add the following declaration inside the *Program* class.

   ```
   private static Messenger _messenger;
   ```

   The *_messenger* member variable provides an instance of the *Messenger* class.

5. Add the following code to the *Main()* method in the *Program* class.

   ```
   static void Main(string[] args)
   {
       _messenger = new Messenger();

       _messenger.OnContactFriendlyNameChange += new
   DMessengerEvents_OnContactFriendlyNameChangeEventHandler(
   _messenger_OnContactFriendlyNameChange);
       _messenger.OnContactPhoneChange += new
   DMessengerEvents_OnContactPhoneChangeEventHandler(
   _messenger_OnContactPhoneChange);
   ```

```csharp
        IMessengerContactAdvanced contact =
            (IMessengerContactAdvanced)_messenger.GetContact(
            "sip:adamb@uc.contoso.com", _messenger.MyServiceId);

        if (contact != null)
        {
            Console.WriteLine("Contact Info for {0}:",
                contact.FriendlyName);
            Console.WriteLine("\tSigninName: {0}",
                contact.SigninName);

            try
            {
                Console.WriteLine("\tPhoneNumber (Work): {0}",
                    contact.get_PhoneNumber(
                    MPHONE_TYPE.MPHONE_TYPE_WORK));
                Console.WriteLine("\tPhoneNumber (Mobile): {0}",
                    contact.get_PhoneNumber(
                    MPHONE_TYPE.MPHONE_TYPE_MOBILE));
                Console.WriteLine("\tPhoneNumber (Home): {0}",
                    contact.get_PhoneNumber(
                    MPHONE_TYPE.MPHONE_TYPE_HOME));
                Console.WriteLine("\tPhoneNumber (Other): {0}",
                    contact.get_PhoneNumber(
                    MPHONE_TYPE.MPHONE_TYPE_CUSTOM));
            }
            catch
            {
                //Your exception logic goes here.
            }
        }

        Console.WriteLine("\nPress Enter key to exit the
            application.\n");
        Console.ReadLine();

            _messenger.OnContactFriendlyNameChange -= new
            DMessengerEvents_OnContactFriendlyNameChangeEventHandler(
            _messenger_OnContactFriendlyNameChange);
          _messenger.OnContactPhoneChange -= new
            DMessengerEvents_OnContactPhoneChangeEventHandler(
            _messenger_OnContactPhoneChange);

        Marshal.ReleaseComObject(_messenger);
        _messenger = null;
        Marshal.ReleaseComObject(_contact);
        _contact = null;
    }
```

The preceding code uses *Messenger.GetContact()* to get an instance of the *IMessengerContactAdvanced* interface of a contact with the SIP URI adamb@uc.contoso.com. Then, the code writes contact information to the console. Also, the *Messenger* class is used to subscribe to event changes to the contact's information.

6. Add the following code to Program.cs to define the event delegates for the contact information.

```
static void _messenger_OnContactPhoneChange(int hr,
    object pContact, MPHONE_TYPE PhoneType, string bstrNumber)
{
    if (hr == 0)
    {
        Console.WriteLine("OnContactPhoneChange for {0}:
            \n\tPhoneType: {1} Number: {2}", contact.FriendlyName,
            PhoneType.ToString(), bstrNumber);
    }
}

static void _messenger_OnContactFriendlyNameChange(int hr,
    object pMContact, string bstrPrevFriendlyName)
{
    IMessengerContactAdvanced contact =
        (IMessengerContactAdvanced)pMContact;

    if (hr == 0)
    {
        Console.WriteLine("OnMyFriendlyNameChange for {0}:
        \n\tPrevFriendlyName: {1} FriendlyName: {2}",
            contact.FriendlyName, bstrPrevFriendlyName,
            contact.FriendlyName);
    }
}
```

In each delegate implementation, note that an instance of *IMessengerContactAdvanced* is passed as an object in the *pMContact* parameter. Cast this *pMContact* to *IMessenger-ContactAdvanced* to use that instance.

7. Run the application by clicking Debug on the Visual Studio menu, and then click Start Debugging.

8. Change the Work phone number for a contact in the local user's contact list.

9. Close the console application and save your work in Visual Studio.

This console application provides the logic to display contact information for a contact and changes to contact information for any contact in the local user's contact list. By running the preceding code, you display the information for the contact with the sign-in name adamb@uc.contoso.com and display changes to the friendly name or phone number for any contact in the user's contact list.

Publishing and Subscribing to Contact Presence

The *IMessengerContactAdvanced* interface provides access to presence information for both the local user and his or her contacts. The *Messenger* class provides events that are raised when presence information changes, which enables your application to subscribe to presence

Chapter 3 Programming a Microsoft Office Communicator Automation API Application

for the local user and his or her contacts. By using both of these features of the API, you can display up-to-date presence information in your application.

When using Office Communicator, you can see elements of presence for the local user, as shown in Figure 3-2.

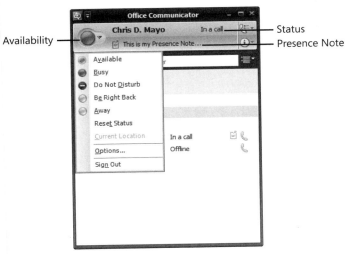

FIGURE 3-2 The user's presence information in Office Communicator.

Similarly, you can see presence information for the user's contacts in the contact list within Office Communicator.

IMessengerContactAdvanced.PresenceProperties Property

Use the *IMessengerContactAdvanced.PresenceProperties* property to access presence information for any contact, including the local user. For example, the following code gets the presence information for the local user and then writes the status to the console.

```
IMessengerContactAdvanced _contact =
    (IMessengerContactAdvanced)
    _messenger.GetContact(_messenger.MySigninName,
    _messenger.MyServiceId);

object[] presenceProps =
    object[])_contact.PresenceProperties;

// Sign In Status.
Console.WriteLine("Local User Status: {0}",
    (MISTATUS)presenceProps[
    (int)PRESENCE_PROPERTY.PRESENCE_PROP_MSTATE]);
```

IMessengerContactAdvanced.PresenceProperties returns an array of objects that holds each of the presence values for a contact. The array can be indexed by using the *PRESENCE_PROPERTY* enumeration. Table 3-2 explains each *PRESENCE_PROPERTY* enumeration value.

TABLE 3-2 *PRESENCE_PROPERTY* Enumeration Values

Element	Explanation
PRESENCE_PROP_MSTATE	Gives the status of the contact as a value in the enumeration *MISTATUS*. Read-only.
PRESENCE_PROP_AVAILABILITY	Gives the availability of the contact as an integer. Read/write for the local user. Read-only for contacts.
PRESENCE_PROP_IS_BLOCKED	Represents whether the contact is blocked by the local user. Boolean value.
PRESENCE_PROP_PRESENCE_NOTE	Presence note as displayed in Office Communicator as a string. Read/write for the local user; read-only for contacts.
PRESENCE_PROP_IS_OOF	Represents whether the contact has set their out of office state. Boolean value. Read-only.
PRESENCE_PROP_TOOL_TIP	Tooltip displayed in Office Communicator for a contact as a string. Read-only.
PRESENCE_PROP_CUSTOM_STATUS_STRING	Custom status string for the contact as a string. Read-only.

IMessengerContactAdvanced.PresenceProperties returns presence information for any contact, but it returns current presence information only for contacts in the local user's contact list. If the contact is not in the user's contact list, calling *IMessengerContactAdvanced. PresenceProperties* returns presence information that is up to two minutes old. To get up-to-date presence information, add the contact to the local user's contact list. Calls to *IMessenger-ContactAdvanced.PresenceProperties* return only presence information based on the level of access granted to the user by the contact.

You can use *IMessengerContactAdvanced.PresenceProperties* to publish presence information for the local user. For example, the following code sets the availability for the signed-in user to *available*.

```
IMessengerContactAdvanced _localUser =
    (IMessengerContactAdvanced)_messenger.GetContact(
    _messenger.MySigninName, _messenger.MyServiceId);

// Set the local user to available (3000).
object[] _presProps = new object[8];

_presProps[(int)PRESENCE_PROPERTY.PRESENCE_PROP_AVAILABILITY]
    = 3000;

_localUser.PresenceProperties = (object)_presProps;
```

Note *IMessengerContactAdvanced.PresenceProperties* can set presence information for the settings that the local user can set from Office Communicator.

Messenger.OnMyStatusChange Event

The *Messenger* class raises the *OnMyStatusChange* event if any element of the local user's presence changes. The following code shows how to subscribe to the *OnMyStatusChange* event by using the *DMessengerEvents_OnMyStatusChangeEventHandler* event handler.

```
Messenger _messenger = new Messenger();

_messenger.OnMyStatusChange += new
    DMessengerEvents_OnMyStatusChangeEventHandler(
    _messenger_OnMyStatusChange);
```
...
```
static void _messenger_OnMyStatusChange(int hr,
    MISTATUS mMyStatus)
{
    // Your code to work with presence goes here...
}
```

Note The *Messenger.OnMyStatusChange* event passes a parameter, *mMyStatus*, of type *MISTATUS* when the event is raised. The *mMyStatus* parameter specifies the presence status of the local user. Because the *Messenger.OnMyStatusChange* event is raised when any element of the local user's presence changes (for example, the local user's presence note), use the *IMessengerContactAdvanced.PresenceProperties* property rather than *mMyStatus* because *PresenceProperties* provides full access to the presence information for a contact. For details, see step 7 in the next "Putting It All Together" section.

Messenger.OnContactStatusChange Event

The *Messenger* class raises the *OnContactStatusChange* event if any element of presence for a contact in the local user's contact list changes. The following code shows how to subscribe to the *OnContactStatusChange* event by using the *DMessengerEvents_OnContactStatusChange-EventHandler* event handler.

```
Messenger _messenger = new Messenger();

_messenger.OnContactStatusChange += new
    DMessengerEvents_OnContactStatusChangeEventHandler(
    _messenger_OnContactStatusChange);
```
...
```
static void _messenger_OnContactStatusChange(object pMContact,
    MISTATUS mStatus)
{
    // Your code to work with presence goes here...
}
```

Note The *MessengerOnContactStatusChange* event fires only for contacts in the local user's contact list.

Tip Like *Messenger.OnMyStatusChange*, *Messenger.OnContactStatusChange* passes a parameter *mStatus* of type *MISTATUS* when the event is raised. Use the *IMessengerContactAdvanced. PresenceProperties* property rather than *mMyStatus* when subscribing to presence information.

Putting It All Together

By using the concepts in this section, you can add local user and contact presence information to your application by using *IMessengerContactAdvanced* and keep that information up to date easily by using the events raised by the *Messenger* class. In this section, you create a console application, use the *Messenger* class and *IMessengerContactAdvanced* interface to display presence information for the local user and a contact, and display changes to presence information for contacts in the local user's contact list, as follows:

1. Start Visual Studio (if it's not already running) and create a new Visual C# Windows console application named ContactPresence.

2. In Solution Explorer, right-click the References node, click Add References, and then, from the COM tab, add references to Microsoft Office Communicator 2007 API Type Library.

3. Open Program.cs and add the following *using* statements.

   ```
   using CommunicatorAPI;
   using System.Runtime.InteropServices;
   ```

4. Add the following declaration inside the *Program* class.

   ```
           private static Messenger _messenger;
   ```

   The *_messenger* member variable provides an instance of the *Messenger* class.

5. Add the following code to the *Main()* method in the *Program* class.

   ```
                   static void Main(string[] args)
                   {
                       _messenger = new Messenger();

                       _messenger.OnContactStatusChange += new
                           DMessengerEvents_OnContactStatusChangeEventHandler(
                           _messenger_OnContactStatusChange);
                       _messenger.OnMyStatusChange += new
                           DMessengerEvents_OnMyStatusChangeEventHandler(
                           _messenger_OnMyStatusChange);
   ```

Chapter 3 Programming a Microsoft Office Communicator Automation API Application

```
        IMessengerContactAdvanced _localUser =
            (IMessengerContactAdvanced)_messenger.GetContact(
            _messenger.MySigninName, _messenger.MyServiceId);

        if (_localUser != null)
        {
            DisplayContactPresence(_localUser);
        }

        // Set the local user to Available (3000) in code.
        object[] _presProps = new object[8];
        _presProps[
            (int)PRESENCE_PROPERTY.PRESENCE_PROP_AVAILABILITY]
            = 3000;

        _localUser.PresenceProperties = (object)_presProps;

        IMessengerContactAdvanced _contact =
            (IMessengerContactAdvanced)_messenger.GetContact(
            "adamb@uc.contoso.com", _messenger.MyServiceId);

        if (_contact != null)
        {
            DisplayContactPresence(_contact);
        }

        Console.WriteLine("\nPress Enter key to exit the
        application.\n");
        Console.ReadLine();

        _messenger.OnContactStatusChange -= new
            DMessengerEvents_OnContactStatusChangeEventHandler(
            _messenger_OnContactStatusChange);
        _messenger.OnMyStatusChange -= new
            DMessengerEvents_OnMyStatusChangeEventHandler(
            _messenger_OnMyStatusChange);

        Marshal.ReleaseComObject(_messenger);
        _messenger = null;
        Marshal.ReleaseComObject(_localUser);
        _localUser = null;
        Marshal.ReleaseComObject(_contact);
        _contact = null;
    }
```

In the preceding code, the *Messenger* class is instantiated and used to subscribe to status event changes for the local user (that is, by using the *OnMyStatusChange* event) and the user's contacts (that is, by using the *OnContactStatusChange* event). Next, you use *Messenger.GetContact()* to get an instance of the local user by using the *IMessenger-ContactAdvanced* interface to display presence information. You use the *IMessenger-ContactAdvanced.PresenceProperties* property to publish presence for the local user. You call *Messenger.GetContact()* again to get an instance of a contact by using the *IMessengerContactAdvanced* interface to display presence information for the contact.

6. Add the following code to Program.cs to define the event delegates for the presence information events.

```
static void _messenger_OnMyStatusChange(int hr,
    MISTATUS mMyStatus)
{
    DisplayContactPresence((IMessengerContactAdvanced)
        _messenger.GetContact(_messenger.MySigninName,
        _messenger.MyServiceId));
}

static void _messenger_OnContactStatusChange(object pMContact,
    MISTATUS mStatus)
{
    DisplayContactPresence((IMessengerContactAdvanced)
        pMContact);
}
```

7. Implement the *DisplayContactPresence* method in the *Program* class by using the following code.

```
static void DisplayContactPresence(
    IMessengerContactAdvanced contact)
{
    object[] _presenceProps = (object[])
        contact.PresenceProperties;

    if (contact.IsSelf)
    {
        Console.WriteLine("Local User Presence Info for {0}:",
            contact.FriendlyName);
    }
    else
    {
        Console.WriteLine("Contact Presence Info for {0}:",
            contact.FriendlyName);
    }

    // Status.
    Console.WriteLine("\tStatus: {0}", (MISTATUS) _presenceProps[
        (int)PRESENCE_PROPERTY.PRESENCE_PROP_MSTATE]);
    Console.WriteLine("\tStatus String: {0}",
        GetStatusString((MISTATUS) _presenceProps[
        (int)PRESENCE_PROPERTY.PRESENCE_PROP_MSTATE]));
    // Status string if status is set to custom.
    Console.WriteLine("\tCustom Status String: {0}",
        _presenceProps[(int)
        PRESENCE_PROPERTY.PRESENCE_PROP_CUSTOM_STATUS_STRING]);

    // Presence or User state.
    Console.WriteLine("\tAvailability: {0}",
        _presenceProps[
        (int)PRESENCE_PROPERTY.PRESENCE_PROP_AVAILABILITY]);
    Console.WriteLine("\tAvailability String: {0}",
        GetAvailabilityString((int) _presenceProps[(int)
        PRESENCE_PROPERTY.PRESENCE_PROP_AVAILABILITY]));
```

```
            // Presence note.
            Console.WriteLine("\tPresence Note: \n'{0}'",
                _presenceProps[(int)
                PRESENCE_PROPERTY.PRESENCE_PROP_PRESENCE_NOTE]);
            // Blocked status.
            Console.WriteLine("\tIs Blocked: {0}", _presenceProps[
                (int)PRESENCE_PROPERTY.PRESENCE_PROP_IS_BLOCKED]);
            // OOF message for contact, if specified.
            Console.WriteLine("\tIs OOF: {0}", _presenceProps[
                (int)PRESENCE_PROPERTY.PRESENCE_PROP_IS_OOF]);
            // Tooltip.
            Console.WriteLine("\tTool Tip: \n'{0}'\n", _presenceProps[
                (int)PRESENCE_PROPERTY.PRESENCE_PROP_TOOL_TIP]);
        }
```

The preceding code uses the *IMessengerContactAdvanced.PresenceProperties* property to retrieve the presence information for a contact and local user.

8. Implement the *GetAvailabilityString* and *GetStatusString* methods by adding the following code to the *Program* class.

```
        static string GetAvailabilityString(int availability)
        {
            switch (availability)
            {
                case 3000:
                    return "Available";
                case 4500:
                    return "Inactive";
                case 6000:
                    return "Busy";
                case 7500:
                    return "Busy-Idle";
                case 9000:
                    return "Do not disturb";
                case 12000:
                    return "Be right back";
                case 15000:
                    return "Away";
                case 18000:
                    return "Offline";
                default:
                    return "";
            }
        }

        static string GetStatusString(MISTATUS mStatus)
        {
            switch (mStatus)
            {
                case MISTATUS.MISTATUS_ALLOW_URGENT_INTERRUPTIONS:
                    return "Urgent interruptions only";
                case MISTATUS.MISTATUS_AWAY:
                    return "Away";
                case MISTATUS.MISTATUS_BE_RIGHT_BACK:
                    return "Be right back";
```

```
                    case MISTATUS.MISTATUS_BUSY:
                        return "Busy";
                    case MISTATUS.MISTATUS_DO_NOT_DISTURB:
                        return "Do no disturb";
                    case MISTATUS.MISTATUS_IDLE:
                        return "Idle";
                    case MISTATUS.MISTATUS_INVISIBLE:
                        return "Invisible";
                    case MISTATUS.MISTATUS_IN_A_CONFERENCE:
                        return "In a conference";
                    case MISTATUS.MISTATUS_IN_A_MEETING:
                        return "In a meeting";
                    case MISTATUS.MISTATUS_LOCAL_CONNECTING_TO_SERVER:
                        return "Connecting to server";
                    case MISTATUS.MISTATUS_LOCAL_DISCONNECTING_FROM_SERVER:
                        return "Disconnecting from server";
                    case MISTATUS.MISTATUS_LOCAL_FINDING_SERVER:
                        return "Finding server";
                    case MISTATUS.MISTATUS_LOCAL_SYNCHRONIZING_WITH_SERVER:
                        return "Synchronizing with server";
                    case MISTATUS.MISTATUS_MAY_BE_AVAILABLE:
                        return "Inactive";
                    case MISTATUS.MISTATUS_OFFLINE:
                        return "Offline";
                    case MISTATUS.MISTATUS_ONLINE:
                        return "Online";
                    case MISTATUS.MISTATUS_ON_THE_PHONE:
                        return "In a call";
                    case MISTATUS.MISTATUS_OUT_OF_OFFICE:
                        return "Out of office";
                    case MISTATUS.MISTATUS_OUT_TO_LUNCH:
                        return "Out to lunch";
                    case MISTATUS.MISTATUS_UNKNOWN:
                        return "Unknown";
                    default:
                        return string.Empty;
                }
            }
```

The preceding code returns the availability and status information of a contact in human-readable form as a string.

9. Run the application by clicking Debug on the Visual Studio menu, and then click Start Debugging.

10. Change the presence information for the local user and one of the local user's contacts.

11. Close the console application and save your work in Visual Studio.

This console application provides the logic to display presence information for the local user and her or his contacts. For example, by running the preceding code, you display presence information for the local user and one of her or his contacts and then change the availability for the local user to *Online*.

Working with the Office Communicator Contact List

The *IMessengerContacts* interface provides access to the user's contact list in Office Communicator 2007 R2, as well as the ability to remove contacts from the contact list. You use the *Messenger* class to add contacts to the contact list.

Messenger.MyContacts Property

A call to the *Messenger.MyContacts* property returns an instance of the *IMessengerContacts* interface. For example, the following code gets an instance of *IMessengerContacts* by using the *Messenger* class.

```
_messenger = new Messenger();

IMessengerContacts _contacts = (IMessengerContacts)
    _messenger.MyContacts;
```

IMessengerContacts.Item() Method

The *IMessengerContacts* interface provides a collection of *IMessengerContactAdvanced* instances, one for each contact in the user's contact list. You use the *IMessengerContacts.Count* property and *IMessengerContacts.Item()* to iterate though this collection, as illustrated in the following code.

```
_messenger = new Messenger();

IMessengerContacts _contacts =
    (IMessengerContacts)_messenger.MyContacts;

IMessengerContactAdvanced _contact;

Console.WriteLine("Contact list for the local user:");

for (int i = 0; i < _contacts.Count; i++)
{
    _contact = (IMessengerContactAdvanced) _contacts.Item(i);
    Console.WriteLine("\t{0}", _contact.FriendlyName);
}
```

Messenger.AddContact() Method

Use the *Messenger.AddContact()* method to add contacts to the local user's contact list. For example, the following code adds a contact with the SIP URI amys@uc.contoso.com.

```
_messenger = new Messenger();

// Adding a contact to the contact list.
_messenger.AddContact(0, "amys@uc.contoso.com");
```

Note In the call to the *Messenger.AddContact()* method, the first parameter, *hwndParent*, is reserved and always set to 0.

IMessengerContacts.Remove() Method

Contacts are removed from the user's contact list by using the *IMessengerContacts.Remove()* method. For example, the following code removes a contact with the SIP URI jamesa@uc.contoso.com from the contact list.

```
Messenger _messenger = new Messenger();

IMessengerContacts _contacts = (IMessengerContacts)_messenger.MyContacts;

// Removing a contact from the contact list.
_contacts.Remove((IMessengerContactAdvanced)
    _messenger.GetContact("jamesa@uc.contoso.com",
    _messenger.MyServiceId));
```

Messenger.OnContactListAdd Event

IMessengerContacts is a static collection representing the local user's contact list when *Messenger.MyContacts* is called. If contacts are added or removed from the contact list, the *IMessengerContacts* instance is not updated. To be notified when a contact is added to the user's contact list, you use the *Messenger.OnContactListAdd* event to receive change notifications. For example, the following code subscribes to this event by using the *DMessengerEvents_OnContactListAddEventHandler* event handler.

```
Messenger _messenger = new Messenger();

_messenger.OnContactListAdd += new
    DMessengerEvents_OnContactListAddEventHandler(
    _messenger_OnContactListAdd);
```

...

```
static void _messenger_OnContactListAdd(int hr,
    object pMContact)
{
    IMessengerContactAdvanced _contact =
        (IMessengerContactAdvanced)pMContact;

    if (hr == 0)
    {
        Console.WriteLine("OnContactListAdd: {0}",
            _contact.FriendlyName);
    }
}
```

MessengerOnContactListRemove Event

To be notified when a contact is removed from the user's contact list, you use the *Messenger. OnContactListRemove* event. For example, the following code uses the *DMessengerEvents_ OnContactListRemoveEventHandler* to subscribe this event.

```
Messenger _messenger = new Messenger();

_messenger.OnContactListRemove += new
    DMessengerEvents_OnContactListRemoveEventHandler(
    _messenger_OnContactListRemove);
```

...

```
static void _messenger_OnContactListRemove(int hr,
    object pMContact)
{
    IMessengerContactAdvanced _contact =
        (IMessengerContactAdvanced) pMContact;

    if (hr == 0)
    {
        Console.WriteLine("OnContactListRemove: {0}",
            _contact.FriendlyName);
    }
}
```

Putting It All Together

Using the concepts in this section, you can display the contact list, add and remove contacts from that list, and subscribe to events raised by changes to the contact list easily by using the *Messenger* class and the *IMessengerContacts* interface. In this section, you create a console application that displays the contact list and changes to the contact list. The code also adds and removes contacts from the contact list to raise these events.

1. Start Visual Studio (if it's not already running) and create a new Visual C# Windows console application named ContactList.

2. In Solution Explorer, right-click the References node, click Add References, and then, from the COM tab, add references to Microsoft Office Communicator 2007 API Type Library.

3. Open Program.cs and add the following *using* statements.

    ```
    using CommunicatorAPI;
    using System.Runtime.InteropServices;
    ```

4. Add the following declaration inside the *Program* class.

    ```
    private static Messenger _messenger;
    ```

    The *_messenger* member variable provides an instance of the *Messenger* class.

5. Add the following code to the *Main()* method in the *Program* class.

```csharp
static void Main(string[] args)
{
    _messenger = new Messenger();

    _messenger.OnContactListAdd += new
        DMessengerEvents_OnContactListAddEventHandler(
        _messenger_OnContactListAdd);
    _messenger.OnContactListRemove += new
        DMessengerEvents_OnContactListRemoveEventHandler(
        _messenger_OnContactListRemove);

    IMessengerContacts contacts = (IMessengerContacts)
        _messenger.MyContacts;
    IMessengerContactAdvanced contact = null;

    Console.WriteLine("Contact list for the local user:");

    for (int i = 0; i < contacts.Count; i++)
    {
        contact = (IMessengerContactAdvanced) contacts.Item(i);
        Console.WriteLine("\t{0}", contact.FriendlyName);
    }

    // Adding a contact to the contact list.
    _messenger.AddContact(0, "amys@uc.contoso.com");

    // Removing a contact from the contact list.
    contacts.Remove((IMessengerContactAdvanced)
        _messenger.GetContact("jamesa@uc.contoso.com",
        _messenger.MyServiceId));

    Console.WriteLine("\nPress Enter key to exit the
        application.\n");
    Console.ReadLine();

    _messenger.OnContactListAdd += new
        DMessengerEvents_OnContactListAddEventHandler(
        _messenger_OnContactListAdd);
    _messenger.OnContactListRemove += new
        DMessengerEvents_OnContactListRemoveEventHandler(
        _messenger_OnContactListRemove);

    Marshal.ReleaseComObject(_messenger);
    _messenger = null;
    Marshal.ReleaseComObject(contacts);
    contacts = null;
    Marshal.ReleaseComObject(contact);
    contact = null;
}
```

The preceding code uses *IMessengerContacts* to write the friendly name of each contact in the contact list to the console. You use *Messenger.AddContact()* to add a contact to the contact list and *IMessengerContacts.Remove()* to remove a contact.

Chapter 3 Programming a Microsoft Office Communicator Automation API Application

6. Add the following code to the *Program* class to write the friendly name of the contact when the contact is added or removed.

```
static void _messenger_OnContactListRemove(int hr,
    object pMContact)
{
    IMessengerContactAdvanced contact =
        (IMessengerContactAdvanced) pMContact;

    if (hr == 0)
    {
        Console.WriteLine("OnContactListRemove: {0}",
            contact.FriendlyName);
    }
}

static void _messenger_OnContactListAdd(int hr,
    object pMContact)
{
    IMessengerContactAdvanced contact =
        (IMessengerContactAdvanced)pMContact;

    if (hr == 0)
    {
        Console.WriteLine("OnContactListAdd: {0}",
            contact.FriendlyName);
    }
}
```

7. Run the application by clicking Debug on the Visual Studio menu, and then click Start Debugging.

8. Verify that the contact list for the local user is written to the console and the contacts specified are added and removed from the contact list.

9. Close the console application and save your work in Visual Studio.

This console application provides the logic to display the contact list and show the results of adding and removing contacts from the contact list.

Starting Conversations

By using the Office Communicator Automation API, you can start conversations on behalf of the local user. Conversation modalities can include IM, voice, and video.

Using the *IMessengerAdvanced* Interface

As described in Chapter 2, "Microsoft Unified Communications APIs Foundation," *IMessengerAdvanced* is an interface implemented by the *Messenger* class.

Part II Office Communicator Automation API

To get a reference to *IMessengerAdvanced*, you need to create an instance of the *Messenger* class and cast it to *IMessengerAdvanced* as shown in the following code.

```
Messenger _messenger = new Messenger();
IMessengerAdvanced _messengerAdv =
    (IMessengerAdvanced)_messenger;
```

IMessengerAdvanced, like all classes and interfaces in the API, is an unmanaged COM type that you access in managed code by using COM Interop. This means that you must release each reference to the *IMessengerAdvanced* interface explicitly. You release the reference by calling the *System.Runtime.InteropServices.Marshal.ReleaseComObject()* method and setting the reference to *NULL*, as shown in the following code.

```
Marshal.ReleaseComObject(_messenger);
_messenger = null;
Marshal.ReleaseComObject(_messengerAdv);
_messengerAdv = null;
```

If you fail to call *Marshal.ReleaseComObject()* and set the reference to *NULL*, your application does not release resources allocated by the API and results in memory leaks.

> **Note** This manner of releasing references applies to all references created by using the Office Communicator Automation API.

IMessengerAdvanced.StartConversation() Method

The *IMessengerAdvanced* interface provides the *StartConversation()* method for creating conversations with one or more contacts using a specific modality. For example, the following code starts an audio conference with the contacts specified in the *sipUris* array and sets the conversation window title to "My Conversation."

```
// An object array of SIP URIs strings for the participants
//   in the conversation.
object[] sipUris = new object[] { "adamb@uc.contoso.com",
    "rl@uc.contoso.com" };

_messengerAdv.StartConversation(
    // The conversation modality (audio in this case).
    CONVERSATION_TYPE.CONVERSATION_TYPE_AUDIO,
    // List of participants.
    sipUris,
    // Not supported.
    null,
    // The conversation window title as as string.
    "My Audio Conversation",
    // Not supported.  Specify "1".
    "1",
    // Not supported. Specify NULL
    null);
```

Figure 3-3 shows the results of this code, an audio conference.

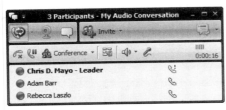

FIGURE 3-3 An audio conference started with *StartConversation()*.

The *sipUris* variable is an object array that specifies the SIP URIs of the participants. By passing a single SIP URI in the array, you create a conversation with the contact corresponding to that SIP URI. By passing multiple SIP URIs, you start a conference with all of the contacts specified. The conversation modality is specified by using the CONVERSATION_TYPE enumeration. Table 3-3 provides an explanation of the values supported in CONVERSATION_TYPE.

TABLE 3-3 *CONVERSATION_TYPE* **Enumeration Values**

Value	Explanation
CONVERSATION_TYPE_IM	Specifies IM communication
CONVERSATION_TYPE_PHONE	Specifies a phone call by using telephone URIs
CONVERSATION_TYPE_LIVEMEETING	Not supported
CONVERSATION_TYPE_AUDIO	Specifies a phone call by using Voice over Internet Protocol (VoIP), Public Switched Telephone Network (PSTN), or both
CONVERSATION_TYPE_VIDEO	Specifies a video conversation
CONVERSATION_TYPE_PSTN	Specifies a phone call by using PSTN

Putting It All Together

By using *IMessengerAdvanced.StartConversation()*, you can start conversations easily on behalf of the local user. In this section, you create a console application that creates an audio conference programmatically.

1. Start Visual Studio (if it's not already running) and create a new Visual C# Windows console application named StartConversation.

2. In Solution Explorer, right-click the References node, click Add References, and then, from the COM tab, add references to Microsoft Office Communicator 2007 API Type Library.

3. Open Program.cs and add the following *using* statements.

   ```
   using CommunicatorAPI;
   using System.Runtime.InteropServices;
   ```

4. Add the following declarations inside the *Program* class.

   ```
   private static Messenger _messenger;
   private static IMessengerAdvanced _messengerAdv;
   ```

   The *_messenger* member variable provides an instance of the *Messenger* class and *_messengerAdv* provides a reference to the *IMessengerAdvanced* interface.

5. Add the following code to the *Main()* method in the *Program* class.

   ```
   static void Main(string[] args)
   {
       _messenger = new Messenger();
       _messengerAdv = (IMessengerAdvanced)_messenger;

       // An object array of SIP URIs strings for the participants
       //   in the conversation.
       object[] sipUris = new object[] { "adamb@uc.contoso.com",
           "rl@uc.contoso.com" };

       _messengerAdv.StartConversation(
           // The conversation modality (audio in this case).
           CONVERSATION_TYPE.CONVERSATION_TYPE_AUDIO,
           // The participants.
           sipUris,
           // Not supported.
           null,
           // The conversation window title as as string.
           "My Audio Conversation",
           // Not supported.  Pass "1".
           "1",
           // Not supported.
           null);

       Console.WriteLine("Press Enter key to exit the
       application.");
       Console.ReadLine();

       Marshal.ReleaseComObject(_messenger);
       _messenger = null;
       Marshal.ReleaseComObject(_messengerAdv);
       _messengerAdv = null;
   }
   ```

6. The previous code creates an instance of *IMessengerAdvanced* and calls *StartConversation()* to create an audio conference with the contacts specified.

7. Run the application by clicking Debug on the Visual Studio menu, and then click Start Debugging.

   Note that the audio conference started with the contacts specified.

8. Close the console application and save your work in Visual Studio.

This console application provides the basic logic for using the *IMessengerAdvanced.StartConversation()* method to start a conversation. For example, Figure 3-4 shows the results of running this code.

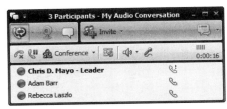

FIGURE 3-4 A conversation started with *IMessengerAdvanced.StartConversation()*.

Summary

This chapter explains the Office Communicator Automation API classes, interfaces, and events that you use to automate Office Communicator. By using this API, you can sign the local user in to and out of Office Communicator, display contact information, publish and subscribe to presence information, work with the contact list, and start conversations programmatically.

Additional Resources

- Microsoft Communicator 2007 SDK (*http://go.microsoft.com/fwlink/?linkid=143206*)
- "COM" Overview (*http://go.microsoft.com/fwlink/?linkid=143207*)
- COM Interoperation with .NET Overview (*http://go.microsoft.com/fwlink/?linkid=143208*)
- Office Communicator 2007: Enhanced Presence Model White Paper (*http://go.microsoft.com/fwlink/?linkid=143209*)
- Office Communicator Help (*http://go.microsoft.com/fwlink/?linkid=143210*)
- Communicator 2007 and Communicator 2007 R2 Automation API Documentation (*http://go.microsoft.com/fwlink/?LinkID=126311*)

Chapter 4
Embedding Contextual Collaboration

This chapter will help you to:

- Implement the end-to-end contextual collaboration scenario by building a sample application that:
    - Displays an application-specific contact list with presence.
    - Starts application-specific conversations.
    - Accepts application-specific conversations in your application.

Introduction to Contextual Collaboration

In Chapter 2, "Microsoft Unified Communications APIs Foundation," we discussed the scenarios that the Microsoft Office Communicator Automation application programming interface (API) enables, including the following:

- **Embedding presence with application-specific contact lists**   Using the API to build contact lists specific to your application and showing Office Communicator presence for those contacts.

- **Enhancing communications**   Adding instant messaging (IM), audio, and video communications directly in your application to allow your users to communicate and collaborate directly from your application.

- **Driving application context-specific communications**   Using the API to integrate data from your application into the conversations that your application starts to provide application-specific context for the conversation.

In this chapter, you learn how to implement these scenarios in a sample application. Implemented together, these scenarios are often called *contextual collaboration*.

> ### Microsoft Office Outlook 2007 and Office Communicator 2007 R2 Integration
>
> *Chris Mayo*
> *Technical Evangelist*
>
> This article looks at the integration of Microsoft Office Communicator 2007 R2 with Office Outlook 2007 as an example of the communication features you can build into an end-to-end contextual collaboration solution with the Office Communicator Automation API. When Office Communicator 2007 R2 is installed on a computer with Outlook 2007, Outlook 2007 automatically provides new integrated communication features.

Figure 4-1 shows an e-mail from Adam Barr to Chris. Notice how the presence of every contact is displayed on the From:, To:, and Cc: lines (in this case, it's just Adam). This is an example of custom contact list integration. The contacts displayed on the From:, To:, and Cc: lines do not need to be added to Chris's Office Communicator contact list.

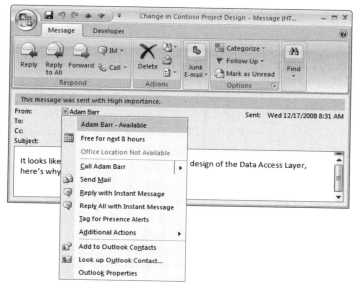

FIGURE 4-1 E-mail in Outlook 2007 with integrated Office Communicator presence.

Right-click the presence icon to display a context menu that is similar to what you see in Office Communicator 2007 R2, as shown in Figure 4-1. This context menu, as well as the IM and Call menu items on the Message tab of the ribbon, can be used to communicate with contacts in this e-mail thread using IM or an audio call.

For example, selecting the Call menu item sends Adam an invitation to an audio call with the subject of the call set to the subject of the e-mail (in this case, "Change in Contoso Project Design"). The IM and Call menu items are examples of providing custom communication features in an application.

When Adam accepts the call, he receives a hyperlink in the IM portion of the conversation that allows him to retrieve the context of the call (in this case, the e-mail about the design change). Clicking that hyperlink opens the e-mail, as shown in Figure 4-2. This hyperlink supplies context to the conversation (why Chris called Adam) and provides an application-specific way for Adam to view that context (opening the e-mail).

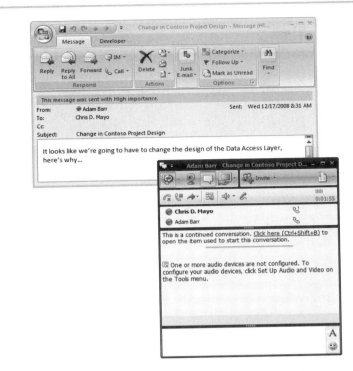
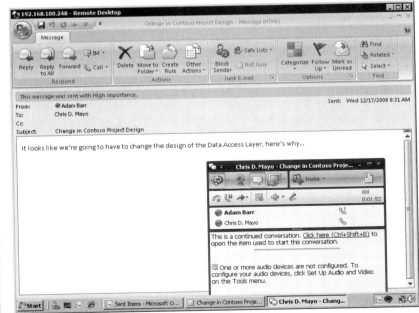

FIGURE 4-2 An active audio call started from Outlook 2007 using e-mail as context.

> This immediately improves the communication experience for users of Outlook 2007. Outlook 2007 users can use presence to determine how to communicate with other contacts and start conversations directly from Outlook 2007. In addition, application context integrated into those conversations allows both parties to start the conversation with the same information. In this example, this process allows Adam to know immediately why Chris is calling.

Scenario

The contextual collaboration scenario defined previously can be implemented in any client application using the Office Communicator Automation API. In this chapter, we implement the contextual collaboration feature using a sample scenario to illustrate this point.

For example, in a retail enterprise, employees in one store often call employees in other stores to check the availability of an item for a customer. To confirm the accuracy of the inter-store inventory system, which is refreshed only every 24 hours, employees call the target store to make sure the item is available and to ask the store to hold that item for the customer before directing the customer to drive to the other store. If both the calling employee and the called employee are running an application with integrated contextual collaboration features, the communication experience can be more efficient. For the employee checking on the inventory, the application can place the call and specify the item of interest within the communication context. When the other employee receives the call, the application can use the context to look up the inventory of the item and display information to help expedite the communication.

Business Value

Implementing the Office Communicator Automation API contextual collaboration scenario adds business value to any client application by increasing ease of communication and associated productivity for information workers. Communication becomes more convenient and more focused when it is seamlessly integrated into the information workers' typical workflow. Communication is also more efficient because all participants in the conversation share the same context within the conversation.

Choice of Technology

The Office Communicator Automation API is a great solution for the contextual collaboration scenario for the following reasons:

- **Developer productivity** Using the Office Communicator Automation API to take advantage of the functionality in Office Communicator 2007 R2 is much faster than building a real-time communications solution from scratch.
- **Proven solution** The Office Communicator Automation API is already in use, providing contextual collaboration features in Outlook 2007 as well as other Microsoft products.
- **Ease of use** Users already using Office Communicator 2007 R2 are familiar with the user interface when integrated in your application.

However, keep in mind that when using the Office Communicator Automation API, your application works in tandem with Office Communicator 2007 R2 by automating it. If you need to integrate communication features directly into your client application (such as hosting the video stream in your application user interface) or you don't want to require Office Communicator to be deployed with your application, you need to use one of the other API options discussed in Chapter 2.

Test Environment

To develop and test this solution, you need the following:

- Microsoft Office Communications Server 2007 R2, deployed as described in Chapter 9, "Preparing the UC Development Environment."
- Two domain accounts, including an account to develop the solution and an account to test the solution, configured for Office Communications Server 2007 R2 and able to sign in to Office Communicator as detailed in Chapter 9.
- A development environment that meets the Office Communicator Automation API requirements, as specified in the "Office Communicator 2007 and Office Communicator 2007 R2 Automation API" documentation at *http://go.microsoft.com/fwlink/?LinkID=126311* in the section titled "Getting Started Using Office Communicator Automation API."

Overall Code Structure

This section walks through the steps required to build a contextual collaboration solution using the Office Communicator Automation API and Microsoft Visual Studio 2008. This application provides a solution for the inventory request scenario described earlier in this chapter by doing the following:

- Displaying application-specific contact lists for employees to call other stores to make inventory requests
- Launching application-specific conversations directly from the application
- Accepting application-specific conversations in the application and displaying the context of the application

Displaying Application-Specific Contact Lists

When building a contextual collaboration solution, you want to display application-specific contact lists populated with contacts particular to the data displayed in your application. For example, you may want to display the presence of employees who are working in a store. You can use the Office Communicator Automation API to display application-specific contact lists and contact presence.

For most projects, embedding the same presence icon and context menu that you see in Office Communicator 2007 R2 directly into your application is the best option because Office Communicator users are already familiar with this interface.

The easiest way to integrate Office Communicator 2007 R2 contact presence information is to use one of the various presence control samples provided by the Office Communicator Automation API team. For example, the WPF Presence Controls for Microsoft Office Communicator 2007—Microsoft Office Communicator 2007 SDK Sample (*http://go.microsoft.com/fwlink/?linkid=143214*) provides Windows Presentation Foundation (WPF) presence controls with full source code.

Displaying User Presence with *MyPersona*

If you want to show the presence of the local user and a custom contact list, you can use the WPF Presence Controls sample to create a WPF application such as the one shown in Figure 4-3. This application uses the *MyPersona* control to show the presence of the local user and the *PersonaList* control to show a custom contact list. The *PersonaList* control is made up of *Persona* controls, which are used to show the presence of a single contact.

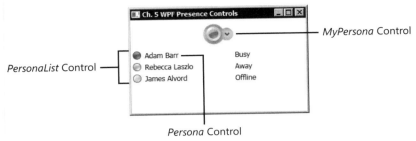

FIGURE 4-3 Showing presence with the *MyPersona*, *PersonaList*, and *Persona* controls.

The *MyPersona* control automatically shows the presence for the local user using the Office Communicator API. The *Persona* and *PersonaList* controls need a specific SIP URI of a contact to show the presence of that contact. For example, the following Microsoft Extensible Application Markup Language (XAML) code defines a *PersonaList* control named *personaList1*.

```
<presence:PersonaList x:Name="personaList1" Grid.Column="0"
     Grid.Row="1"
     ShowContextMenu="True"
     ShowDisplayName="True"
     ShowAvailability="True"
     ShowToolTip="True"
     ShowDetailedToolTipText="True"/>
```

The following code populates *personaList1* with the contacts Adam, Renee, and James.

```
// Create a list of contacts specific to your application.
//  Note: They don't have to be contacts in the
//    local user's contact list.
List<string> sipUris = new List<String>()
{
    "adamb@uc.contoso.com",
    "rl@uc.contoso.com",
    "jamesa@uc.contoso.com"
};

// Add the list of contacts to the SipUris property
//   to populate the control.
personaList1.SipUris = sipUris;
```

Displaying Application-Specific Context Menus with *PersonaList*

When you right-click a contact, the *Persona* control displays a context menu similar to the context menu displayed by Office Communicator, as shown in Figure 4-4. Note that the context menu also includes a custom communication menu item named Inventory Request. The *Persona* control supports creating custom menu items for application-specific communication features.

FIGURE 4-4 The *Persona* control showing a contact's context menu.

The custom context menu item is added to a *PersonaList* control, *personaList1*, by using the following code.

```
        // Create a new menu item for your application specific
        //   communication feature.
        List<MenuItem> customItems = new List<MenuItem>();

        MenuItem customMenuItem1 = new MenuItem();
        customMenuItem1.Header = "Inventory Request";
        customMenuItem1.Name = "inventoryRequest";

        customItems.Add(customMenuItem1);

        personaList1.CustomMenuItemList = customItems;

        personaList1.CustomMenuItemClicked +=
            new EventHandler<CustomMenuItemClickedEventArgs>
                (personaList1_CustomMenuItemClicked);
    }

    void personaList1_CustomMenuItemClicked(object sender,
        CustomMenuItemClickedEventArgs e)
    {
        // Your code goes here...
    }
```

Putting It All Together: Displaying an Application-Specific Contact List with Contact Presence

Using the concepts from the preceding sections, you can quickly build an application that displays an application-specific contact list with contact presence. In this section, you create a WPF application with these features using the WPF Presence Controls.

1. Download and install the Office Communicator 2007 Software Development Kit (SDK) located at *http://go.microsoft.com/fwlink/?LinkID=85980*.

2. Download, install, and build the WPF Presence Controls for Microsoft Office Communicator 2007—Microsoft Office Communicator 2007 SDK Sample located at *http://go.microsoft.com/fwlink/?linkid=143214*.

3. Start Visual Studio 2008, and create a new Microsoft Visual C# Windows WPF application named ContextualCollaboration.

4. Right-click the References node in Solution Explorer, select Add Reference, click the COM tab, and select the Microsoft Office Communicator 2007 API Type Library in the Component Name list box. Click OK.

5. Right-click the References node in Solution Explorer, select Add References, click the Browse tab, and browse to the WPFMOCPresenceControls.dll file in the install directory of the WPF Presence Controls Sample.

6. Open Window1.xaml, click the XAML tab, and then add the bold text that is shown in the following code example.

```xaml
<Window x:Class="Ch04_01_ContextualCollaboration.Window1"
    xmlns="http://schemas.microsoft.com/winfx/2006/xaml/presentation"
    xmlns:x="http://schemas.microsoft.com/winfx/2006/xaml"
    xmlns:presence="clr-namespace:Microsoft.Samples.Office.UnifiedCommunications.PresenceControls;assembly=WPFMOCPresenceControls"
    Title="Contextual Collaboration" Height="300" Width="300">
    <Grid ShowGridLines="False">
        <Grid.ColumnDefinitions>
            <ColumnDefinition Width="*" />
        </Grid.ColumnDefinitions>
        <Grid.RowDefinitions>
            <RowDefinition Height="50" />
            <RowDefinition Height="100" />
            <RowDefinition Height="*" />
        </Grid.RowDefinitions>

        <presence:MyPersona x:Name="myPersona" Grid.Column="0"
            Grid.Row="0"/>
        <presence:PersonaList x:Name="personaList1" Grid.Column="0"
            Grid.Row="1"
            ShowContextMenu="True"
            ShowDisplayName="True"
            ShowAvailability="True"
            ShowToolTip="True"
            ShowDetailedToolTipText="True"/>
    </Grid>
</Window>
```

7. The XAML in the previous code example defines a *MyPersona* control named *myPersona* for showing the presence of the signed-in user and a *PersonaList* control named *personaList1* for showing a custom list of contacts.

8. Right-click the design surface of Window1.xaml, and select View Code.

9. Add the following lines of code to the existing *using* statements to bring the Office Communicator, InteropServices, and WPF Presence Controls sample into scope.

```
using CommunicatorAPI;
using System.Runtime.InteropServices;
using Microsoft.Samples.Office.UnifiedCommunications.PresenceBase;
```

10. To populate the contact list and add a custom menu item for making application-specific calls about an inventory request, enter the following code in the constructor of Window1.

```
public Window1()
{
    InitializeComponent();

    // Create a list of contacts specific to your application.
    //   Note: They don't have to be contacts in the
    //   local user's contact list.
    List<string> sipUris = new List<String>()
    {
        "adamb@uc.contoso.com",
        "rl@uc.contoso.com",
        "jamesa@uc.contoso.com"
    };

    // Add the list of contacts to the SipUris property
    //   to populate the control.
    personaList1.SipUris = sipUris;

    // Create a new menu item for your application specific
    //   communication feature.
    List<MenuItem> customItems = new List<MenuItem>();

    MenuItem customMenuItem1 = new MenuItem();
    customMenuItem1.Header = "Inventory Request";
    customMenuItem1.Name = "inventoryRequest";

    customItems.Add(customMenuItem1);

    personaList1.CustomMenuItemList = customItems;

    personaList1.CustomMenuItemClicked +=
        new EventHandler<CustomMenuItemClickedEventArgs>
            (personaList1_CustomMenuItemClicked);
}
```

11. Define the event handler in the *Window1* class for the Inventory Request custom menu item, as shown in the following code example. This event handler triggers when the user clicks this menu item.

```
void personaList1_CustomMenuItemClicked(object sender,
    CustomMenuItemClickedEventArgs e)
{
    // Your code goes here...
}
```

12. To run the application, click Debug, and then click Start Debugging.

13. Change the presence of the local user in Office Communicator. Log on to Office Communicator on another computer using one of the accounts in the application-specific contact list shown previously, and change the presence of the contact in Office Communicator. Note the changes in presence shown in the *MyPersona* and *PersonaList* controls.

14. Right-click a contact in the application contact list and click Call, and then click Communicator Call. Note the call being placed between the local user and the contact.

15. Close the ContextualCollaboration application, and save your work in Visual Studio.

ContextualCollaboration shows how easy it is to use the WPF Presence Controls sample to embed presence and communication features in an application. Following the steps described previously produces a WPF application, as shown in Figure 4-4. Note the context application-specific menu item named Inventory Request.

When the Office Communicator Call menu item is selected, a call is placed with the contact, Adam Barr, as shown here.

Starting Application-Specific Conversations

Chapter 3, "Programming a Microsoft Office Communicator Automation API Application," shows that you can start conversations for the local user programmatically using the *IMessengerAdvanced* interface. These conversations can include IM, audio, and video modalities. The *IMessengerAdvanced* interface also provides the ability to send application data in the IM portion of a conversation to create application-specific communication features.

IMessengerAdvanced.StartConversation()

As discussed in Chapter 3, *IMessengerAdvanced* provides the *StartConversation()* method for creating conversations with one or more contacts using a specific modality. For example, the following code starts an audio conference with the contacts specified in the *sipUris* array and creates a conversation window named "My Audio Conversation."

```
Messenger _messenger = new Messenger();
IMessengerAdvanced _messengerAdv =
    (IMessengerAdvanced)_messenger;


// An object array of SIP URIs strings for the participants
//  in the conversation.
object[] sipUris = new object[] { "adamb@uc.contoso.com",
    "rl@uc.contoso.com" };

object obj = _messengerAdv.StartConversation(
    // The conversation modality (audio in this case).
    CONVERSATION_TYPE.CONVERSATION_TYPE_AUDIO,
    // The participants.
    sipUris,
    // Not supported.
    null,
    // The conversation window title as as string.
    "My Audio Conversation",
    // Not supported.  Specify "1".
    "1",
    // Not supported. Specify NULL
    null);
```

Figure 4-5 shows the result of this code, which initiates an audio conference using Office Communicator.

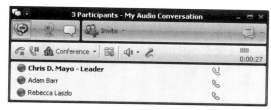

FIGURE 4-5 Starting an audio conference using *StartConversation()*.

Notice that *StartConversation()* returns an object. This object is a window handle (HWND) of the conversation window created. You can use this value to refer to the conversation window after it's created.

Receiving Notification of New Conversations

Conversations using the Office Communicator Automation API are started asynchronously. When calling *StartConversation()*, the conversation invitations are sent to the participants, but the conversation window is not actually created until the first invitation is accepted.

When the conversation window is created, the *Messenger* class raises the *OnIMWindowCreated* event. This event is raised for all conversations, not just IM conversations as the name implies. The following code shows how to subscribe to the *OnIMWindowCreated* event using the *DMessengerEvents_OnIMWindowCreatedEventHandler* event handler.

```
Messenger _messenger = new Messenger();

_messenger.OnIMWindowCreated += new
    DMessengerEvents_OnIMWindowCreatedEventHandler(
    _messenger_OnIMWindowCreated);

...

void _messenger_OnIMWindowCreated(object pIMWindow)
{
    IMessengerConversationWndAdvanced newConversation =
        (IMessengerConversationWndAdvanced)pIMWindow;

    // Code to run when the conversation is created
}
```

The *DMessengerEvents_OnIMWindowCreatedEventHandler* supports a single parameter, *pIMWindow*, of the *object* type. This parameter is an instance of *IMessengerConversationWndAdvanced*, which represents the conversation created. *IMessengerConversationWndAdvanced* is an important interface because it provides access to the conversation after it's created and allows you to manipulate the conversation (such as sending IM to the conversation window). If you want to refer to the conversation window after the *OnIMWindowCreated* event is raised, store this object instance in a variable in your code.

Receiving Notification of Destroyed Conversations

A conversation window is destroyed when the local user closes the conversation window or when all participants leave the conversation. The *Messenger* class raises the *OnIMWindowDestroyed* event when the conversation window is destroyed. The following code subscribes to the *OnIMWindowDestroyed* event through the *DMessengerEvents_OnIMWindowDestroyedEventHandler* event handler.

```
Messenger _messenger = new Messenger();

_messenger.OnIMWindowDestroyed -= new
    DMessengerEvents_OnIMWindowDestroyedEventHandler(
    _messenger_OnIMWindowDestroyed);

...
```

```
static void _messenger_OnIMWindowDestroyed(object pIMWindow)
{
    // Code to run when the conversation is closed...
}
```

Like *OnIMWindowCreated*, *OnIMWindowDestroyed* applies to all conversation types, not just IM conversations. A single parameter, *pIMWindow*, is passed of the *object* type, which is an instance of *IMessengerConversationWndAdvanced*. Use this event to release any references you have of the conversation window that has been destroyed.

> **Note** When *OnIMWindowDestroyed* is raised, the *IMessengerConversationWndAdvanced* instance passed is no longer a valid reference to the conversation window because the conversation window has been destroyed. If you want to compare the reference passed to *OnIMWindowDestroyed* to an instance of *IMessengerConversationWndAdvanced* from the *OnIMWindowCreated* event, you have to compare the object references by using *Object.ReferenceEquals()*.

Sending IM Text to Conversations

The *IMessengerConversationWndAdvanced* interface provides the *SendText()* method for sending IM text to a conversation window. You can also use this method to transmit application-specific metadata within the conversation. For example, you can send IM to a newly created conversation window using the following code.

```
static void _messenger_OnIMWindowCreated(object pIMWindow)
{
    IMessengerConversationWndAdvanced newConversation =
        (IMessengerConversationWndAdvanced)pIMWindow;

    newConversation.SendText("Some IM text…");
}
```

SendText() takes a single parameter—the string to be sent to the IM conversation window.

Putting It All Together: Sending Application IM Text to Your Conversation

Using the concepts from the preceding sections, you can build an application quickly that starts a conversation and sends some application metadata in the IM portion of the conversation. In this section, you build on the ContextualCollaboration sample code shown earlier in this chapter by performing the following steps to add the communication features:

1. Start Visual Studio 2008 (if it's not already running) and open the ContextualCollaboration.csproj project.

2. Open Window1.xaml.cs and add the following declarations to the *Window1* class.

   ```
   private Messenger _messenger;
   private IMessengerAdvanced _messengerAdv;

   // Store the HWND of conversation created
   //   with StartConversation().
   private int _myConversationHWND;
   // Store reference to conversation
   //   created with StartConversation().
   private IMessengerConversationWndAdvanced _myConversation;
   // Used to store context to pass in call to SendText().
   private string _myContext;
   ```

   The variable, *_myConversationHWND*, holds the value returned from the call to *IMessengerAdvanced.StartConversation()*. The *_myConversation* variable refers to the conversation window after it is created. Finally, *_myContext* is used to store the application metadata passed to the conversation after it is created.

3. Add the code to the constructor of the *Window1* class to instantiate the variables *_messenger* and *_messengerAdv* and subscribe to the conversation events.

   ```
   public Window1() // Constructor for class Window1
   {
       InitializeComponent();

       ...

       _messenger = new Messenger();
       _messengerAdv = (IMessengerAdvanced)_messenger;

       _messenger.OnIMWindowCreated += new
           DMessengerEvents_OnIMWindowCreatedEventHandler(
           _messenger_OnIMWindowCreated);
       _messenger.OnIMWindowDestroyed += new
           DMessengerEvents_OnIMWindowDestroyedEventHandler(
           _messenger_OnIMWindowDestroyed);
   }
   ```

4. Add the following code to start a conversation when the Inventory Request menu item is clicked.

   ```
   void personaList1_CustomMenuItemClicked(object sender,
   CustomMenuItemClickedEventArgs e)
   {
       _myContext =
       "<InventoryRequest><Item>12345</Item></InventoryRequest>";

       object obj = _messengerAdv.StartConversation(
           // The conversation modality (audio in this case).
           CONVERSATION_TYPE.CONVERSATION_TYPE_AUDIO,
           // The participants selected by the local user.
           e.SipUri,
   ```

```
                // Not supported.
                null,
                // The conversation window title as as string.
                "Inventory Request: Item 12345",
                // Not supported.  Pass "1".
                "1",
                // Not supported.
                null);

            _myConversationHWND = int.Parse(obj.ToString());
        }
```

Note that the conversation window handle returned from the call to *StartConversation()* is stored in the *_myConversationHWND* variable so that the conversation window can later be identified when the event handler, *OnIMWindowCreated*, is called to notify your application that the conversation window was created.

5. Add the following code to implement *OnIMWindowCreated* in the *Window1* class to send application-specific data, *_myContext* (defined in step 4), when the conversation window is created.

```
        void _messenger_OnIMWindowCreated(object pIMWindow)
        {
            IMessengerConversationWndAdvanced newConversation =
                (IMessengerConversationWndAdvanced) pIMWindow;

            // Is the conversation the one I started by calling
            //    StartConversation()?
            if (newConversation.HWND == _myConversationHWND)
            {
                // Get a reference to the conversation if needed
                //    later to add/remove contacts, send IM, retreive
                //    history...
                _myConversation = newConversation;
                // Send the application data to the IM conversation.
                _myConversation.SendText(_myContext);
            }
        }
```

In the previous code, if the window handle of the new conversation (provided by the *IMessengerConversationWndAdvanced.HWND* property) matches the window handle of the conversation created by *StartConversation()* (stored in the *_myConversationHWND* variable), the code uses the *IMessengerConversationWndAdvanced.SendText()* method to send the application data to the IM conversation.

6. Next, add code to implement the *OnIMWindowDestroyed* event handler in the *Window1* class to release the reference to the conversation window when the conversation window is destroyed.

```
        void _messenger_OnIMWindowDestroyed(object pIMWindow)
        {
            // When the conversation is destroyed, compare window
```

```
            //  destroyed to conversation window reference.
            //  Note: When OnIMWindowDestroyed is called the
            //  pIMWindows is no longer a valid COM object.
            //  The only way you can refer it is as an object,
            //  hence the use of ReferenceEquals.
            if (object.ReferenceEquals((object)_myConversation,
                pIMWindow))
            {
                Marshal.ReleaseComObject(_myConversation);
                _myConversation = null;

                _myConversationHWND = 0;
            }
        }
```

In the previous code, *object.ReferenceEquals()* is used to compare the window destroyed (passed in the *pIMWindow* parameter) to the *_myConversation* variable. If the references match, the conversation window destroyed is the conversation started by your application. Therefore, the reference should be freed using *MarshalReleaseComObject()*.

7. Add the code in bold to the *Window1* class to implement the *IDisposable* interface so Office Communicator Automation API references can be freed when the window is released.

```
    public partial class Window1 : Window, IDisposable
```

8. Implement the *IDisposable* interface by adding the following code to the *Dispose()* method in the *Window1* class.

```
        public void Dispose()
        {
            _messenger.OnIMWindowCreated -= new
                DMessengerEvents_OnIMWindowCreatedEventHandler(
                _messenger_OnIMWindowCreated);
            _messenger.OnIMWindowDestroyed -= new
                DMessengerEvents_OnIMWindowDestroyedEventHandler(
                _messenger_OnIMWindowDestroyed);

            Marshal.ReleaseComObject(_messenger);
            _messenger = null;
            Marshal.ReleaseComObject(_messengerAdv);
            _messengerAdv = null;

            if (_myConversation != null)
            {
                Marshal.ReleaseComObject(_myConversation);
                _myConversation = null;
            }
        }
```

The previous code uses *Marshal.ReleaseComObject()* to release references to Office Communicator Automation API classes and interfaces.

9. Run the application by clicking Debug on the Visual Studio menu, and then click Start Debugging.

10. Select an online contact in the contact list, and right-click and select the Inventory Request menu item.

11. A conversation is started, and the contents of the _myContext_ variable are transmitted.

12. Close the ContextualCollaboration application, and save your work in Visual Studio.

ContextualCollaboration shows how easy it is to use the Office Communicator Automation API to add an application-specific communication feature to your application. Following the steps in the previous procedure produces a conversation like the one shown in Figure 4-6.

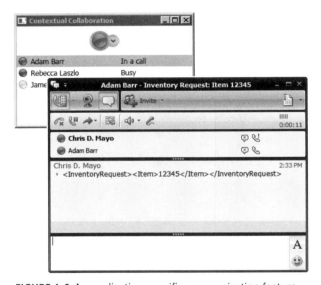

FIGURE 4-6 An application-specific communication feature.

Accepting Application-Specific Conversations

The previous section demonstrated how to start a new application-specific conversation. This section demonstrates how to accept incoming conversations and process application-specific data.

Retrieving Text from IM Conversations

IMessengerConversationWndAdvanced provides the *History* property for retrieving the IM conversation from the conversation window. This property can be used to accept application-specific data when a conversation is started. Due to the asynchronous nature of the API, the *History* method doesn't return the IM message from a conversation when the conversation window is first created, which is when the *OnIMWindowCreated* event handler is triggered. Your application must spawn a new thread to poll the *History* property.

Putting It All Together: Receiving Application-Specific Conversations with *IMessengerAdvanced*

Using the concepts from the preceding section, you can add the ability to accept application-specific conversations from other users. In this section, you build on the ContextualCollaboration example by adding this feature, as follows:

1. Start Visual Studio 2008 (if it's not already running) and open the ContextualCollaboration.csproj project.

2. Open Window1.xaml to define a list box to show the item number of incoming inventory requests (application-specific conversations) by adding the code in bold text to the following code example.

   ```
   <Window …
       Title="Contextual Collaboration" Height="300" Width="300" >
       <Grid ShowGridLines="False">
           …
           <ListBox Name="lbConversations" DockPanel.Dock="Top"
               Grid.Column="0" Grid.Row="2"/>
       </Grid>
   </Window>
   ```

3. Open Window1.xaml.cs and add the following declarations to the *Window1* class.

   ```
   // Timer used to poll IM history from application specific conversations.
   private Timer _historyTimer;
   // Counter to track number of polls for incoming messages.
   private int _historyTicks;
   ```

   The *_historyTimer* variable is used to poll for the IM conversation history using a new thread. The *_historyTicks* variable is used to stop polling if the IM conversation history is not returned after a reasonable amount of time has passed.

4. From the following code example, add the code in bold to the *_messenger_OnIMWindowCreated* method.

   ```
   void _messenger_OnIMWindowCreated(object pIMWindow)
   {
       IMessengerConversationWndAdvanced newConversation =
           (IMessengerConversationWndAdvanced)
           pIMWindow;

       // Is the conversation the one I started by calling
       //    StartConversation()?
       if (newConversation.HWND == _myConversationHWND)
       {
           …
       }
   ```

```
            else // incoming new conversation
            {
                //When a new conversation is created, start a timer
                //  to poll for History() for one second.
                _historyTimer = new Timer(
                        new TimerCallback(ConversationHistory_Tick),
                        newConversation,
                        TimeSpan.FromSeconds(0),
                        TimeSpan.FromSeconds(1)
                        );
            }
        }
```

In the previous code, if the newly created conversation window was not started by our application (where *IMessengerConversationWndAdvanced.HWND* does not equal *_myConversationHWND*), you can assume that this incoming new conversation could be an application-specific conversation sent from another user. A *System.Threading. Timer* is created and assigned to *_historyTimer* to poll for IM conversation text.

5. Implement the *TimerCallback* method inside the *Window1* class.

```
        private void ConversationHistory_Tick(object state)
        {
            IMessengerConversationWndAdvanced conversationWnd =
                (IMessengerConversationWndAdvanced)state;

            try
            {
                // Check if the conversation has
                //  application specific data.
                if ((conversationWnd.History != null))
                {
                    // Turn off the timer.
                    _historyTimer.Change(Timeout.Infinite,
                        Timeout.Infinite);
                    _historyTimer.Dispose();

                    DisplayApplicationContext(conversationWnd.History);
                }
                else
                {
                    _historyTicks += 1;
                    // If polling has not found History within
                    //  1 sec, assume it's not an application
                    //   specific conversation and stop polling.
                    if (_historyTicks > 1000)
                    {
                        // Shut down the timer.
                        _historyTimer.Change(Timeout.Infinite,
                            Timeout.Infinite);
                        _historyTimer.Dispose();
                    }
                }
            }
        }
```

```
        catch
        {
            // In the event that an exception is thrown trying
            //  to access the convesatoin window, turn off the
            //  timer to stop polling.
            _historyTimer.Change(Timeout.Infinite,
                Timeout.Infinite);
            _historyTimer.Dispose();
        }
    }
```

In the previous code, the *IMessengerConversationWndAdvanced* instance from *OnIMWindowCreated* is passed as state to the callback (in the *state* parameter). If *IMessengerConversationWndAdvanced.History* returns a value, *Timer* is disabled, and *History* is passed to a method named *DisplayApplicationContext* to determine whether it actually contains application-specific data. If *IMessengerConversatonWndAdvanced.History* does not return a value, a counter named *_historyTicks* is incremented. If *_historyTicks* is greater than 1,000 (1,000 milliseconds [ms] because the *Timer* "ticks" every 1 ms), it can be assumed that the conversation was not started by the remote user's application, and *_historyTimer* can be disabled and polling stopped.

6. Implement the *DisplayApplicationContext* method in the *Window1* class.

```
private void DisplayApplicationContext(string context)
{
    try
    {
        // Process context to identify if the context is of
        //  interest to your application.
        if (context.Contains("InventoryRequest"))
        {
            // When you pull History from a
            //  IMessengerConversationAdvanced, it's HTML.
            //  Strip off everything so it only contains
            //  the text passed to SendText().
            string contextBody = context.Replace("&lt;", "<");
            contextBody = contextBody.Replace("&gt;", ">");

            int start = contextBody.IndexOf(
                "<InventoryRequest>", 0);
            int end = contextBody.IndexOf(
                "</InventoryRequest>", start) + 19;

            contextBody = contextBody.Substring(start,
                end - start);

            XmlDocument doc = new XmlDocument();
            doc.InnerXml = contextBody;
            string item = doc.DocumentElement.InnerText;

            Dispatcher.Invoke(DispatcherPriority.Normal,
                new Action(() =>
```

```
                {
                    lbConversations.Items.Add(
                        string.Format(
                        "Inventory Request: Item# {0}", item));

                }));
            }
        }
        catch
        {
            // Handle string manipulation error handling here...
        }
    }
```

In the previous code, the *DisplayApplicationContext* method takes the contents of the *IMessengerConversationWndAdvanced.History* property passed to it and checks for application-specific conversations by looking for the InventoryRequest tag in the message string. If application-specific conversation data is found, any formatting is removed and the contents are added to the *lbConversations* list box.

Testing the Application

To test your application, you copy it to another computer running Office Communicator 2007 R2 using the following procedure:

1. Copy the contents of the ContextualCollaboration\Bin\Debug folder to the computer of one of the contacts listed in the application-specific contact list and run ContextualCollaboration.exe.

2. Run the application on the computer where you wrote the application by clicking Debug and then clicking Start Debugging.

3. Select and right-click an online contact in the contact list, and select the Inventory Request menu item.

    Note that the conversation started on one computer and is received on the other computer, as shown by the item number "12345" displayed in the application.

4. Close the application on both computers, and save your work in Visual Studio.

The ContextualCollaboration example shows how easy it is to use the Office Communicator Automation API to transmit metadata to communicate between two or more remote applications. Following the steps described previously produces the results shown in Figure 4-7.

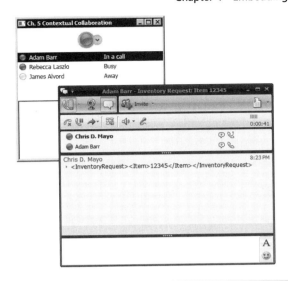
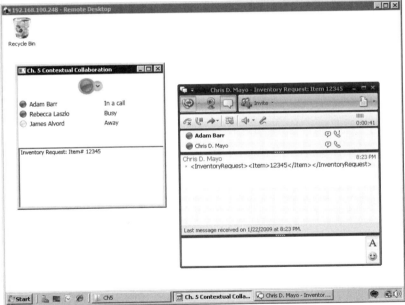

FIGURE 4-7 An application-specific conversation started by Chris and accepted by Adam.

Summary

This chapter defined a contextual collaboration solution that uses the Office Communicator Automation API to automate Office Communicator 2007 R2 so that it sends and receives application-specific messages between remote applications. The Office Communicator Automation API makes it possible to embed the local user's presence and contact list quickly and provides the ability to initiate and accept conversations to transmit application-specific metadata.

Additional Resources

- Microsoft Unified Communications Managed API 2.0 SDK (32 bit) (*http://go.microsoft.com/fwlink/?LinkID=139195*)
- "Unified Communications Managed API 2.0 Core SDK" documentation (*http://go.microsoft.com/fwlink/?LinkID=126312*)
- Microsoft Communicator 2007 SDK (*http://go.microsoft.com/fwlink/?LinkID=141199*)
- "Communicator 2007 and Communicator 2007 R2 Automation API" documentation (*http://go.microsoft.com/fwlink/?LinkID=126311*)
- WPF Presence Controls for Microsoft Office Communicator 2007—Microsoft Office Communicator 2007 SDK Sample (*http://go.microsoft.com/fwlink/?linkid=143214*)

Part III
Unified Communications Managed API Workflow

Part III covers the Unified Communications Managed API (UCMA) Workflow API, as shown in the following figure. The UCMA Workflow API provides specialized Windows Workflow activities to help you quickly build instant messaging and voice-based workflow applications using the visual designer in Visual Studio. Because of the abstraction provided by the UCMA Workflow API, you do not have to work with the more complex underlying UCMA for large portions of your application.

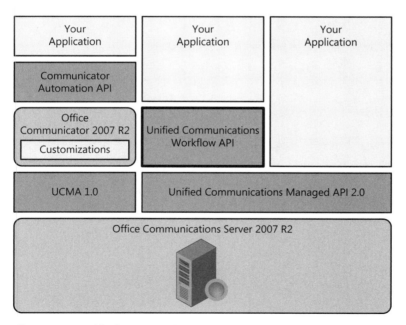

Chapter 5, "Unified Communications Managed API (UCMA) Workflow," explains the API in detail, and Chapter 6, "Business Process Communication," walks through an example of creating a business process communications application using UCMA Workflow Activities.

Chapter 5
Unified Communications Managed API (UCMA) Workflow

This chapter will help you to:

- Understand how the Unified Communications Managed API (UCMA) Workflow Activities provide a way to develop applications that can interact with the user over the phone or instant messaging (IM) channel as new interfaces.

- Understand how UCMA Workflow enables faster development of applications by providing prepackaged custom activities for common tasks.

- Understand how UCMA Workflow enables intelligent routing decisions by providing a way to query for presence information.

UCMA Workflow

UCMA Workflow consists of a few Workflow Runtime services and a set of custom Workflow Activities that provide Unified Communications (UC) functionality. UCMA Workflow builds on Windows Workflow Foundation, which is part of Microsoft .NET Framework 3.0 and .NET Framework 3.5.

Before jumping into the details of this API, it is important to understand how the activities and services relate to the application. Figure 5-1 shows how the UCMA Workflow Runtime services and Workflow Activities are related to each other.

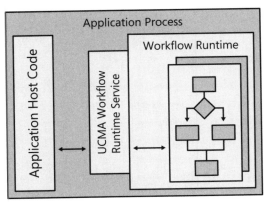

FIGURE 5-1 Anatomy of an application using the UCMA Workflow.

A UCMA Workflow application consists of workflow components and nonworkflow components. In Figure 5-1, workflow components are depicted by blocks designated as "UCMA Workflow Runtime Service" and "Workflow Runtime." The nonworkflow component is represented by the block designated as "Application Host Code." Within the application host code, the UCMA Core is used to handle the connection to Microsoft Office Communications Server 2007 R2. This means that the application provisioning required to run a UCMA Workflow application as a trusted service is the same as the application provisioning required to run a UCMA Core application. (For details about configuring and provisioning a UCMA Workflow application, see Chapter 9, "Preparing the UC Development Environment.") The UCMA Workflow application must create the Workflow Runtime and add the necessary UCMA Workflow Runtime services. The UCMA Workflow Runtime services are used to pass data from the application host code to the workflow instances running under the Workflow Runtime.

Using Project Templates

To use the UCMA Workflow, you must install the UCMA 2.0 Software Development Kit (SDK). (For details on how to install the UCMA 2.0 SDK, see the "Configuring UCMA Core" section in Chapter 9.) This SDK comes with project templates that are specific to the UCMA Workflow and make it easier to get started with developing UCMA Workflow applications. The templates are available for both C# and Microsoft Visual Basic. The following project templates are supported in the UCMA SDK:

- **Inbound Sequential Workflow Console Application**   This project template is the starting point for creating UC-enabled workflow applications that handle incoming phone calls or instant messages.

- **Outbound Sequential Workflow Console Application**   This project template is the starting point for creating UC-enabled workflow applications that create outgoing phone calls or instant messages.

In both cases, the host process is a console application by default, but you can easily change it to any suitable application host (for example, a Windows service).

After installing the UCMA 2.0 SDK, these project templates are available in Microsoft Visual Studio 2008 under the Communications Workflow node in the Project Types pane of the New Project dialog box. This is shown in Figure 5-2.

Chapter 5 Unified Communications Managed API (UCMA) Workflow

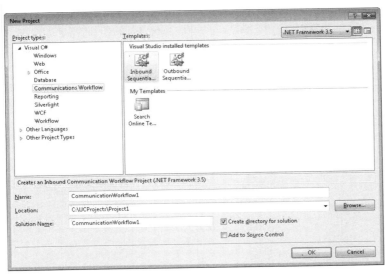

FIGURE 5-2 New project templates for the Communications Workflow.

The previous example shows the location of a C# project template. The Visual Basic template is located in the same relative location under the Visual Basic node.

Selecting a Workflow Language

After you create the new UCMA Workflow project by clicking OK in the New Project dialog box, the Select Language dialog box appears. This dialog box lists all of the language packs installed on the machine. If no language packs are installed, the user cannot create a project using this template. Language packs are required for speech recognition and for playing a message to the user (that is, text-to-speech message).

Using Workflow Designer

Visual Studio 2008 supports Workflow Designer to allow the workflow part of a UCMA Workflow application to be constructed by dragging and dropping appropriate activities from the Visual Studio Toolbox. The designer surface also helps you visualize the flow logic of the application. The designer surface is also known as the *designer canvas*.

As shown in Figure 5-3, the toolbox in the left panel lists all of the UCMA Workflow Activities under the Unified Communications Workflow tab. The designer canvas in the right panel is the sequential workflow as generated by the Inbound Sequential Workflow Console Application project template. As with other workflow activities, you can drag UCMA Workflow Activities onto the designer canvas.

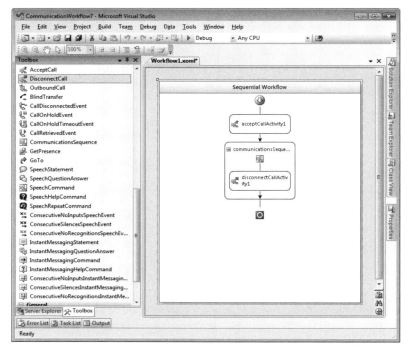

FIGURE 5-3 The Unified Communications Workflow Toolbox and designer canvas.

Workflow Runtime Services

UCMA Workflow Activities need custom workflow services to interact with the non-workflow world. These are added to the Workflow Runtime for the UCMA Workflow Activities to execute properly. Two services are included in the UCMA Workflow: *CommunicationsWorkflowRuntimeService* and *TrackingDataWorkflowRuntimeService*.

CommunicationsWorkflowRuntimeService The Workflow Runtime is the execution environment in which workflow instances execute. The *CommunicationsWorkflowRuntimeService* is required for UCMA Workflow Activities to execute properly; UCMA Workflow Activities fail to execute and throw exceptions at run time if you do not add this service to the Workflow Runtime. The application uses this service to pass objects (for example, phone calls and IM calls) or data (for example, the language to use) to the workflow. In UCMA, IM connections are also referred to as a call.

This service can be added to the Workflow Runtime, as shown in the *InitializeWorkflow* method of the Program.cs file, which is automatically generated by the project templates.

```
WorkflowRuntime _wRuntime = new WorkflowRuntime();
_wRuntime.AddService(new CommunicationsWorkflowRuntimeService());
```

This service provides the following functionality:

- Add phone or IM calls to a particular workflow instance.

    In a typical scenario, the application receives a call. The application creates an instance of a workflow to handle the incoming call or IM by associating the call or IM with a workflow instance, as shown in the following code example.

    ```
    // Create an instance of the workflow class called Workflow1
    WorkflowInstance _wInstance = _wRuntime.CreateWorkflow(typeof(Workflow1));
    // Get the CommunicationsWorkflowRuntimeService from the Workflow Runtime

    CommunicationsWorkflowRuntimeService _cWFRService =
    (CommunicationsWorkflowRuntimeService)
    _wRuntime.GetService(typeof(CommunicationsWorkflowRuntimeService));

    // Pass the call to the CommunicationsWorkflowRuntimeService service and associate it
    with the workflow instance created

    _cWFRService.EnqueueCall(_wInstance.InstanceId, call);
    ```

- Retrieve call(s) associated with a particular workflow instance.

    When the workflow instance executes, the calls associated with the workflow instance can be retrieved by calling either of the following two methods.

    ```
    _cWFRService.DequeueCall(/* specify Workflow Instance ID */);
    ```

    or

    ```
    _cWFRService.GetWorkflowCalls(/* specify Workflow Instance ID */);
    ```

    Workflow Instance ID is a property of the *Workflow* instance.

    The *AcceptCall* activity uses the *DequeueCall* method to accept the call. Custom activities can also use these methods to retrieve the call as needed.

    As the name suggests, *DequeueCall* removes the call from the service. However, *GetWorkflowCalls* does not remove the calls from the service.

    DequeueCall is useful in cases in which each phone call is handled by a new instance of the workflow instance. For example, in a help desk application, a new instance of the application's workflow is created for each incoming call to walk the caller through a dialog with the automated system. *GetWorkflowCalls* is useful in cases in which the application logic might depend on multiple calls being received. For example, the application might need to connect two calls. In this case, you use the *GetWorkflowCalls* method to get both calls before connecting them.

- The *GetWorkFlowCulture* and *SetWorkFlowCulture* methods are used to retrieve or set the language culture of the workflow instance.

Because many UCMA Workflow Activities involve interacting with the user, you are required to specify a language. This language is used to ensure that the correct speech recognition engine is used to recognize user input and that the appropriate synthesis engine is used for voice output. The language of the workflow instance can be set as follows.

```
// Create an instance of the workflow, Workflow1 is the workflow class name
WorkflowInstance _wInstance = _wRuntime.CreateWorkflow(typeof(Workflow1));

// Get the CommunicationsWorkflowRuntimeService from the Workflow Runtime

CommunicationsWorkflowRuntimeService _cWFRService =
(CommunicationsWorkflowRuntimeService)
_wRuntime.GetService(typeof(CommunicationsWorkflowRuntimeService));

// Set the workflow instance language

_cWFRService.SetWorkflowCulture(_wInstance.InstanceId, new CultureInfo("en-US"));
```

- The *GetEndpoint* and *SetEndpoint* methods are used to retrieve and set (respectively) the endpoint to be used in a workflow instance.

    The UCMA Workflow Activities *OutboundCall* and *GetPresence* require an endpoint (as described in the "Endpoints" section in Chapter 7, "Structure of a UCMA Application") to execute. An endpoint is associated with a workflow instance in the same way that a call is associated with a workflow instance. The activities in a workflow instance automatically assume the context of the endpoint associated with the workflow. You can set the endpoint that you want the workflow instance to use as follows.

```
// Definition of an endpoint
private static ApplicationEndpoint _endpoint;

// Create an instance of the workflow class called Workflow1

WorkflowInstance _wInstance = _wRuntime.CreateWorkflow(typeof(Workflow1));

// Get the CommunicationsWorkflowRuntimeService from the Workflow Runtime

CommunicationsWorkflowRuntimeService _cWFRService =
(CommunicationsWorkflowRuntimeService)
_wRuntime.GetService(typeof(CommunicationsWorkflowRuntimeService));

// Specify the endpoint for the workflow instance
    _cWFRService.SetEndpoint(_wInstance.InstanceId, _endpoint);
```

TrackingDataWorkflowRuntimeService   You use *TrackingDataWorkflowRuntimeService* to track values from the *MainPrompt* and *RecognitionResult* properties of certain activities. It is represented by the *TrackingDataWorkflowRuntimeService* object of the UCMA Workflow. For example, when the *SpeechStatementActivity* property, *IsDataTrackingEnabled*, is set to *True*, the *SpeechStatement* activity stores the value of its *Main* prompt in this service. This data is stored in the service for the lifetime of the workflow instance in which the activity is running. This service is optional and is applicable only for the following activities:

- *SpeechStatementActivity*
- *SpeechQuestionAnswerActivity*
- *InstantMessagingStatementActivity*
- *InstantMessagingQuestionAnswerActivity*

This functionality is useful in the cases when the caller has to be transferred to a human operator or agent. This service contains a record of the dialog between the user and the automated system, which can then be sent to the agent to provide context about the caller. This prevents the agent from asking the caller the same questions that the automated system already asked. All of the data for a particular workflow instance is deleted when the workflow instance terminates.

You add this service in the *InitializeWorkflow* method in the Program.cs file. The Program.cs file is generated automatically by the project templates.

```
WorkflowRuntime _wRuntime = new WorkflowRuntime();
_wRuntime.AddService(new TrackingDataWorkflowRuntimeService());
```

Using the *TrackingDataWorkflowRuntimeService*, the developer can achieve the following:

- Add property values to track during the lifetime of the workflow instance.

    The *SpeechStatement*, *SpeechQuestionAnswer*, *InstantMessagingStatement*, and *InstantMessagingQuestionAnswer* activities automatically add property values to this service when the *IsDataTrackingEnabled* property on these activities is set to *True*. The *MainPrompt* property for all of these activities is tracked. In addition, the *RecognitionResult* property is tracked for *SpeechQuestionAnswerActivity* and *InstantMessagingQuestionAnswerActivity*.

    Before custom activities can add property values to this service, they must obtain a handle to the service object. In a workflow instance, any object that implements the *IServiceProvider* interface can access *TrackingDataWorkflowRuntimeService*. *ActivityExecutionContext* is an example of such an object. You can access *ActivityExecutionContext* by overriding the *Execute* method of *SequentialWorkflowActivity*. You can retrieve the *TrackingDataWorkflowRuntimeService* as follows.

    ```
    /// <summary>
    /// Override method of the Sequential Workflow Activity
    /// </summary>
    ```

```
protected override ActivityExecutionStatus Execute(Activity
ExecutionContext executionContext)

{

        TrackingDataWorkflowRuntimeService _tdRuntimeService =
(TrackingDataWorkflowRuntimeService)(executionContext.GetService
(typeof(TrackingDataWorkflowRuntimeService)));

    return base.Execute(executionContext);

}
```

To add new properties to the *TrackingDataWorkflowRuntimeService* to be tracked, use the *AddTrackingData(Guid, ActivityTrackingData)* method. The *ActivityTrackingData* object represents the name of the activity and the value of that property.

- Retrieve data tracked for a particular workflow instance.

    You can retrieve data that is being tracked by using the *GetTrackingData(Guid)* method from *TrackingDataWorkflowRuntimeService*. This returns a collection of type *ActivityTrackingData*. Each *ActivityTrackingData* object represents the name of an activity, the name of the property to be tracked for that activity, and the value of that property.

General Activities

The two activities in the UCMA Workflow, which are listed in Table 5-1, are important because they define the flow of the work.

TABLE 5-1 General UCMA Workflow Activities

Activity	Description
CommunicationsSequenceActivity	An activity that provides infrastructure for most UCMA Workflow Activities to execute. You can use this for scoping and as a container activity.
GotoActivity	An activity that moves the execution of the workflow to the target activity.

CommunicationsSequenceActivity

CommunicationsSequenceActivity provides a base container specific for executing most UCMA Workflow Activities. *CommunicationsSequenceActivity* provides the following benefits:

- Acts as a container for other activities
- Provides custom views for dialog functionality
- Enables the use of *GotoActivity*
- Provides a call to its child activities to execute

Acting as a Container Activities can be dropped inside a *CommunicationsSequenceActivity* to take advantage of the functionality defined in *CommunicationsSequenceActivity*. *CommunicationsSequenceActivity* is derived from a Windows Workflow Sequence activity. You can expand and collapse this activity. You can collapse this activity to save screen space so that you can work on other areas of the workflow.

Providing custom views for dialog functionality *CommunicationsSequenceActivity* provides two custom views, *Commands* and *CommunicationsEvents*, as shown in Figure 5-4. You use these views to enable commands and handle communication events, which are important to implement any useful dialog. For more information about how to use these views, see the "Command Activities," "Call Control Communications Event Activities," and "Dialog Communications Event Activities" sections later in this chapter.

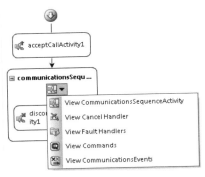

FIGURE 5-4 *CommunicationsSequenceActivity* views.

Note You can also navigate to these views by right-clicking *CommunicationsSequenceActivity* and selecting View Commands or View CommunicationsEvents.

You can use *CommunicationsSequenceActivity* to define a scope of *CommunicationEvents* activities. For example, a *CommunicationEvent* defined in a *CommunicationsSequenceActivity* is not visible outside of that scope.

For more information about how to use *CommunicationsSequenceActivity* for scoping communications events, see the "What Are Communication Events?" sidebar later in this chapter.

Enabling the use of the *Goto* activity After a Workflow Runtime executes an activity, that activity cannot be executed again in the same workflow instance. As a result, special logic is required in constructs like the *While* activity, which requires you to execute the same activity more than once. You achieve this by cloning, which is a Windows Workflow construct. Before executing the activity, a clone of the activity is made and the cloned activity is executed. This preserves the original activity to be executed later (at which point, it is cloned again). For more information about activity execution context and cloning, see the "Understanding the Activity Execution Context" article at http://msdn.microsoft.com/en-us/library/aa349099.aspx.

Because the *Goto* activity by its nature can cause its target activity to be executed again, the target activity has to be cloned. The *CommunicationsSequence* activity clones each of its children before executing them. As a result, these activities are available to be executed again.

Providing a call to its children activities to execute Because UCMA Workflow Activities perform actions on a call whether it is a phone call or an IM call, *CommunicationsSequenceActivity* defines the call to be operated on by its children activities. For example, the *SpeechStatement* activity needs to know which call on which to play the message, and the *InstantMessagingStatement* activity needs to know which IM session to reply to. Determining which phone or IM call is associated with an activity is accomplished by the *CallProvider* activity. The *AcceptCall* activity and the *OutboundCall* activity expose the *CallProvider* object as a property. Table 5-2 lists the properties defined by the *CallProvider* object, which are required to establish a dialog with the user.

TABLE 5-2 **Properties of the *CallProvider* object**

Property	Description
Call	The call to be used
DtmfRecognizer	The engine used to recognize dual-tone multifrequency (DTMF) input against DTMF grammars for the call
RecognitionConnector	The object to help in media flow during recognition for the call
SpeechRecognizer	The engine to recognize speech or IM input against speech grammars for the call
SynthesisConnector	The object to help in media flow voice synthesis for the call
Synthesizer	The engine to synthesize the voice output for the call
ToneController	The object that identifies the DTMF tones

Binding to a CallProvider *property* The *CallProvider* property of *CommunicationsSequenceActivity* can be linked to the *CallProvider* property that is exposed by one of the following activities: *AcceptCall* or *OutboundCall*. You perform this linking by using *ActivityBind* so that the value of the property is available only at run time. For information about *ActivityBind*, see the "Using Dependency Properties" MSDN article at http://msdn.microsoft.com/en-us/library/ms734499.aspx. This linking ensures that all children activities of *CommunicationsSequenceActivity* manipulate the call provided by the *CallProvider* on *CommunicationsSequenceActivity*. Figure 5-5 demonstrates how a *CallProvider* property is created and passed on to UCMA Workflow Activities by using *CommunicationsSequenceActivity*.

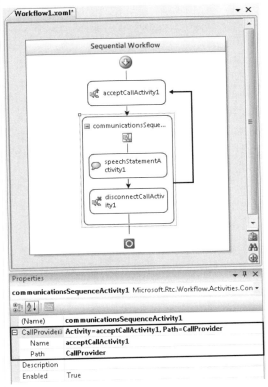

FIGURE 5-5 *CommunicationsSequenceActivity* binding to the *CallProvider* property of the *AcceptCall* activity.

The following list describes the activities in Figure 5-5:

- The *AcceptCall* activity named *acceptCallActivity1* creates a *CallProvider* object that is exposed publicly as a property of the same name.

- The *CallProvider* property of the *CommunicationsSequenceActivity* named *communicationsSequenceActivity1* is bound to the *CallProvider* property of the activity, *acceptCallActivity1*. This is denoted by the line running from *CommunicationsSequence-Activity* to the *acceptCall* activity.

- At run time, the *SpeechStatementActivity* named *speechStatementActivity1* uses the call associated with its parent *CommunicationsSequenceActivity* (in this case, *communicationsSequenceActivity1*). Because the *CallProvider* property of *communicationsSequenceActivity1* is bound to the *CallProvider* property created by *acceptCallActivity1*, *speechStatementActivity1* plays the message on the same call that was accepted by *acceptCallActivity1*. Similarly, *disconnectCallActivity1* terminates the call accepted by *acceptCallActivity1*, which it obtained from its parent *CommunicationsSequenceActivity*.

Using **CallProvider** *to support multiple calls*   The same concept can be extended to have multiple calls in a workflow, as shown in Figure 5-6. In this example, there are two *CommunicationsSequenceActivity* activities. The *CallProvider* property of *parentActivity1* is bound to the *CallProvider* property of *acceptCallActivity1* (as shown by the dotted arrow) and the *CallProvider* property of *parentActivity2* is bound to the *CallProvider* property of *outboundCallActivity1* (as shown by the solid arrow). All of the children of *parentActivity1* (that is, *speechStatementActivity1* and *disconnectCallActivity1*) perform operations on the call bound to *acceptCallActivity1*. All of the children of *parentActivity2* (that is, *speechStatementActivity2* and *disconnectCallActivity2*) perform operations on the call established by *outboundCallActivity1*.

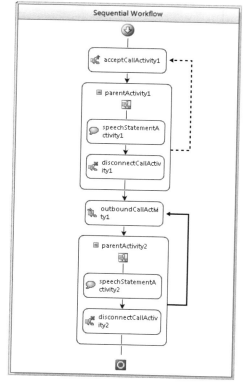

FIGURE 5-6 Two *CommunicationsSequenceActivity* activities binding to different *CallProvider* properties.

Using **CallProvider** *in nested* **CommunicationsSequenceActivity**   The same binding concept can then define which call to use in nested *CommunicationsSequence* activities. In the workflow shown in Figure 5-7, all of the children of *parentActivity2* use the call placed by *outboundCallActivity1* even though *parentActivity2* is nested inside *parentActivity1*. This occurs because the *CallProvider* property of *parentActivity2* is bound to the *CallProvider* property of *outboundCallActivity1*.

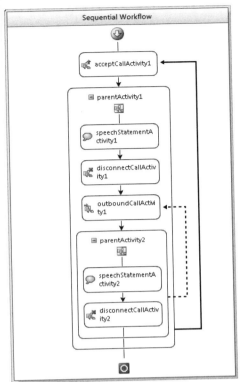

FIGURE 5-7 Binding to different *CallProvider* properties within nested *CommunicationsSequence* activities.

To figure out which call is used with a particular activity (that is, Activity A), just search its parent tree to find the first *CommunicationsSequenceActivity* whose *CallProvider* property is bound to a *CallProvider* property of another activity (for example, an *AcceptCall* activity, an *Outbound Call* activity, or a custom *CallProvider* activity). It is possible that a *CommunicationsSequenceActivity* in a parent tree does not have its *CallProvider* activity bound to anything. In this case, continue through the tree until you reach a *CommunicationsSequenceActivity* in the tree that has the *CallProvider* property set.

The *Goto* Activity

The *Goto* activity moves the execution to a different activity in the workflow. A good dialog design involves jumps in execution. For example, if the caller wants to return to the beginning of the dialog tree, the dialog execution should move to the start of the application. The *Goto* activity facilitates such dialog designs.

To specify a target for a *Goto* activity, set the *TargetActivityName* property of the *Goto* activity as shown in Figure 5-8. You can open the panel in Figure 5-8 by clicking the ... button in the window of the *TargetActivityName* property. In the following example, the valid target activities (that is, *speechStatementActivity1*, *sequenceActivity1*, and *disconnectCallActivity1*) are enabled for selection and invalid targets are disabled for selection.

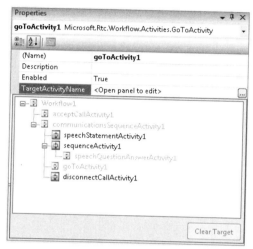

FIGURE 5-8 Valid targets for the *TargetActivityName* property of *Goto*.

Call Control Activities

Four activities are available to perform call control actions, as shown in Table 5-3. Some of these activities are valid for both phone calls and IM calls, but some are specific only to phone calls.

TABLE 5-3 Call Control Activities

Activity	Description
AcceptCall	Accepts the incoming call associated with the workflow instance. This activity accepts both phone calls and IM calls.
DisconnectCall	Disconnects the call specified by the *CallProvider* property of *CommunicationsSequenceActivity* to which it belongs. This activity disconnects both phone and IM calls.
OutboundCall	Places an outbound call using the endpoint associated with the workflow instance. This can create either a phone call or an IM call.
BlindTransfer	Transfers the current call to the specified target. The current call is defined by the *CallProvider* property of *CommunicationsSequenceActivity* to which the *BlindTransfer* belongs. This activity can perform a blind transfer on phone calls, but not IM calls.

The *AcceptCall* Activity

The *AcceptCall* activity accepts either a phone call or an IM call. If the call is not incoming, *AcceptCall* throws an *InvalidOperation* exception. The *AcceptCall* is associated with the workflow instance to which it belongs. The call is retrieved from the *CommunicationsWorkflow-RuntimeService* using the *DequeueCall* API.

The *AcceptCall* activity also creates a *CallProvider* object that is exposed as a property. Because an *AcceptCall* activity can accept one call, one *CallProvider* object is created for that call. Table 5-4 describes the properties of the *AcceptCall* activity.

TABLE 5-4 Properties of *AcceptCallActivity*

Property	Description
CallProvider	Exposes an object that wraps other objects needed to support a dialog (for example, *Call, SpeechSynthesizer,* or *SpeechRecognizer*). This is available after the *AcceptCall* activity executes.
ContentIds	A list of *ContentIds* to be passed to the UCMA as part of accepting the call. You need to set this property before the *AcceptCall* activity executes.
Headers	Defines Session Initiation Protocol (SIP) headers (referred to as *SignalingHeaders*) that are passed when accepting the call. You can specify the *SignalingHeaders* by using this property before the *AcceptCall* activity executes.

The *DisconnectCall* Activity

The *DisconnectCall* activity disconnects a phone call or an IM call that is exposed by the *CallProvider* property of the parent *CommunicationsSequenceActivity*. If the parent *CommunicationsSequenceActivity* does not have a value specified for *CallProvider*, it tries to get it from the grandparent *CommunicationsSequenceActivity*. It follows the tree upward until it finds a *CommunicationsSequenceActivity* that has the specified *CallProvider* value.

If the call is already disconnected, *DisconnectCall* does not throw any exceptions. It is likely that the caller will end the call before the *DisconnectCall* activity executes. *DisconnectCall* also cleans up all of the *CallProvider* objects associated with the call. Table 5-5 lists a property and description of the *DisconnectCall* activity.

TABLE 5-5 Property of *DisconnectCallActivity*

Property	Description
Headers	Defines SIP headers (that is, *SignalingHeaders*) passed when disconnecting the call. You can specify the *SignalingHeaders* by using this property before the *DisconnectCall* activity executes.

The *OutboundCall* Activity

The *OutboundCall* activity creates a new outbound call. The type of call created by this activity depends on the value of the *CallType* property.

The *OutboundCall* activity also creates a list of objects needed for supporting a dialog to be used by other UCMA Workflow Activities. These objects are wrapped as a *CallProvider* object, which is exposed as a property on *OutboundCall*. Because only one call is created by the *OutboundCall* activity, only one *CallProvider* object is created per call. Table 5-6 describes the properties of the *OutboundCall* activity.

TABLE 5-6 Properties of *OutboundCallActivity*

Property	Description
CalledParty	The party to which the call is being placed. The value should be a valid SIP Uniform Resource Identifier (URI). Examples: sip:someone@contoso.com or tel:1112223333
CallProvider	Exposes an object that contains all of the objects needed to support a dialog (for example, *Call*, *SpeechSynthesizer*, or *SpeechRecognizer*). This is available after the *OutboundCall* activity executes.
CallType	The type of the call created by the *OutboundCall* activity. Possible values are *AudioVideoCall* or *InstantMessagingCall*.
CustomMimeParts	A list of custom Multipurpose Internet Mail Extensions (MIME) part descriptions.
Headers	A list of *SignalingHeaders* to be passed to the UCMA as part of accepting the call. You need to set this property before the *AcceptCall* activity executes.

The *BlindTransfer* Activity

The *BlindTransfer* activity transfers the call to a specified party without waiting for the transfer to succeed or fail. This is valid only for phone calls, as described in Table 5-7.

TABLE 5-7 Properties of *BlindTransferActivity*

Property	Description
CalledParty	The party the call is being transferred to. The value should be a valid SIP URI. Examples: sip:someone@contoso.com or tel:1112223333

Dialog Activities

Dialog activities are the backbone of the interaction between the user and an application. Using these activities, an application can play or send messages to the user and recognize their responses.

The UCMA Workflow has four activities that enable interaction with the user. Each activity is valid for either a phone call or an IM call, as described in Table 5-8.

TABLE 5-8 Dialog Activities

Activity	Description
InstantMessagingStatement	Sends an instant message to the user.
InstantMessagingQuestionAnswer	Asks a question by sending an instant message to the user and recognizes the user's text response.
SpeechQuestionAnswer	Asks the user a question and recognizes user response via a phone call. You can use either a recorded .wav file or a synthesized text-to-speech message. Both speech and DTMF inputs from the user can be recognized.
SpeechStatement	Plays a message to the user via a phone call. You can use either a recorded .wav file or a synthesized text-to-speech message.

The *SpeechStatement* Activity

The *SpeechStatement* activity plays a specified message to the user. This activity can execute only in an audio-video (A/V) call. The message can be a prerecorded .wav file or a synthesized message using a text-to-speech engine, as described in Table 5-9.

TABLE 5-9 Properties of *SpeechStatementActivity*

Property	Description
IsDataTrackingEnabled	When this property is set to *True*, if *TrackingDataWorkflowRuntimeService* has been added to the Workflow Runtime, it stores the value of *MainPrompt*.
MainPrompt	The prompt that the activity plays to the user when it executes. This is of the type *Microsoft.Speech.Synthesis.PromptBuilder*.

Playing a message A message can be played to the user by setting the prompt on the *SpeechStatement* activity. You can set the prompt on the *SpeechStatement* activity either at the designer level or in code.

At design time, you can set static prompts in the Properties window, as shown in Figure 5-9.

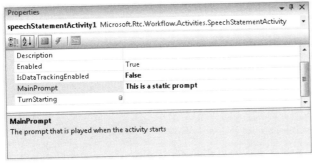

FIGURE 5-9 Setting static prompts for the *SpeechStatement* activity.

Alternatively, you can set a dynamic prompt in code by using the *TurnStarting* event handler. For details, see the "Turn and the *TurnStarting* Event" sidebar. The following code example is a sample event handler.

```
private void speechStatementActivity1_TurnStarting(object sender,
Microsoft.Rtc.Workflow.Activities.SpeechTurnStartingEventArgs e)

{}
```

SpeechTurnStartingEventArgs exposes the prompt that is played and the type of the prompt. The prompt type is not relevant to *SpeechStatement* because it has only one type (*Main*). You can use this event handler to change the prompt dynamically on this activity.

> **Turn and the *TurnStarting* Event**
>
> A turn is a unit of dialog between two parties. In the case of a *SpeechStatement* activity, turn corresponds to the activity playing the message and the user receiving it. *TurnStarting* is an event that is raised just before the *SpeechStatement* activity is ready to play the message. *TurnStarting* event handler can be registered like any other .NET event handler. Visual Studio provides an easy way to do this in the property window of the selected activity.

The *SpeechQuestionAnswer* Activity

The *SpeechQuestionAnswer* activity plays a message to the user and recognizes the user's response. The user can respond via speech or DTMF. This activity can execute only in an A/V call. The message can be a prerecorded .wav file or a synthesized message using a text-to-speech engine. The *SpeechQuestionAnswer* activity properties are described in Table 5-10.

TABLE 5-10 Properties of *SpeechQuestionAnswerActivity*

Property	Description
CanBargeIn	When set to *True*, the user can interrupt the question with an answer. If it is set to *False*, the user's response is discarded until the question finishes playing.
CompleteTimeout	The length of silence following user speech before the speech recognizer finalizes a result.
DtmfGrammars	A collection of grammar files for recognizing the user's DTMF input.
ExpectedDtmfInputs	An array of strings that are valid for the user's DTMF input.
ExpectedSpeechInputs	An array of strings that are valid for the user's speech input.
Grammars	A collection of grammar files for recognizing the user's speech input.
IncompleteTimeout	The length of silent time after which recognition ends. This value applies when the speech prior to the silence is an incomplete match of all active grammars.

TABLE 5-10 Properties of *SpeechQuestionAnswerActivity*

Property	Description
InitialSilenceTimeout	The length of time in which the user has to start responding or the activity treats the user's response as silence. The default is 3 seconds.
IsDataTrackingEnabled	When set to *True*, if *TrackingDataWorkflowRuntimeService* has been added to the Workflow Runtime, it stores the value of *MainPrompt* that is played and the corresponding recognition result.
MainPrompt	The prompt that the activity plays to the user when this activity executes. This is of the type *Microsoft.Speech.Synthesis.PromptBuilder*.
PreFlushDtmf	When set to *True*, all of the DTMF digits in the buffer are discarded and the users cannot type ahead their DTMF responses. When set to *False*, users can type ahead their DTMF responses.
Prompts	A collection of all of the prompts available for this activity.
RecognitionResult	The result of the recognition of the user's speech or DTMF response.

Asking a question You can ask the user a question by setting the prompt on the *SpeechQuestionAnswer* activity. The prompt on the activity can be set in the same way as you would with the *SpeechStatement* activity, either at the designer level or in code.

At design time, you can set static prompts in the Properties window, as shown in Figure 5-10.

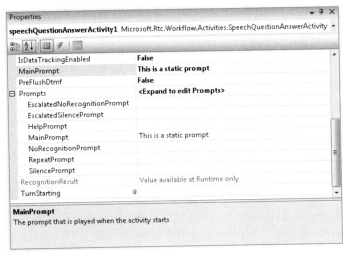

FIGURE 5-10 Setting static prompts for the *SpeechQuestionAnswer* activity.

Alternatively, you can set a dynamic prompt in code by using the *TurnStarting* event handler. For details, see the "Turn and the *TurnStarting* Event" sidebar earlier in this chapter. The following code example is a sample event handler.

```
private void speechQuestionAnswerActivity1_TurnStarting(object sender,
Microsoft.Rtc.Workflow.Activities.SpeechTurnStartingEventArgs e)
{}
```

SpeechTurnStartingEventArgs exposes the prompt that is played and the type of the prompt. You can use this event handler to change the prompt on this activity dynamically.

> ## Prompt Types
>
> The *SpeechQuestionAnswer* and *InstantMessagingQuestionAnswer* activities have many types of prompts. Table 5-11 describes what each prompt type is used for.
>
> **TABLE 5-11 Prompt Types**
>
> | Prompt Type | Description |
> | --- | --- |
> | Acknowledge | This prompt is played before the user input is sent for recognition. This is available only for the *InstantMessagingQuestionAnswer* activity. |
> | EscalatedNoRecognition | This prompt is played by the activity when the user's response is not recognized a second time. This is not mandatory. This is available only for the *SpeechQuestionAnswerActivity* and *InstantMessagingQuestionAnswer* activities. |
> | EscalatedSilence | This prompt is played by the activity when the user does not respond in the time specified in the *InitialSilenceTimeout*. This is not mandatory. This is available only for the *SpeechQuestionAnswerActivity* and *InstantMessagingQuestionAnswer* activities. |
> | Help | This prompt is played by the activity when the user response triggers a help command in that scope. For details, see the "What Is a Command?" sidebar later in this chapter. This is not mandatory. This is available only for the *SpeechQuestionAnswerActivity* and *InstantMessagingQuestionAnswer* activities. |
> | Main | This is the first prompt that the activity plays. This is mandatory. If it is not set, the activity throws an exception. This is available only for the *SpeechStatement, SpeechQuestionAnswerActivity, InstantMessagingStatement*, and *InstantMessagingQuestionAnswer* activities. |
> | NoRecognition | This prompt is played by the activity the first time the user's response is not recognized. This is not mandatory. This is available only for the *SpeechQuestionAnswerActivity* and *InstantMessagingQuestionAnswer* activities. |
> | Repeat | This prompt is played by the activity when the user response triggers a repeat command in that scope. For details, see the "What Is a Command?" sidebar later in this chapter. This is not mandatory. This is available only for *SpeechQuestionAnswer*Activity. |
> | Silence | This prompt is played by the activity the first time the user does not respond in the time specified in *InitialSilenceTimeout*. This is not mandatory. This is available only for the *SpeechQuestionAnswerActivity* and *InstantMessagingQuestionAnswer* activities. |

Prompt Fallback Logic

As mentioned in Table 5-11, all prompts except *Main* are optional. However, there are conditions in which one of these nonoptional prompts needs to be played. For example, if a user's response cannot be recognized but the *NoRecognition* prompt has not been specified, then a fallback prompt is played. The fallback prompt that is played depends on the prompt that was supposed to be played. Prompt types and their fallback prompt types are listed in Table 5-12.

TABLE 5-12 Prompt Types and the Associated Fallback Prompt Types

Prompt Type	Fallback Prompt Type
Main	Not applicable; this prompt is mandatory.
Acknowledge	No fallback. If it is not set, then this prompt is not played.
EscalatedNoRecognition	NoRecognition
EscalatedSilence	Silence
Help	Main
NoRecognition	Help
Repeat	Main
Silence	Help

For example, suppose that you have set only the *Main* prompt for a *SpeechQuestionAnswer* activity. If the user's response is not recognized, ideally the *NoRecognition* prompt is played. However, because the *NoRecognition* prompt is not specified, *SpeechQuestionAnswer* tries to play the *Help* prompt. Again, because the *Help* prompt is also not specified (that is, only the *Main* prompt is specified), the *Main* prompt, which is the fallback prompt for the *Help* prompt, is played.

Recognizing user responses *SpeechQuestionAnswer* needs to know a set of expected user responses to recognize. *SpeechQuestionAnswer* has four properties that enable the developer to specify this set. The following description uses speech as the example. The process for recognizing DTMF input is very similar.

Specifying expected inputs *ExpectedSpeechInputs*, a property on *SpeechQuestionAnswer*, is an array of strings that the developer can specify. At run time, this array is converted to a grammar so that, if the user responds with anything in this set, it can be recognized. You can specify the set of valid user responses by using the String Collection Editor window from the Properties window, as shown in Figure 5-11. This is useful in scenarios in which the list of expected inputs is finite; for example, a list of pizza sizes or yes/no responses. Similarly, you can use *ExpectedDtmfInputs* for DTMF inputs.

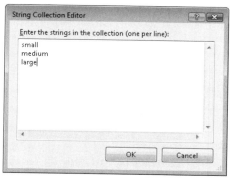

FIGURE 5-11 A String Collection Editor window opened from the *ExpectedSpeechInputs* property of *SpeechQuestionAnswer*.

Specifying the grammar file   You can use a grammar file to list the expected inputs. A grammar file is a more flexible and powerful way of specifying the set of expected inputs (for example, dates and numbers). These grammar files have to follow a certain schema, as described by the Speech Recognition Grammar Specification (SRGS).

Given a grammar file, the file can be attached to *SpeechQuestionAnswer* by using the *Grammars* property. This is a collection of grammar files used to recognize user input. In the Properties window, you can specify multiple grammar files by separating them with a semi-colon. Similarly, you can specify DTMF grammars by using the *DTMFGrammars* property.

> ### What Is a Grammar?
>
> A grammar is intended for use by speech recognizers so that developers can specify the words and patterns of words for which you want the speech recognizer to listen. For example, the grammar file should contain all of the words or pattern of words that the developer deems a valid response to the question, "What size pizza would you like?"

InstantMessagingStatementActivity

InstantMessagingStatementActivity sends a specified message to the user. This activity can execute only on IM calls. Table 5-13 lists *InstantMessagingStatementActivity* properties.

TABLE 5-13 Properties of *InstantMessagingStatementActivity*

Property	Description
IsDataTrackingEnabled	When set to *True*, and if *TrackingDataWorkflowRuntimeService* has been added to the Workflow Runtime, it stores the value of *MainPrompt*.
MainPrompt	The message that the activity sends to the user when this activity executes. This is of the type *string*.

Setting the prompt   You can set the prompt on *InstantMessagingStatementActivity* either at the designer level or in code.

At design time, you can set static prompts in the Properties window, as shown in Figure 5-12.

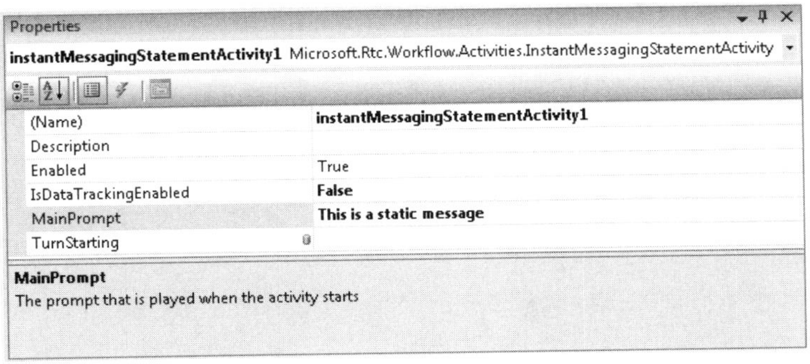

FIGURE 5-12 Setting a static prompt for *InstantMessagingStatementActivity*.

Alternatively, you can set a dynamic prompt in code by using the *TurnStarting* event handler. For details, see the "Turn and the *TurnStarting* Event" sidebar earlier in this chapter. The following code example is a sample event handler.

```
private void instantMessagingStatementActivity1_TurnStarting(object sender,
Microsoft.Rtc.Workflow.Activities.InstantMessagingTurnStartingEventArgs e)
{}
```

InstantMessagingTurnStartingEventArgs exposes the prompt that is played and the type of the prompt. The type of prompt is not relevant to *InstantMessagingStatement* because it has only one type (*Main*). You can use this event handler to change the prompt on this activity dynamically.

InstantMessagingQuestionAnswerActivity

InstantMessagingQuestionAnswerActivity sends a message to the user and recognizes the user's response. This activity can execute only on IM calls. Table 5-14 lists the *InstantMessagingQuestionAnswerActivity* properties.

TABLE 5-14 Properties of *InstantMessagingQuestionAnswerActivity*

Property	Description
ExpectedInputs	An array of strings that are valid for the user's IM input.
Grammars	A collection of grammar files for recognizing a user's IM input.
InitialSilenceTimeout	The length of time in which the user has to start responding or the activity treats the user's response as silence. The default is 30 seconds.

TABLE 5-14 Properties of *InstantMessagingQuestionAnswerActivity*

Property	Description
IsDataTrackingEnabled	When set to *True*, and if *TrackingDataWorkflowRuntimeService* has been added to the Workflow Runtime, it stores the value of *MainPrompt* that is played and the corresponding recognition result.
MainPrompt	The message that the activity sends to the user when this activity executes. This is of the type *string*.
Prompts	A collection of all of the prompts available for this activity.
RecognitionResult	The result of the recognition of the user's IM response.

Asking an IM question   You can ask the user an IM question by setting the prompt on *InstantMessagingQuestionAnswerActivity*. The prompt on the activity can be set in the same way as you would with *InstantMessagingStatementActivity*, either at the designer level or in code.

At design time, you can set static prompts in the Properties window, as shown in Figure 5-13.

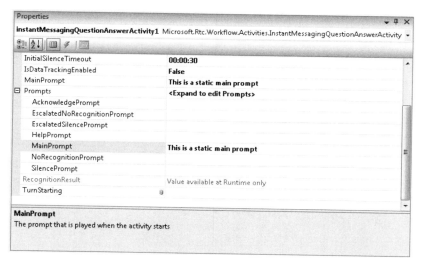

FIGURE 5-13 Setting static prompts for *InstantMessagingQuestionAnswerActivity*.

Alternatively, you can set a dynamic prompt in code by using the *TurnStarting* event handler. For details, see the "Turn and the *TurnStarting* Event" sidebar earlier in this chapter. The following code example is a sample event handler.

```
private void instantMessagingQuestionAnswerActivity1_TurnStarting(object sender,
    Microsoft.Rtc.Workflow.Activities.InstantMessagingTurnStartingEventArgs e)
{}
```

InstantMessagingTurnStartingEventArgs exposes the prompt that is played and the type of the prompt. You can use this event handler to change the prompt on this activity dynamically.

Recognizing the user's text response   Similar to *SpeechQuestionAnswer*, *InstantMessagingQuestionAnswerActivity* needs to have a set of expected user text responses to recognize. *InstantMessagingQuestionAnswerActivity* has two properties that enable the developer to specify this set.

Specifying expected inputs   *ExpectedInputs*, a property on *InstantMessagingQuestionAnswerActivity*, is an array of strings that the developer can specify. This is exactly the same as the *ExpectedSpeechInputs* or *ExpectedDTMFInputs* property of *SpeechQuestionAnswer*.

Specifying the grammar file   You can use a grammar file to list the expected inputs. Even though these files follow a certain schema as described by the SRGS, you can use it to recognize IM input.

Given a grammar file, the file can be attached to *InstantMessagingQuestionAnswerActivity* by using the *Grammars* property in the exact way that you would with *SpeechQuestionAnswer*.

Command Activities

Five activities are available to enable developers to add commands to the dialog interaction with the user. Each activity is valid for either a phone call or an IM call. Command activities and descriptions are listed in Table 5-15.

TABLE 5-15 Different Types of Command Activities

Activity	Description
InstantMessagingCommand	A generic command that is triggered based on the attached grammar file. Valid only for IM calls.
InstantMessagingHelpCommand	A command used for scenarios to provide help to the user. It is triggered based on the attached grammar file. Valid only for IM calls.
SpeechCommand	A generic command that is triggered based on the attached speech or DTMF grammar file. Valid only for phone calls.
SpeechHelpCommand	A command used for scenarios to provide help to the user. It is triggered based on the attached speech or DTMF grammar file. Valid only for phone calls.
SpeechRepeatCommand	A command used for scenarios to repeat the question back to the user. It is triggered based on the attached speech or DTMF grammar file. Valid only for phone calls.

What Is a Command?

A command can be considered a valid user response that may not be a direct answer to the question asked. For example, answers to the question, "Which size pizza would you like?" could include "Small" or "Large." However, the following responses are also valid, even though they do not answer the question:

- What sizes do you have?
- Do you have a family-size pizza?
- What was that again?
- I think I have changed my mind. I don't want pizza anymore.

The first two responses ask for more information (that is, requesting help). The third response asks for the question to be repeated. The last response just states that the user does not want any pizza. These responses can be called commands because the user is commanding that more information be provided or that the dialog be ended.

Commands are an integral part of a good dialog user interface when interacting with an automated system, be it over the phone or via IM. They allow users to ask for more information (that is, Help), replay the question again (that is, Repeat), or perform more generic actions, such as restarting. It increases the chance that users understand what is expected of them so that they can provide the correct input.

SpeechCommandActivity

SpeechCommandActivity is a command activity that you can use for any generic command when interacting via a phone call. You can specify this activity for a scope as defined by a *CommunicationsSequence* activity. Anytime *SpeechQuestionAnswer* is active in that scope, the *SpeechCommand* activity, if defined, is also active in that scope.

For example, a *SpeechQuestionAnswer* activity is defined to ask for a pizza size, as shown in Figure 5-14.

Also, *SpeechCommandActivity* is defined in that scope as well. You can add *SpeechCommandActivity* in the command view of *CommunicationsSequenceActivity* only. You can open the *Command* view by right-clicking *CommunicationsSequenceActivity* and then clicking View Commands. A snapshot of a *SpeechCommandActivity* is added in the *Command* view, as shown in Figure 5-15. The *SpeechCommandActivity* is named *operatorCommand*.

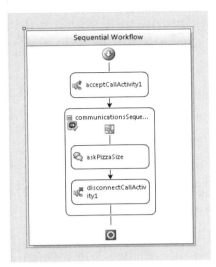

FIGURE 5-14 A simple workflow with a *SpeechQuestionAnswer* activity named *askPizzaSize*.

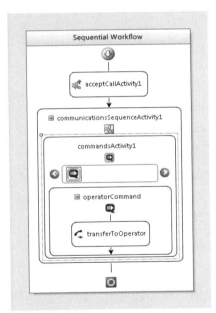

FIGURE 5-15 A *Command* view of *communicationsSequenceActivity1* showing the *SpeechCommand* named *operatorCommand*.

In this particular example, when the *askPizzaSize* activity is executing, *operatorCommand* is also active. If the user's response matches the grammar defined in the *operatorCommand* grammar, the *operatorCommand* activity executes all of its children. In this case, the *transferToOperator* activity, which is a *Blind Transfer* activity, executes. Table 5-16 lists the properties of *SpeechCommandActivity*.

TABLE 5-16 Properties of *SpeechCommandActivity*

Property	Description
DtmfGrammars	A collection of grammar files for recognizing the user's DTMF input for this command
ExpectedDtmfInputs	An array of strings that are valid for the user's DTMF input for this command
ExpectedSpeechInputs	An array of strings that are valid for the user's speech input for this command
Grammars	A collection of grammar files for recognizing the user's speech input for this command
RecognitionResult	The result of the recognition of the user's speech or DTMF response if matched by the command grammar

SpeechHelpCommandActivity

SpeechHelpCommandActivity is a custom speech command activity. It is added to the *Commands* view of a *CommunicationsSequenceActivity* in the same way as for *SpeechCommandActivity*. Like *SpeechCommandActivity*, it is also active during all of the *SpeechQuestionAnswer* activities in the scope that this command is defined within. However, when the user input is matched with any of the grammars defined for *SpeechHelpCommand-Activity*, the *Help* prompt of *SpeechQuestionAnswerActivity* is played automatically.

For example, a *SpeechQuestionAnswerActivity* named *askPizzaSize*, shown previously in Figure 5-14, is defined with the *MainPrompt* and *HelpPrompt* sets, as shown in Figure 5-16. Also imagine a *SpeechHelpCommandActivity* called *helpCommand* is defined in the *Commands* view in the same way *operatorCommand* was defined.

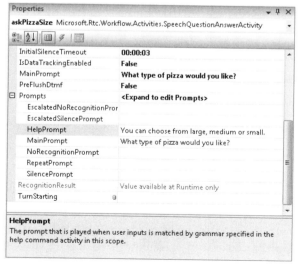

FIGURE 5-16 The *MainPrompt* and *HelpPrompt* set for the *askPizzaSize* activity.

When the *askPizzaSize* activity executes, it plays its *MainPrompt*, "What size pizza would you like?" If the user's response is matched with the *helpCommand* grammar, the *askPizzaSize* activity automatically plays its *HelpPrompt*, "You can choose from large, medium, or small." If the *HelpPrompt* is not defined, the prompt fallback logic is used.

SpeechRepeatCommandActivity

SpeechRepeatCommandActivity behaves exactly the same way as *SpeechHelpCommandActivity* except when the user input is matched with any of the grammars defined for *SpeechRepeatCommandActivity*. In this case, the *Repeat* prompt of *SpeechQuestionAnswerActivity* is played automatically. If the *Repeat* prompt is not defined, the prompt fallback logic is used.

InstantMessagingCommandActivity

InstantMessagingCommandActivity is the IM version of *SpeechCommandActivity*. You can use it to implement generic commands for IM dialog interactions, and it is active for IM calls when *InstantMessagingQuestionAnswerActivity* is executing. By replacing all speech activities with their equivalent IM activities in the example given earlier in this chapter for *SpeechCommandActivity*, you can easily create a sample application for *InstantMessagingCommandActivity*. Table 5-17 lists the properties of *InstantMessagingCommandActivity*.

TABLE 5-17 Properties of *InstantMessagingCommandActivity*

Property	Description
ExpectedInputs	An array of strings that are valid for the user's IM input for this command
Grammars	A collection of grammars for recognizing the user's IM input for this command
RecognitionResult	The result of the recognition of the user's IM response if matched by the command grammar

InstantMessagingHelpCommandActivity

InstantMessagingHelpCommandActivity is a custom IM command activity. It behaves the same way as *SpeechHelpCommandActivity* except for IM dialogs using *InstantMessagingQuestionAnswerActivity*.

Call Control Communications Event Activities

Four call control communication event activities available in the UCMA Workflow allow for the handling of call control communication events. One activity is valid for both phone calls and IM calls, while others are specific to either a phone call or an IM call, as described in Table 5-18.

TABLE 5-18 **Call Control Communications Event Activities**

Activity	Description
CallDisconnectedEvent	An event when the call (phone or IM) is disconnected.
CallOnHoldEvent	An event when the call is put on hold. Valid only for phone calls.
CallOnHoldTimeoutEvent	An event when the call is on hold for a specified amount of time. Valid only for phone calls.
CallRetrievedEvent	An event when the call that was on hold is retrieved. Valid only for phone calls.

What Are Communication Events?

There are two different types of communication events. A call control communication event occurs on the call that is used for communication. A dialog communication event is an event that occurs as part of the dialog process. They are distinguished from call control communication events because call control communication events are based on the events on the call, whereas dialog communication events are related only to the dialog between the two parties. For example, if a pizza counter attendant asks a user, "What size pizza would you like?" but the user does not say anything, the attendant is going to ask the question again. After a while (that is, after the attendant has asked the same question *n* number of times), the attendant realizes that the user is either not interested in getting pizza or does not understand the question. The attendant can then take specific action. This realization by the pizza counter attendant can be called a dialog event.

Dialog events are an integral part of a good dialog user interface. They allow the system to realize that the user is not making progress with the automated system. The system can then decide to either transfer the call to an agent or disconnect the call and play or send a (hopefully) useful message to the user if no agents are available.

CallDisconnectedEventActivity

CallDisconnectedEventActivity is an event handler activity for a "call disconnected" event, whether a phone call or an IM call. You can specify this activity for a scope, as defined by a *CommunicationsSequenceActivity*. This activity can be added only in the *CommunicationEvents* view of *CommunicationsSequenceActivity*. You can open the *CommunicationsEvents* view of *CommunicationsSequenceActivity* by right-clicking *CommunicationsSequenceActivity* and then clicking View Communication Events on the context menu. An example of a *CommunicationsSequenceActivity* with a *CallDisconnectedEventActivity* added in its *CommunicationsEvents* view is shown in Figure 5-17.

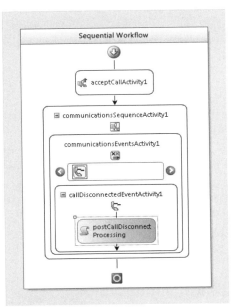

FIGURE 5-17 The *CommunicationsEvents* view of *communicationsSequenceActivity1* showing *callDisconnectedEventActivity1*.

In the preceding example, a code activity, *postCallDisconnectedProcessing*, is executed when the call is disconnected.

When *CallDisconnectedEventActivity* is executed, it cancels all of the other activities that are executing in *CommunicationsSequenceActivity*. For example, if a *SpeechQuestionAnswerActivity* is executing when the call is disconnected, *CallDisconnectedEventActivity* in that scope cancels *SpeechQuestionAnswerActivity*. If no such *CallDisconnectedEventActivity* is present in the scope, the activities that need a call to execute (in this example, *SpeechQuestionAnswerActivity*) throw an *InvalidOperationException* because they cannot execute without a call. Table 5-19 lists the property of *CallDisconnectedEventActivity*.

TABLE 5-19 Property of *CallDisconnectedEventActivity*

Property	Description
CallStateTransitionReason	Exposes the reason why the call was disconnected

CallOnHoldEventActivity

CallOnHoldEventActivity is an event handler activity for a "call put on hold." This works only for a phone call because IM calls cannot be put on hold. This activity can be specified for a scope as defined by a *CommunicationsSequenceActivity*. This activity can be added only in the *CommunicationEvents* view of *CommunicationsSequenceActivity*.

> **Note** The speech dialog activities, like *SpeechQuestionAnswerActivity* and *SpeechStatementActivity*, intrinsically handle calls being put on hold. *CallOnHoldEventActivity* is used if the application wants to perform extra processing when the call is put on hold.

CallOnHoldTimeoutEventActivity

CallOnHoldTimeoutEventActivity is an event handler activity for a "call on hold for a certain amount of time." Its property and description are shown in Table 5-20. Due to the nature of this event, it can occur only after the call has been put on hold. The amount of time after which this event occurs can be set using a property on the activity. This works only for a phone call because IM calls cannot be put on hold. This activity can be specified for a scope as defined by a *CommunicationsSequenceActivity*. This activity can be added only in the *CommunicationEvents* view of *CommunicationsSequenceActivity*.

TABLE 5-20 Property of *CallOnHoldTimeoutEventActivity*

Property	Description
CallOnHoldTimeout	The length of time for which the call is on hold before this activity is executed

CallRetrievedEventActivity

CallRetrievedEventActivity is an event handler activity for a "call being retrieved." Due to the nature of this event, it can occur only after the call has been put on hold. This works only for a phone call because IM calls cannot be put on hold or retrieved. You can specify this activity for a scope as defined by *CommunicationsSequenceActivity*. This activity can be added only in the *CommunicationEvents* view of *CommunicationsSequenceActivity*.

> **Note** The speech dialog activities, like *SpeechQuestionAnswer* and *SpeechStatement*, intrinsically handle calls being retrieved. The *CallRetrievedEventActivity* is used if the application wants to perform extra processing when the call is retrieved.

Dialog Communications Event Activities

Six dialog communication event activities are available that allow for handling dialog events. Each activity is valid for either a phone call or an IM call, as described in Table 5-21.

TABLE 5-21 **Dialog Communications Event Types**

Activity	Description
ConsecutiveNoInputsInstantMessagingEvent	An event when the application fails to understand a user response or the user fails to respond *n* times in a row. Valid only for IM calls.
ConsecutiveNoInputsSpeechEvent	An event when the application fails to understand a user response or the user fails to respond *n* times in a row. Valid only for phone calls.
ConsecutiveNoRecognitionsInstantMessagingEvent	An event when the application fails to understand a user response *n* times in a row. Valid only for IM calls.
ConsecutiveNoRecognitionsSpeechEvent	An event when the application fails to understand a user response *n* times in a row. Valid only for phone calls.
ConsecutiveSilencesInstantMessagingEvent	An event when the user fails to respond *n* times in a row. Valid only for IM calls.
ConsecutiveSilencesSpeechEvent	An event when the user fails to respond *n* times in a row. Valid only for phone calls.

ConsecutiveNoInputsSpeechEventActivity

ConsecutiveNoInputsSpeechEventActivity is an event handler activity for a user providing an invalid input *n* times in a row. Its property is listed in Table 5-22. By default, the value of *n* is 3. An invalid input is when the user either does not provide any input (that is, remains silent) or the user's input is not recognized. This works only for speech dialogs over a phone call. This activity can be specified for a scope as defined by a *CommunicationsSequenceActivity*. You add this activity in the *CommunicationEvents* view of *CommunicationsSequenceActivity*.

TABLE 5-22 **Property of** *ConsecutiveNoInputsSpeechEventActivity*

Property	Description
MaximumNoInputs	The number of consecutive times that invalid user input is received before this activity is executed. The default value is 3.

ConsecutiveSilencesSpeechEventActivity

ConsecutiveSilencesSpeechEventActivity is an event handler activity for a user remaining silent *n* times in a row. Its property is listed in Table 5-23. By default, the value of *n* is 3. This works only for speech dialogs over a phone call. This activity can be specified for a scope as defined by a *CommunicationsSequenceActivity*. You add this activity in the *CommunicationEvents* view of *CommunicationsSequenceActivity*.

TABLE 5-23 Property of *ConsecutiveSilencesSpeechEventActivity*

Property	Description
MaximumSilences	The number of consecutive times that the user remains silent before this activity is executed. The default value is 3.

ConsecutiveNoRecognitionsSpeechEventActivity

ConsecutiveNoRecognitionsSpeechEventActivity is an event handler activity for user input not being recognized *n* times in a row. Its property is listed in Table 5-24. By default, the value of *n* is 3. This works only for speech dialogs over a phone call. This activity can be specified for a scope as defined by a *CommunicationsSequenceActivity*. You add this activity in the *CommunicationEvents* view of *CommunicationsSequenceActivity*.

TABLE 5-24 Property of *ConsecutiveNoRecognitionsSpeechEventActivity*

Property	Description
MaximumNoRecognitions	The number of consecutive times that user input is not recognized before this activity is executed. The default value is 3.

ConsecutiveNoInputsInstantMessagingEventActivity

ConsecutiveNoInputsInstantMessagingEventActivity is an event handler activity for a user providing an invalid input *n* times in a row. Its property is listed in Table 5-25. By default, the value of *n* is 3. An invalid input is when the user either does not provide any input (that is, does not respond) or the user's input is not recognized. This works only for IM dialogs over an IM call. You can specify this activity for a scope as defined by a *CommunicationsSequenceActivity*. You add this activity in the *CommunicationEvents* view of *CommunicationsSequenceActivity*.

TABLE 5-25 Property of *ConsecutiveNoInputsInstantMessagingEventActivity*

Property	Description
MaximumNoInputs	The number of consecutive times that invalid user input is received before this activity is executed. The default value is 3.

ConsecutiveSilencesInstantMessagingEventActivity

ConsecutiveSilencesInstantMessagingEventActivity is an event handler activity for a user not responding *n* times in a row. Its property is described in Table 5-26. By default, the value of *n* is 3. This works for only for IM dialogs over an IM call. You can specify this activity for a scope as defined by a *CommunicationsSequenceActivity*. You add this activity in the *CommunicationEvents* view of *CommunicationsSequenceActivity*.

TABLE 5-26 Property of *ConsecutiveSilencesInstantMessagingEventActivity*

Property	Description
MaximumSilences	The number of consecutive times that the user does not respond before this activity is executed. The default value is 3.

ConsecutiveNoRecognitionsInstantMessagingEventActivity

ConsecutiveNoRecognitionsInstantMessagingEventActivity is an event handler activity for user input not being recognized *n* times in a row. Its property is described in Table 5-27. By default, the value of *n* is 3. This works only for IM dialogs over an IM call. You can specify this activity for a scope as defined by a *CommunicationsSequenceActivity*. You add this activity in the *CommunicationEvents* view of *CommunicationsSequenceActivity*.

TABLE 5-27 Property of *ConsecutiveNoRecognitionsInstantMessagingEventActivity*

Property	Description
MaximumNoRecognitions	The number of consecutive times that user input is not recognized before this activity is executed. The default value is 3.

Presence-Related Activity

Presence is one of the most important value propositions of any UC platform. The UCMA Workflow ships with an activity related to presence, as described in Table 5-28.

TABLE 5-28 Presence-Related Activity

Activity	Description
GetPresence	Queries the presence information for user(s)

GetPresenceActivity

GetPresenceActivity queries the presence information of a user or a group of users. In addition to the presence state, it returns the current activity of the user (for example, in a meeting, in a call) and also out-of-office message if set. Its properties are described in Table 5-29. The out-of-office message is published if the user has set it in Microsoft Office Outlook. If the user has not set an out-of-office message, this value is *NULL*.

TABLE 5-29 Properties of *GetPresenceActivity*

Property	Description
Targets	A list of *RealTimeAddress* objects that represents the users whose presence will be queried.
Results	A dictionary representing the *RealTimeAddress* and its presence query result. The *RealTimeAddress* is used as a key in the dictionary. The *PresenceResult* object is the value representing the result of the query.

PresenceResult

PresenceResult is the object that represents the result of the presence query performed by the *GetPresence* activity. Its properties are described in Table 5-30.

TABLE 5-30 Properties of *PresenceResult*

Property	Description
CurrentState	A string representing the current activity of the user (for example, "In a meeting", "In a call").
OofNote	A string representing the out-of-office message as published by the user. If the user has not published an out-of-office message, this value is *NULL*.
PresenceStatus	An *enum* value listing the presence status of the user. It is of type *Microsoft.Rtc.Collaboration.Presence.PresenceAvailability*.

Summary

The UCMA Workflow Activities provide a way to develop applications that can interact with the user over the phone or an IM channel as new interfaces. It enables faster development of such applications by providing prepackaged custom activities for common tasks (for example, playing or sending a message to the user or asking the user a question and recognizing user input). It also enables intelligent routing decisions by providing a way to query for presence information. Finally, because this API does not dictate which application to use, developers are free to use any application host that they choose.

Additional Resources

- "Getting Started with Workflow Foundation (WF)" (*http://msdn.microsoft.com/en-us/netframework/aa663328.aspx*)
- Microsoft Unified Communications Managed API 2.0 SDK (32 bit) (*http://go.microsoft.com/fwlink/?LinkID=140790*)
- Microsoft Unified Communications Managed API 2.0 SDK (64 bit) (*http://go.microsoft.com/fwlink/?LinkID=139195*)
- "Unified Communications Managed API 2.0 Workflow SDK Documentation" (*http://go.microsoft.com/fwlink/?LinkID=133578&clcid=0x409*)
- "Understanding the Activity Execution Context" (*http://msdn.microsoft.com/en-us/library/aa349099.aspx*)
- "Using Dependency Properties" (*http://msdn.microsoft.com/en-us/library/ms734499.aspx*)

Chapter 6
Business Process Communication

This chapter will help you to:

- Understand how to implement a workflow application.
- Understand the advantages of using a workflow application.
- See how to support both phone and instant message modalities in the same application.

Scenario

Expense reporting and approval is a business process common to most enterprises. These processes have a deadline, and human intervention is required to complete them. For example, the typical approval process requires that a supervisor review expense reports and sign off on them before they are sent to the reimbursement department. Failure to meet payment deadlines can result in either penalty fees (for example, for credit cards) or extra processing.

This scenario can be enhanced by monitoring unapproved expense reports, and as the deadline approaches, alerting the approver with the information necessary to take effective and timely action.

This chapter covers the step-by-step implementation of a Unified Communications Managed API (UCMA) Workflow application that does the following:

- Queries for the presence of the approver to determine the best communication channel
- Notifies the approver of pending tasks
- Provides different communication channel options to complete pending tasks based on the approver's availability
- Informs a delegate of pending tasks if the approver is out of the office

Business Value

By using multiple communications channels and presence information, the business process can make better decisions, help reduce costs to the enterprise, and increase productivity. For example, if the approver is out of the office and the pending tasks are important, the delegate is asked to complete the task. This ensures that important tasks do not become critical due to lack of action.

The value added to the business process derives from the following:

- Reduced human latency in automated business processes
- Increased productivity due to choosing the best communication channel
- Cost reduction created by increasing the process efficiency due to presence-based intelligent routing

Choice of Technology

The UCMA Workflow provides the following benefits that make it ideal in the preceding scenarios:

- An interactive application is easy to create by dragging activities onto the workflow canvas.
- Custom activities provide the infrastructure to interact with users over a phone call or instant messaging (IM) call.
- Custom activities also provide constructs to enable a more natural dialog with the user.
- It can be integrated to use the benefits of Microsoft Office Communications Server 2007 R2.

Overall Code Structure

The rest of this chapter lists and explains the steps for building an expense reporting and approval application using UCMA Workflow Activities and Microsoft Visual Studio 2008. This application provides a solution to the preceding scenario.

Test Environment

To build and test this solution, you need the following:

- Office Communications Server 2007 R2 installed and configured on your domain.
- At least two users with Microsoft Office Communicator 2007 R2 installed who are able to sign in to the Office Communicator.
- Your development environment must match the UCMA Workflow specified in the UCMA Software Development Kit (SDK) documentation at *http://go.microsoft.com/fwlink/?LinkID=133578*.

Building the Application

The following sections break down the application-building process into smaller tasks. For each task, there is a list of steps to be followed.

Task 1: Create a New Communication Workflow Project

The Outbound Sequential Workflow Console Application project template, which is part of the UCMA SDK, provides a good starting point for building this application because it automatically generates code to integrate with Office Communications Server 2007 R2.

To create a project using this template, perform the following steps:

1. Start Visual Studio as an administrator. In Windows Vista, you may have to do this by right-clicking the Microsoft Visual Studio 2008 icon and then clicking Run As Administrator.
2. On a computer running Windows Vista, click Continue when the User Account Control prompts to allow this operation.
3. On the Visual Studio menu, click File, New Project to open the New Project dialog box.
4. In the Project Types tree, click Communications Workflow.
5. Select Outbound Sequential Workflow Console Application and then click OK.
6. In the Select Language dialog box, choose the language type and then click OK.

Table 6-1 lists the files generated by this project template.

TABLE 6-1 Outbound Sequential Workflow Console Application Template Files

File Name	Description
Program.cs	Defines a console application similar to the one generated by the Console Application project template that comes with Visual Studio 2008. This file, which is generated automatically, contains all of the code needed to integrate with Office Communications Server 2007 R2.
Workflow1.xoml	Defines the semantics of the application's workflow as viewed in the designer canvas. This file, in combination with the Workflow1.xoml.cs file, defines the workflow.
Workflow1.xoml.cs	Contains the code-besides for the workflow. This file, in combination with the Workflow1.xoml file, defines the workflow.

Task 2: Configure the Application to Connect to Office Communications Server

You need to modify the automatically generated code in the Program.cs file to connect to your specific server running Office Communications Server.

To deploy a UCMA Workflow application, the application must sign in to Office Communications Server. Because an application is not a user, the administrator cannot use an Active Directory Domain Services (AD DS) user account. Instead, the administrator needs to create an AD DS *Contact* object to represent the application and specify the application as a trusted service. You need to gather the information listed in Table 6-2 before you modify the Program.cs file. You can obtain this information from your Office Communications Server administrator, who creates the AD DS *Contact* object and specifies your application as a trusted service.

TABLE 6-2 Information Needed to Update Program.cs

Name	Description	Action to Be Taken
Application Port	Port that the application listens on.	Update the value in the constructor of *platformSettings* in the *Initialize* method.
Application URI	The Session Initiation Protocol (SIP) Uniform Resource Identifier (URI) assigned to the *Contact* object that represents your applications, to which users place calls or send IM messages.	Update the value of the variable *applicationUri* in the *Initialize* method.
Certificate	The certificate that authenticates the application machine running Office Communications Server. This certificate must be trusted by servers running Office Communications Server.	Install the certificate in the Local Computer certificate store on the computer where the application is installed.
FQDN	The fully qualified domain name (FQDN) for the server running Office Communications Server on which the *Contact* object to be created is homed.	Update the value of the variable *ocsFqdn* in the *Initialize* method.
GRUU	The Globally Routable User Agent URI (GRUU) of the trusted service added to Office Communications Server for this application.	Update the value of the variable *gruu* in the *Initialize* method.
SIP Port	Port number on which your server running Office Communications Server listens for SIP connections. The default is 5061.	Update the value of the variable *ocsTlsPort* if the value is not the default.

More Info For details, see Chapter 9, "Preparing the UC Development Environment."

Task 3: Allow User Input to the Workflow Instance

In this scenario, a specific event triggers the business workflow. Examples of such events are "Approval deadline in three days" or "Going over the expense limit requires signoff." In this example, you use console input to trigger the workflow. You can replace this easily with a handler for any event you choose.

1. Update the *Main* method with the following code.

   ```
   /// <summary>
   /// Main
   /// </summary>
   private static void Main()
   {
      bool _continue = true;
     string _approverUri = string.Empty;
     string _reportTitle = string.Empty;
     string _amount = string.Empty;
     string _filedBy = string.Empty;
     string _onlineDelegateUri = string.Empty;

      Initialize();
      while (_continue)
      {
            Console.Write("Enter the approver's SIP URI or 'exit' to quit: ");
            _approverUri = Console.ReadLine();

            if(_approverUri != "exit")
            {
                   Console.Write("Enter the expense report title: ");
                   _reportTitle = Console.ReadLine();
                   Console.Write("Enter the expense report amount: ");
                   _amount = Console.ReadLine();
                   Console.Write("Enter the name of person filing the report: ");
                   _filedBy = Console.ReadLine();
                   Console.Write("Enter the SIP URI of an online delegate: ");
                   _ onlineDelegateUri = Console.ReadLine();
                   StartWorkflow(_approverUri, _reportTitle, _amount,
   _filedBy, _onlineDelegateUri);
   }
            else
            {
                       _continue = false;
            }
      }

      Cleanup();
   }
   ```

   When you run this application, the application prompts the user to enter the SIP address of the approver, an expense report name, an expense amount, the name of the person filing the report, and the SIP address of a delegate that is online. If the user types **Exit,** the application closes. After the user enters the required input, the application starts the workflow by calling the *StartWorkflow()* method. In a real-world implementation, this information would be provided by the process that triggers the workflow.

2. Modify the signature of the *StartWorkflow* method as follows.

   ```
   private static void StartWorkflow(string approverUri, string reportTitle, string
   amount, string filedBy, string onlineDelegate)
   ```

3. Add the following lines at the beginning of the *StartWorkflow* method to pass the approver URI, expense report details, and online delegate URI to the workflow instance.

```
Dictionary<string, object> namedArgs = new Dictionary<string, object>();
namedArgs.Add("ApproverSipAddress", new RealTimeAddress("sip:" + approverUri));
namedArgs.Add("ReportTitle", reportTitle);
namedArgs.Add("Amount", amount);
namedArgs.Add("FiledBy", filedBy);
namedArgs.Add("OnlineDelegate", new RealTimeAddress("sip:" + onlineDelegate));
WorkflowInstance workflowInstance = _workflowRuntime.CreateWorkflow(typeof(Workflow1),
namedArgs);
```

Note You may have to add *using Microsoft.Rtc.Signaling;* and *using System.Collections .Generic;* to the top of Program.cs if they are not already in the file.

4. To make the data passed into the *Workflow* instance accessible, public properties are defined in the workflow. In Workflow1.xoml.cs, add the following lines of code inside the definition of the *Workflow1* class as follows.

```
public RealTimeAddress ApproverSipAddress { get; set; }
public string ReportTitle {get; set; }
public string Amount {get; set; }
public string FiledBy {get; set; }
public RealTimeAddress OnlineDelegate {get; set; }
```

Task 4: Get the Approver's Presence Information

By using the presence information of the approver, the application can use more efficient logic to get the expense report approved. If the approver is online, the application can contact the approver directly. If the approver is offline, the application can contact the designated delegate. The following steps query for the presence of the approver:

1. The workflow file (that is, Workflow1.xoml) is prepopulated with activities to start an outbound call. However, in this scenario you need to get the approver's presence information before making an outbound call. To start from a clean state, delete all of the activities from the designer canvas.

2. Drag a *CommunicationsSequence* activity from the Unified Communications Workflow tab onto the designer canvas over the text *Drop Activities To Create A Sequential Workflow* and, in the Property pane, rename it **overallContainer**. This activity is needed to define a dialog with the user.

3. Drag a *Code* activity from the Windows Workflow v3.0 tab onto the designer canvas inside the *overallContainer* activity and, in the Property pane, rename it **setParameters**. You use this activity to set any parameters needed to execute the workflow, such as access data passed into the workflow (see Tasks 4, 6, 7.2.1, and 7.3.1).

4. Drag a *GetPresence* activity from the Unified Communications Workflow tab onto the designer canvas after the *setParameters* activity and rename it **getApproverPresence**. You use this activity to get the presence information of the approver. The workflow should look like the following:

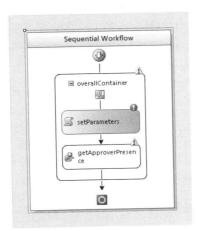

5. Double-click the *setParameters* activity to generate the code handler for that activity. Update the handler method with the following code to set the *Target* property of the *getApproverPresence* activity with the value of *ApproverSipAddress*.

```
private void setParameters_ExecuteCode(object sender, EventArgs e)
{
    /// Set the target of getApproverPresence activity
    this.getApproverPresence.Targets.Clear();
    this.getApproverPresence.Targets.Add(ApproverSipAddress);
}
```

Task 5: Implement Branching Logic Based on the Approver's Presence

The objective is to have the expense report reviewed and then approved or rejected. Therefore, the application logic follows different branches based on the presence of the approver. In cases in which the approver's presence status is Offline, Do Not Disturb, or Out of Office, the application determines that the approver is not available and contacts its delegate. For all other presence states, the approver is considered to be available to review the request.

The following steps implement branching logic based on the approver's presence:

1. Drag an *IfElse* activity from the Windows Workflow Foundation v3.0 pane onto the designer canvas after the *getApproverPresence* activity.

2. In the Property pane, select the *ifElseBranchActivity1* branch and rename it **cantBeContactedBranch**. You use *cantBeContactedBranch* to define the business logic for the situation in which the approver cannot be reached.

3. In the Property pane, select the *ifElseBranchActivity2* branch and rename it **canBeContactedBranch**. You use *canBeContactedBranch* to define the business logic for the situation in which the approver can be reached.

4. Select *cantBeContactedBranch* and change the value of the *Condition* property from *(None)* to *Code Condition*.

5. The *cantBeContactedBranch* branch executes only if a specified condition, which is defined in step 6, is *True*. To specify this condition, click the plus sign (+) to expand the *Condition* property. Type the method name **CheckIfUserCantBeContacted**, and then press Enter. This takes you to the automatically generated method stub. This method defines the logic to check whether this condition is met.

6. Update the *CheckIfUserCantBeContacted* method with the following code.

```
private void CheckIfUserCantBeContacted(object sender, ConditionalEventArgs e)
{
   if(this.getApproverPresence.Results[ApproverSipAddress].CurrentState.ToLower().
Contains("out of office") ||
this.getApproverPresence.Results[ApproverSipAddress].PresenceStatus ==
Microsoft.Rtc.Collaboration.Presence.PresenceAvailability.Offline ||
this.getApproverPresence.Results[ApproverSipAddress].PresenceStatus ==
Microsoft.Rtc.Collaboration.Presence.PresenceAvailability.DoNotDisturb)
   {
            // Set the value of e.Result so that ifElse executes this branch
            e.Result = true;
   }
   else
   {
            // Set the value of e.Result so that ifElse does not execute this branch
            e.Result = false;
   }
}
```

You use the *getApproverPresence* activity to obtain the approver's presence status. If the approver is unavailable, *cantBeContactedBranch* is executed. Otherwise, *canBeContactedBranch* is executed. The *Results* property of the *getApproverPresence* activity contains the result of the presence query. The *getApproverPresense.Results* activity is a *Dictionary* object. To obtain the approver's presence, the approver's SIP URI is specified as a key.

At this point, the workflow should look like the following:

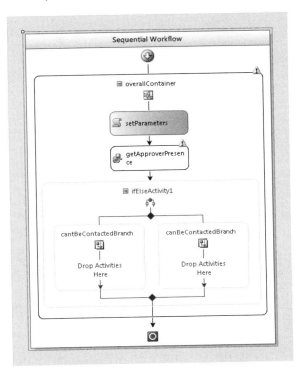

Task 6: Update *cantBeContactedBranch*

Now you update *cantBeContactedBranch*, which determines the actions taken when the approver cannot be contacted. In this situation, the workflow contacts the delegate instead.

1. Drag a *Code* activity inside *cantBeContactedBranch* and change its name from *codeActivity1* to **findDelegate**. You use this *Code* activity to find the delegate of the approver.

2. Double-click the *findDelegate* activity to generate the method stub, and update the method with the following code. The approver's SIP address is replaced with the delegate's SIP address.

   ```
   private void findDelegate_ExecuteCode(object sender, EventArgs e)
   {
   ApproverSipAddress = OnlineDelegate;
   }
   ```

3. Drag a *Goto* activity onto the designer canvas below the *findDelegate* code activity, and then change its name from *gotoActivity1* to **gotoPresenceQuery**.

4. In the Property pane of the *gotoPresenceQuery* activity, open the *TargetActivityName* pane by clicking the ellipsis (...) button at the end of the field.

5. Double-click the *setParameters* code activity to set it as the target activity.

   At the end of these steps, the workflow should look like the following:

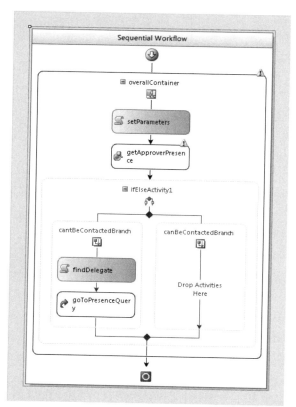

If the approver is unavailable, the workflow returns to the "restart from the presence query" step with the delegate of the primary approver.

Task 7: Update *canBeContactedBranch*

Now you update *canBeContactedBranch*, which determines the actions to be taken when the approver can be contacted.

Task 7.1: Create Branches for Different Modalities

Before contacting the approver, the application needs to determine the appropriate modality to use. If the approver's presence is set to Away, Be Right Back, or Offline, the approver is contacted by phone. If the approver is in a phone call or her presence status is set to Available or Busy, the approver is contacted by IM.

The following steps implement this logic:

1. Drag an *IfElse* activity from the Windows Workflow Foundation v3.0 tab onto the designer canvas inside *canBeContactedBranch*.

2. In the Property pane, change the name of the first branch from *ifelseBranchActivity1* to **PhoneMode**. *PhoneMode* defines the path to use to call the approver.

3. Click the *PhoneMode* branch and change the value of the *Condition* property from *(None)* to *Declarative Rule Condition*.

4. Click the plus sign (+) next to the *Condition* property to expand it. Type **PhoneModeCondition** as the value for *ConditionName*.

5. Click the ellipsis (…) button at the end of the *Expression* field to open the Select Condition dialog box. Select PhoneModeCondition from the list and click Edit… to open the Rule Condition Editor dialog box.

6. Add the following declarative rule in the Rule Condition Editor dialog box and then click OK. This ensures that *canBeContactedBranch* is executed if the approver is set to Away or Be Right Back.

   ```
   this.getApproverPresence.Results[ApproverSipAddress].PresenceStatus == Microsoft.Rtc.
   Collaboration.Presence.PresenceAvailability.Away ||
   this.getApproverPresence.Results[ApproverSipAddress].PresenceStatus == Microsoft.Rtc.
   Collaboration.Presence.PresenceAvailability.BeRightBack
   ```

7. In the Property pane, change the name of the second branch from *ifelseBranchActivity2* to **IMMode**. No condition is required to be set for the *IMMode* branch because this branch is executed only if the conditions for the *PhoneMode* branch are not satisfied.

The workflow should look like the following:

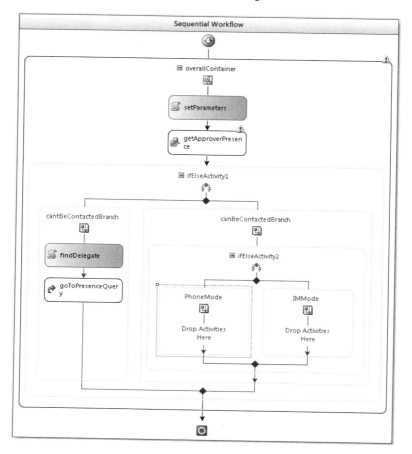

Task 7.2: Contact the Approver by Phone

The following tasks define the logic for calling and interacting with the approver by phone:

1. Place an outbound phone call to the approver.
2. Bind the outbound phone call to a *CommunicationsSequenceActivity*.
3. Play an introductory message.
4. Ask the approver about the action he or she wants to take and recognize speech or dual-tone multifrequency (DTMF) input.
5. Branch the responses based on the approver's input.
6. Define *Approve* and *Decline* actions.

Task 7.2.1: Place an Outbound Phone Call to the Approver

1. Drag an *OutboundCall* activity inside the *PhoneMode* branch and change its name from *outboundCallActivity1* to **startPhoneCall**. The *startPhoneCall* activity initiates the phone call with the approver.

2. Set the *CallType* property to *AudioVideoCall*.

3. Open the Workflow1.xoml.cs file and then add the following code to the *setParameters_ExecuteCode* method. This sets up the *startPhoneCall* activity to place an outbound phone call to the approver's SIP URI.

   ```
   this.startPhoneCall.CalledParty = ApproverSipAddress;
   ```

Task 7.2.2: Bind the Outbound Phone Call to a *CommunicationsSequenceActivity*

In the UCMA Workflow, each *CommunicationsSequenceActivity* must know on which call it is executing. You achieve this by binding the *CallProvider* property of *CommunicationsSequenceActivity* to the *CallProvider* property of an *AcceptCall* activity or an *OutboundCall* activity.

1. Drag a *CommunicationsSequence* activity after the *startPhoneCall* activity and change its name from *communicationsSequenceActivity1* to **speechDialog**. The *speechDialog* activity contains the logic flow of the dialog with the approver over the phone.

2. In the *speechDialog* Property pane, click the ellipsis (…) button at the end of the field of the *CallProvider* property to open the Bind CallProvider To An Activity's Property dialog box, as shown here.

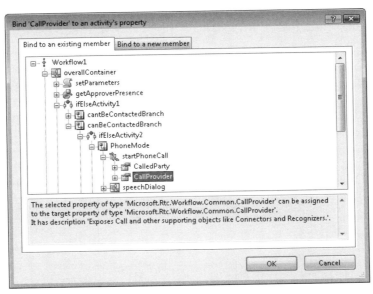

3. Expand the *overallcontainer*, *ifElseActivity1*, and *canBeContactedBranch* nodes in the tree to reach the *CallProvider* node, which is a property of the *startPhoneCall* activity, as shown previously. Click on the *CallProvider* node to select it.

4. Click OK to close the Bind dialog box. This binds the *CallProvider* property of the *speechDialog* activity with the *CallProvider* property of the *startPhoneCall* activity.

   The workflow should look like the following:

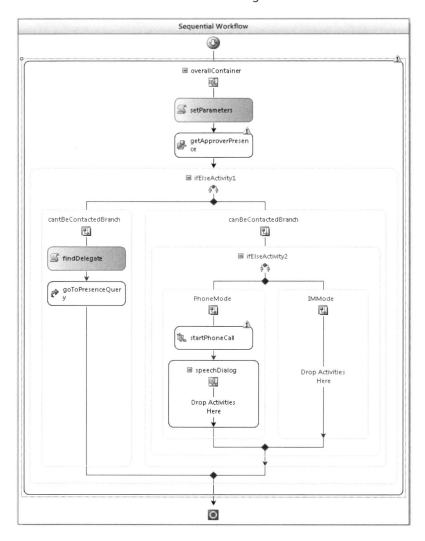

Task 7.2.3: Play an Introductory Message

1. Add a *SpeechStatementActivity* inside the *speechDialog* activity and change its name from *speechStatementActivity1* to **introMessage**. The *introMessage* activity plays the introduction message to the approver.

2. In the *MainPrompt* property of the *introMessage* activity, type the message **Hello, this is the expense report alerting system.**

Task 7.2.4: Prompt the Approver for an Action and Recognize Speech or DTMF Input

1. Add a *SpeechQuestionAnswerActivity* after the *introMessage* activity and change its name from *speechQuestionAnswerActivity1* to **askForActionsQuestion**. The *askForActionsQuestion* activity prompts the approver for input regarding the pending expense report.

2. Type **askForActionsQuestion_TurnStarting** as the value of the *TurnStarting* property and then press Enter to generate the stub for the *askForActionsQuestion_TurnStarting* event handler automatically. The *askForActionsQuestion_TurnStarting* method dynamically generates the prompt that the *askForActionsQuestion* activity plays.

3. Update this event handler stub with the following code.

    ```
    private void askForActionsQuestion_TurnStarting(object sender, Microsoft.Rtc.Workflow.
    Activities.SpeechTurnStartingEventArgs e)
    {
        this.askForActionsQuestion.MainPrompt.SetText(string.Format("Report Title Name: 
    {0}, Filed By: {1}, Amount: {2}. Please say approve or decline", ReportTitle, FiledBy, 
    Amount));
    }
    ```

    At this point, you must specify a list of valid speech and DTMF inputs for the question asked by the *askForActionsQuestion* activity. Define the valid speech and DTMF responses for this question as follows:

4. At the end of the *ExpectedSpeechInputs* property of the *askForActionsQuestion* activity, click the ellipsis (...) button to open the String Collection Editor dialog box.

5. Type **approve** and **decline** on separate lines. You can type additional synonyms on separate lines as well.

6. Click OK.

7. At the end of the *ExpectedDTMFInputs* property, click the ellipsis (...) button to open the String Collection Editor dialog box.

8. Type **1** and **2** on separate lines. The DTMF value *1* corresponds to the *approve* speech input and the DTMF value *2* corresponds to the *decline* speech input, as specified in step 5.

9. Click OK.

Task 7.2.5: Branch the Responses Based on the Approver's Action

1. Drag an *IfElse* activity from the Windows Workflow Foundation v3.0 tab onto the designer canvas after the *askForActionsQuestion* activity.

2. In the Property pane, change the name of the first branch from *ifElseBranchActivity1* to **approveBranch**. You use *approveBranch* to define the logic for when the user chooses to approve the pending expense report.

3. Click *approveBranch* and then change the value of the *Condition* property from *(None)* to *Declarative Rule Condition*.

4. Click the plus sign (+) next to the *Condition* property to expand it. Type **ApproveCondition** as the value for *ConditionName*.

5. At the end of the *Expression* field, click the ellipsis (…) button to open the Select Condition dialog box. Select ApproveCondition from the list and click Edit… to open the Rule Condition Editor dialog box.

6. In the Rule Condition Editor dialog box, add the following declarative rule. This condition ensures that the branch is executed if the approver wants to approve the expense.

   ```
   this.askForActionsQuestion.RecognitionResult.Text.ToLower().Contains("approve") ||
   this.askForActionsQuestion.RecognitionResult.Text.ToLower().Contains("1")
   ```

7. In the Property pane, change the name of *ifElseBranchActivity2* to **declineBranch**. There is no need to set the condition for *declineBranch* because it is executed only if the condition specified in step 6 is not met.

   Now, you can use *approveBranch* to interface with the expense report approval system to approve the expense in question. You can use *declineBranch* to interface with the expense report approval system to decline the expense in question.

The *PhoneMode* branch that defines the speech dialog should look like the following:

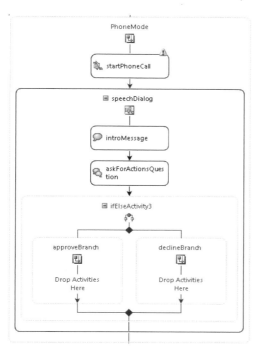

Task 7.2.6: Define Approve and Decline Actions

1. Drag a *Code* activity inside *approveBranch* and change its name from *codeActivity1* to **ApproveExpense**. You use *ApproveExpense* to approve the expense.

2. Drag a *Code* activity inside *declineBranch* and change its name from *codeActivity1* to **DeclineExpense**. You use *DeclineExpense* to decline the expense.

3. Double-click these *Code* activities to generate the code handlers. You can use these handlers to integrate with the expense reporting system to approve or decline the expense.

4. For this example, update the handlers as follows.

```
private void ApproveExpense_ExecuteCode(object sender, EventArgs e)
{
// Update this to integrate with expense report management system to approve  expense
    Console.WriteLine("Expense Approved");
}

private void DeclineExpense_ExecuteCode(object sender, EventArgs e)
{
// Update this to integrate with expense report management system to decline expense
    Console.WriteLine("Expense Declined");
}
```

The phone dialog branch should look like the following:

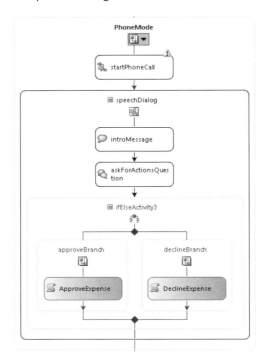

5. Drag a *SpeechStatementActivity* after the *ApproveExpense* code activity inside *approveBranch* and change its name from *speechStatementActivity1* to **ExpenseApprovedMessage**. You use *ExpenseApprovedMessage* to play a message to the approver that notifies her or him that the approval process succeeded.

6. Update the *MainPrompt* property with the message **The expense report has been approved.**

7. Drag a *SpeechStatementActivity* after the *DeclineExpense* code activity inside *declineBranch* and change its name from *speechStatementActivity1* to **ExpenseDeclinedMessage**. You use *ExpenseDeclinedMessage* to play a message to the approver that notifies her or him that the decline process finished.

8. Update the *MainPrompt* property with the message **The expense report has been declined.**

9. Drag a *SpeechStatementActivity* after the *ifElseActivity3* activity and change its name from *speechStatementActivity1* to **GoodbyeMessage**. You use *GoodbyeMessage* to play a goodbye message.

10. Update the *MainPrompt* property with the message **Goodbye.**

The speech dialog branch should look like the following:

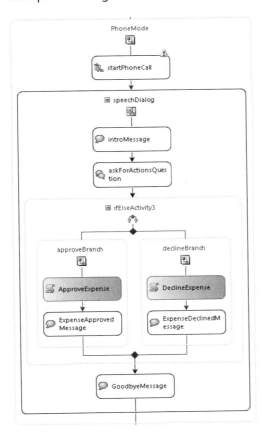

Task 7.3: Contact the Approver by IM

The following tasks define the logic for interacting with the approver by IM:

1. Place an outbound IM call to the approver.
2. Bind the outbound IM call to a *CommunicationsSequenceActivity*.
3. Send an introductory IM message.
4. Prompt the approver for an action and recognize IM input.
5. Branch the responses based on the approver's input.
6. Define approve and decline actions.

Task 7.3.1: Place an Outbound IM Call to the Approver

1. Drag an *OutboundCallActivity* inside the *IMMode* branch and change its name from *outboundCallActivity1* to **startIMSession**. You use *startIMSession* to start an outbound IM session with the approver.

2. Set the *CallType* property of *startIMSession* to *InstantMessagingCall*.

3. Open the Workflow1.xoml.cs file and then add the following code to the *setParameters_ExecuteCode* method to set the SIP URI to match the approver's SIP URI.

   ```
   this.startIMSession.CalledParty = ApproverSipAddress;
   ```

Task 7.3.2: Bind the Outbound IM Call with a *CommunicationsSequenceActivity*

As mentioned during Task 7.2, the *CallProvider* property of *CommunicationsSequenceActivity* must be bound with the *CallProvider* property of the *OutboundCall* activity.

1. Drag a *CommunicationsSequenceActivity* after the *startIMSession* and change its name from *communicationsSequenceActivity1* to **imDialog**. *imDialog* defines the IM dialog that interacts with the approver.

2. In the *imDialog* Property pane, click the ellipsis (...) button at the end of the field next to the *CallProvider* property to open the Bind dialog box, as shown here.

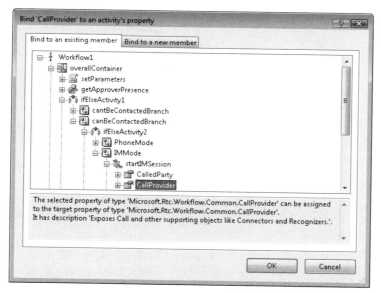

3. Expand the tree in the Bind dialog box until you reach the properties of *startIMSession*.

4. Click the *CallProvider* property of *startIMSession* and then click OK.

The *speechDialog* activity is collapsed in all screenshots while you develop the IM dialog. The workflow should look like the following:

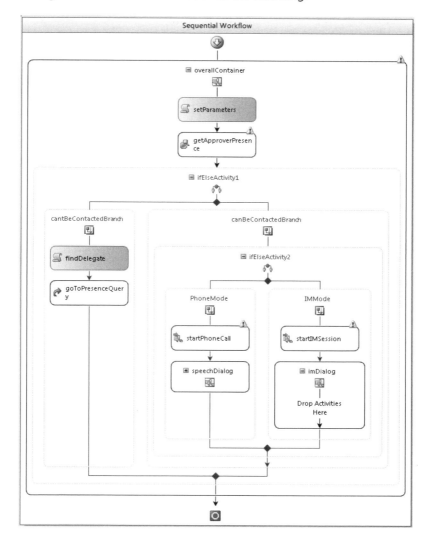

Task 7.3.3: Send an Introductory IM Message

1. Add an *InstantMessagingStatementActivity* inside *imDialog* and change its name from *instantMessagingStatementActivity1* to **introIMMessage**.

2. Update the *MainPrompt* property with the message **Hello, this is the expense report alerting system.**

Task 7.3.4: Prompt the Approver for an Action and Recognize IM Input

1. Add an *InstantMessagingQuestionAnswerActivity* after the *introIMMessage* activity and change its name from *instantMessagingStatementActivity1* to **askForActionsIMQuestion**.

2. Type **askForActionsIMQuestion_TurnStarting** as the value of the property *TurnStarting* and then press Enter to generate the stub for the *TurnStarting* event handler automatically.

3. Update the *TurnStarting* event handler stub with the following code.

```
private void askForActionsIMQuestion_TurnStarting(object sender, Microsoft.Rtc.
Workflow.Activities.InstantMessagingTurnStartingEventArgs e)
{
   this.askForActionsIMQuestion.MainPrompt = string.Format("Report Title Name: {0},
Filed By: {1}, Amount: {2}. Please enter approve or decline", ReportTitle, FiledBy,
Amount);
}
```

You must specify a list of valid text inputs for the question asked by the *askForActionsIMQuestion* activity. Specify the following text inputs as valid answers.

4. At the end of the *ExpectedInputs* property, click the ellipsis (...) button to open the String Collection Editor dialog box.

5. Type **approve** or **decline** in separate lines in the dialog box. You can type additional synonyms on separate lines as well.

6. Click OK.

Task 7.3.5: Branch the Responses Based on the Approver's Input

1. Drag an *IfElse* activity from the Windows Workflow Foundation v3.0 tab onto the designer canvas after the *askForActionsIMQuestion* activity.

2. In the Property pane, change the name of the *ifElseBranchActivity1* branch to **approveIMBranch**. You use *approveIMBranch* to define the logic when the user approves the pending expense report.

3. Click *approveIMBranch* and change the *Condition* property from *(None)* to *Declarative Rule Condition*.

4. Click the plus sign (+) next to the *Condition* property to expand it, and then type **ApproveIMCondition** as the value of *ConditionName*.

5. At the end of the *Expression* field, click the ellipsis (...) button to open the Select Condition dialog box. Select ApproveIMCondition from the list and click Edit... to open the Rule Condition Editor dialog box.

6. In the Rule Condition Editor dialog box, add the following declarative rule. This condition ensures that the branch is executed if the approver wants to approve the expense.

   ```
   this.askForActionsIMQuestion.RecognitionResult.Text.ToLower().Contains("approve")
   ```

7. In the Property pane, change the name of the *ifElseBranchActivity2* branch to **declineIMBranch**. There is no need to set the condition for this branch because it is executed only if the condition specified in step 6 is not met.

   Now, you can use *approveIMBranch* to interface with the expense report approval system to approve the expense in question. You can use *declineIMBranch* to interface with the expense report approval system to decline the expense in question.

   The workflow should now look like the following:

   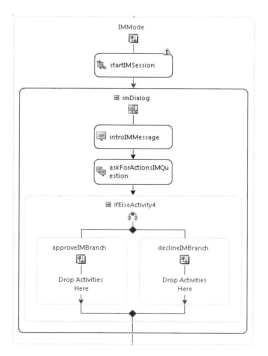

Task 7.3.6: Define Approve and Decline Actions

1. Drag a *Code* activity inside *approveIMBranch* and change its name from *codeActivity1* to **ApproveIMExpense**. You use *ApproveIMExpense* to approve the expense.

2. Drag another *Code* activity inside *declineIMBranch* and change its name from *codeActivity1* to **DeclineIMExpense**. You use *DeclineIMExpense* to reject the expense.

3. Select *ApproveIMExpense* and, in the Property pane, set the *ExecuteCode* value to *ApproveExpense_ExecuteCode* as follows:

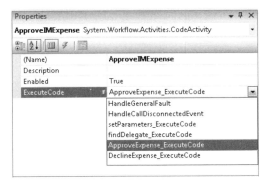

4. Take the same action for *DeclineIMExpense* and set its *ExecuteCode* value to *DeclineExpense_ExecuteCode*.

   By performing these steps, you ensure that the approval or decline actions through speech or IM execute the same code. The IM dialog branch should now look like the following:

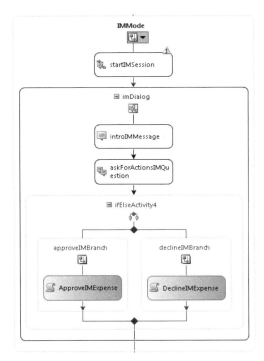

5. Drag an *InstantMessagingStatementActivity* after *ApproveIMExpense* and change its name from *instantMessagingStatementActivity1* to **ExpenseApprovedIMMessage**. You

use *ExpenseApprovedIMMessage* to send a message to the approver notifying him or her that the approval process succeeded.

6. Update the *MainPrompt* property with the message **The expense report has been approved.**

7. Drag an *InstantMessagingStatementActivity* after *DeclineIMExpense* and change its name from *instantMessagingStatementActivity1* to **ExpenseDeclinedIMMessage**. You use *ExpenseDeclinedIMMessage* to play a message to the approver notifying him or her that the decline process succeeded.

8. Update the *MainPrompt* property with the message **The expense report has been declined.**

9. Drag an *InstantMessagingStatementActivity* after *ifElseActivity4* and change its name from *instantMessagingStatementActivity1* to **GoodbyeIMMessage**. You use *GoodbyeIMMessage* to play a goodbye message.

10. Update the *MainPrompt* property with the value **Goodbye.**

The IM dialog branch should now look like the following:

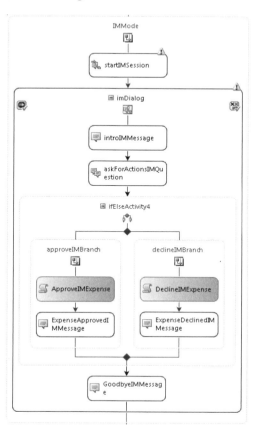

Task 7.4: Add Commands to the Dialog

In this task, you enhance the user interface by providing a more natural dialog flow. By performing the following steps, you add commands for help and repeat to the speech dialog and add a command for help to the IM dialog.

Task 7.4.1: Add Commands to the Speech Dialog

1. Right-click the *speechDialog* activity and then click View Commands to open the *Commands* view.

2. Drag *SpeechHelpCommandActivity* onto the *Commands* view. You use *SpeechHelpCommandActivity* to provide help to users when they request it.

3. At the end of the *ExpectedSpeechInputs* property, click the ellipsis (…) button to open the String Collection Editor dialog box and specify the list of text inputs that you want this command to recognize.

4. In the String Collection Editor dialog box, type **help**. You can type other synonyms on separate lines as well.

5. Click OK.

6. Drag a *SpeechRepeatCommand* onto the *Commands* view. You use *SpeechRepeatCommand* to repeat the question to the user when requested.

7. At the end of the *ExpectedSpeechInputs* property, click the ellipsis (…) button to open the String Collection Editor dialog box and specify the list of text inputs that you want this command to recognize.

8. In the Array Builder dialog box, type **repeat**. You can type other synonyms on separate lines as well.

9. Click OK.

    At this point, the speech dialog has been updated with speech help and speech repeat commands. Follow the next steps to define how you want the workflow to respond when the user triggers any of these commands. The following steps define the application behavior:

10. Right-click the *speechDialog* activity and click View CommunicationsSequenceActivity to open the *Main* view of the *speechDialog* activity.

11. Select *askForActionsQuestion* activity and, in the Property pane, expand the *Prompts* property.

12. Update the *HelpPrompt* property with the message **Please say Approve or press 1 to approve the expense report, or say Decline or press 2 to decline the expense report.**

13. Update the *RepeatPrompt* property in the *askForActionsQuestion_TurnStarting* method in the Workflow1.xoml.cs file by adding the following code.

    ```
    this.askForActionsQuestion.Prompts.RepeatPrompt.SetText(string.Format("Report Title
    Name: {0}, Filed By: {1}, Amount: {2}. Please say approve or decline", ReportTitle,
    FiledBy, Amount));
    ```

Task 7.4.2: Add Commands to the IM Dialog

1. Right-click the *imDialog* activity and click View Commands to open the *Commands* view.

2. Drag an *InstantMessagingHelpCommand* onto the *Commands* view. You use *InstantMessagingHelpCommand* to provide help to users when they request it.

3. At the end of the *ExpectedInputs* property, click the ellipsis (...) button to open the String Collection Editor dialog box and specify the list of text inputs that you want to recognize for this command.

4. In the String Collection Editor dialog box, type **help**. You can type other synonyms on separate lines as well.

5. Click OK.

    At this point, you have updated the IM dialog with an IM help command. The following steps define how the workflow should respond if the user triggers any of these commands. The following steps define the application behavior:

6. Right-click the *imDialog* activity and click View CommunicationsSequenceActivity to open the *Main* view of the *imDialog* activity.

7. Select *askForActionsIMQuestion* and, in the Property pane, expand the *Prompts* property.

8. Update the *HelpPrompt* property with the message **Please enter Approve to approve the expense report, or enter Decline to decline the expense report.**

Task 7.5: Disconnecting the Call

After the phone or IM dialog completes, the application must disconnect the call. To do this, drag a *DisconnectCall* activity after the *ifElseActivity1* activity.

Task 7.6: Add Events to the Dialog

In this task, you add handlers to the phone and IM dialog in case the user has difficulty responding or the application fails to understand user input. Similar to commands, these handlers improve the user dialog interface by either giving users better options (such as DTMF) or transferring to a live agent. In this example, the dialog event handlers disconnect the call. Follow these steps to add dialog event handlers to the speech and IM dialog of the application.

Task 7.6.1: Add Events to the Speech Dialog

1. Right-click the *speechDialog* activity and click View CommunicationsEvents to open the *Events* view.

2. Drag a *ConsecutiveSilencesSpeechEvent* activity onto the text *Drop A CommunicationsEvent Here*. By default, it is named *consecutiveSilencesSpeechEventActivity1*, and it is executed when the user stays silent in response to a question three times in a row.

3. Drag a *Goto* activity inside *consecutiveSilencesSpeechEventActivity1* and onto the text *Drop Activities Here*. The default name for this activity is *gotoActivity1*.

4. In the Property pane, set the *TargetActivityName* property of *gotoActivity1* that you added in step 3 to *disconnectCallActivity1*.

5. Drag a *ConsecutiveNoRecognitionsSpeechEvent* activity in the view next to *consecutiveSilencesSpeechEventActivity1*. By default, it is named *consecutiveNoRecognitionsSpeechEventActivity1*, and it is executed when the user's speech or DTMF response is not recognized three times in a row.

6. Drag a *Goto* activity inside *consecutiveNoRecognitionsSpeechEventActivity1* and onto the text *Drop Activities Here*.

7. In the Property pane, set the *TargetActivityName* of *gotoActivity2* that you added in step 6 to *disconnectCallActivity1*.

8. Drag a *ConsecutiveNoInputsSpeechEvent* activity into the view next to *consecutiveNoRecognitionsSpeechEventActivity1*. By default, it is named *consecutiveNoInputsSpeechEventActivity1*, and it is executed when the user stays silent or his or her speech or DTMF response is not recognized three times in a row.

9. Drag a *Goto* activity inside *consecutiveNoInputsSpeechEventActivity1* and onto the text *Drop Activities Here*.

10. In the Property pane, set the *TargetActivityName* of *gotoActivity3* that you added in step 9 to *disconnectCallActivity1*.

Task 7.6.2: Add Events to the IM Dialog

1. Right-click the *imDialog* activity and click View CommunicationsEvents to open the *Events* view.

2. Drag a *ConsecutiveSilencesInstantMessagingEvent* activity in the view onto the text *Drop A CommunicationsEvent Here*. By default, it is named *consecutiveSilencesInstantMessagingEventActivity1,* and it is executed when the user does not respond to a question three times in a row.

3. Drag a *Goto* activity inside *consecutiveSilencesInstantMessagingEventActivity1* and onto the text *Drop Activities Here*.

4. In the Property pane, set the *TargetActivityName* of *gotoActivity4* that you added in step 3 to *disconnectCallActivity1*.

5. Drag a *ConsecutiveNoRecognitionsInstantMessagingEvent* activity in the view next to *consecutiveSilencesInstantMessagingEventActivity1*. By default, it is named *consecutiveNoRecognitionsInstantMessagingEventActivity1*, and it is executed when the user's IM response is not recognized three times in a row.

6. Drag a *Goto* activity inside *consecutiveNoRecognitionsInstantMessagingEventActivity1* and onto the text *Drop Activities Here*.

7. In the Property pane, set the *TargetActivityName* of *gotoActivity5* that you added in step 6 to *disconnectCallActivity1*.

8. Drag a *ConsecutiveNoInputsInstantMessagingEvent* activity in the view next to *consecutiveNoRecognitionsInstantMessagingEventActivity1*. By default, it is named *consecutiveNoInputsInstantMessagingEventActivity1*, and it is executed when the user does not respond or her or his IM response is not recognized three times in a row.

9. Drag a *Goto* Activity inside *consecutiveNoInputsInstantMessagingEventActivity1* and onto the text *Drop Activities Here*.

10. In the Property pane, set the *TargetActivityName* of *gotoActivity6* that you added in step 9 to *disconnectCallActivity1*.

Task 7.7: Add Call Events

Now, you need to update the application to support disconnection of a phone call or an IM call. This is necessary to prepare the application to handle a user disconnecting the phone call or IM in the middle of a dialog. Follow these steps to add this support:

1. Right-click the *overallContainer* activity and click View CommunicationsEvents to open the *Events* view.

2. Drag a *CallDisconnectedEventActivity* activity in the view onto the text *Drop a CommunicationsEvent Here* and then, in the Property pane, change its name from *callDisconnectedEventActivity1* to **handleCallDisconnect**.

By doing this, you ensure that the application can handle the situation when the user disconnects the call. If you want the application to take specific action on a disconnected call, you can do this by dragging activities inside the *handlerCallDisconnect* activity. An example of this type of action is logging the call length.

Task 8: Running the Application

To run the application, you need the following:

1. A user account (for example, approver@contoso.com) for the approver logged on to Office Communications Server from Office Communicator.

2. A user account (for example, delegate@contoso.com) for the approver's delegate logged on to Office Communications Server from Office Communicator on a different computer.

Follow these steps to run the application:

1. Verify that the presence status of the approver account is set to Available.

2. In Visual Studio, press F5.

3. When the console window opens and the application prompts with "Enter the approver's SIP URI or Exit to close the application:," type the SIP URI of the approver.

4. When the application prompts "Enter the Expense report Title:", type the expense report title.

5. When the application prompts "Enter the Expense report amount:", type the expense report amount.

6. When the application prompts "Enter the name of the person filing the report:", type the name of the person filing the report.

7. When the application prompts "Enter the SIP URI of an online delegate:", type the SIP URI of the online delegate.

8. Click the incoming call notification on the approver machine when the application tries to connect to the approver. Start interacting with the application once the call is established.

To test the branch that finds the delegate, perform the following steps:

1. Verify that the approver presence is set to Offline.

2. In Visual Studio, press F5.

3. When the application prompts "Enter the approver's SIP URI or Exit to close the application", type the SIP URI of the approver, who is offline.

4. When the application prompts "Enter the expense report title", type the expense report title.

5. When the application prompts "Enter the expense report amount", type the expense report amount.

6. When the application prompts "Enter the name of the person filing the report", type the name of the person filing the report.

7. When the application prompts "Enter the SIP URI of an online delegate", type the SIP URI of the online delegate.

8. The application contacts the delegate account specified in step 7.

Summary

This chapter discussed how to implement an application using the UCMA Workflow for business process communications and information access from different modalities, such as phones or IM. By using Visual Studio 2008, you can add communication activities to an application easily by dragging them onto the designer canvas and setting them up to interact with the user over a new interface (for example, phone and IM). The scenario explored in this chapter, an expense report approval process, demonstrates how you can use the UCMA Workflow to enable your applications for Unified Communications quickly.

Additional Resources

- "Unified Communications Managed API 2.0 Workflow SDK Documentation" (*http://go.microsoft.com/fwlink/?LinkID=133578*)

Part IV
Unified Communications Managed API

Part IV covers the Unified Communications Managed API (UCMA), as shown in the following illustration. This API is completely written in managed code using C# and has a small footprint. It provides a powerful API that can easily scale to thousands of connections to Microsoft Office Communications Server. In addition, you will find that it is very versatile.

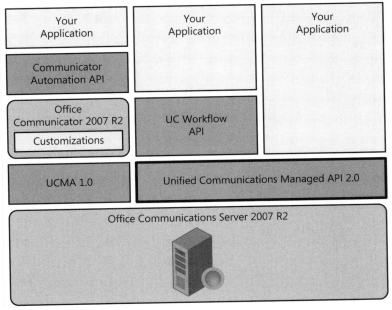

Chapter 7, "Structure of a UCMA Application," explains the API in detail, and Chapter 8, "Publishing Custom Presence with UCMA," describes an example of how to extend the Office Communications Server Enhanced Presence.

Chapter 7
Structure of a UCMA Application

This chapter will help you to:

- Understand the general structure of a Unified Communications Managed API (UCMA) Core application.
- Understand how to create a UCMA Core application.
- Understand the structure of the *CollaborationPlatform* class.
- Understand how to create and control multimodal calls that include both instant messaging (IM) and audio.
- Understand how to create, schedule, and participate in multiparty conferences.
- Understand how to publish and subscribe to presence information.

Creating a UCMA Application

As its name indicates, Unified Communications Managed API (UCMA) is a Microsoft .NET Framework–managed API. As such, this API benefits from all of the features that the .NET Framework provides. Because this API is highly scalable, it is ideal for building server applications. It can support thousands of endpoints and concurrent client connections. Applications built using UCMA can be load-balanced for high availability. Many of the services in Microsoft Office Communications Server use the UCMA stack. This chapter explains the anatomy of the UCMA.

Creating a UCMA application involves the following programming steps:

1. Instantiate the *CollaborationPlatform* object that is responsible for managing connections between the local endpoint and the server (or other endpoints), dispatching messages to endpoints, or providing other services. For a server or middle-tier application, the process involves creating an instance of the *ServerPlatformSettings* class that is initialized with information provided from the provisioning process. For a client application, this process involves creating an instance of the *ClientPlatformSettings* class to specify the connection transport.

2. Create one or more instances of *ApplicationEndpoint* or *UserEndpoint* depending on the application requirements. Typically, you use *ApplicationEndpoint* for an application that provides services to other clients. This application runs under its own security context and does not require any user credentials. An application that uses *UserEndpoint* executes in the security context of the corresponding user. If the application is not configured as trusted by Office Communications Server, it requires the user credentials.

3. Publish and receive presence information for the application or a user. To publish presence, you need to use the *LocalOwnerPresence* property of an *endpoint* instance. To receive presence, you need to use the *RemotePresence* property of an *endpoint* instance. An application can receive the presence information on an "ongoing" (subscription) or "on-demand" (query) basis.

4. Create conversations between the local participant and a remote participant, and handle calls between the participants. Every conversation is bound to an established endpoint when the conversation is created. To handle incoming calls, the application must register type-specific call handlers with the local endpoint by calling the *RegisterForIncomingCall* method on the *endpoint* instance.

5. Schedule and join a conference involving more than two participants. In a conference, participants converse with each other indirectly through appropriate multipoint control units (MCUs). For a UCMA application, MCUs for instant messaging (IM) and audio/video (A/V) calls are supported. A two-party conversation is changed into a conference when more participants are invited to join. This process is known as escalating a conversation into a conference.

Either for presence or conversations, the operations of the UCMA application consist of asynchronous processing of messages to and from the Office Communications Server or to and from any remote endpoints involved in active sessions of the application.

When deployed as server-based applications that support large numbers of connections, the UCMA applications must be trusted by Office Communications Server and the application host computer must be configured to support Mutual Transport Layer Security (MTLS). This process is referred to as application provisioning and must be performed before you can run any trusted UCMA application. For details, see the "Configuring UCMA Core" section in Chapter 9, "Preparing the UC Development Environment."

CollaborationPlatform

The application uses the *CollaborationPlatform* class to manage the connection between Office Communications Server and the application. A UCMA application can instantiate this class as either a server platform or a client platform, depending on the choice of the input parameter provided to the *CollaborationPlatform* constructor. For the server platform, the class constructor takes an instance of the *ServerPlatformSettings* class. For the client platform, the constructor takes an instance of the *ClientPlatformSettings* class. You use the *ClientPlatformSettings* class to instantiate a number of endpoints (for example, to emulate a large number of clients to stress-test an application). You use the *ServerPlatformSettings* class to create server-based applications that provide services to many clients. Query/response Web robots are examples of these types of applications. When an application uses the *ServerPlatformSettings* class to configure the *CollaborationPlatform* object, the UCMA library configures and manages a pool of multiple connections to Office Communications Server that is designed to maximize the flow of data between the application and the Office Communications Server. This provides for large-scale message processing in the application. Every UCMA application must create, configure, and start a *CollaborationPlatform* instance before it proceeds to calling other UCMA features.

Creating the Collaboration Platform for a Server Application

Listing 7-1 shows how to create and configure a *CollaborationPlatform* class for use by a server application.

LISTING 7-1 Creating and Configuring a *CollaborationPlatform* Class

```
private static void PreparePlatformAndEndpoint()
{
    //Load the local machine's certificate
    X509Certificate2 cert = GetLocalCertificate();

    //create a ServerPlatformSettings instance to
    //initialize a CollaborationPlatform instance
    ServerPlatformSettings platformSettings =
        new ServerPlatformSettings(
        ConfigurationManager.AppSettings["ocsUserAgentUCMALab1"],
        Dns.GetHostEntry("localhost").HostName,
        Int32.Parse(ConfigurationManager.AppSettings["appLocalPort"]),
        ConfigurationManager.AppSettings["appGruu"],
        cert);

    //Collaboration platform initialization
    _collabPlatform = new CollaborationPlatform(platformSettings);

    //Now that the CollaborationPlatform is configured, call BeginStartup.
    _collabPlatform.BeginStartup(EndCollabStartup, _collabPlatform);
}
```

In Listing 7-1, the parameters, which are used as the *CollaborationPlatform* settings and passed to instantiate the *ServerPlatformSettings* class, are read from the application configuration file associated with the application by using the *ConfigurationManager* class defined in the .NET Framework. This file is created as a part of creating the application. By centralizing these settings in the application configuration file, you provide a single source of configuration information for managing them as the application is moved from the development environment to the testing and deployment phases.

Starting the Collaboration Platform

Starting the newly created platform is an asynchronous process and involves calling the *BeginStartup* method on the *CollaborationPlatform* class to start the process and calling the *CollaborationPlatform.EndStartup* method when the startup process finishes. The

BeginStartup method takes two input parameters. The first parameter is an instance of the callback method to be invoked when the startup operation completes. The second parameter is the input parameter to the callback method. The instance of the *CollaborationPlatform* (*_collabPlatform*) is not ready for use until this startup process finishes successfully. To determine the status of the platform startup, a UCMA application must implement the callback routine to call *CollaborationPlatform.EndStartup* and to verify the status of the operation. The following code example shows this asynchronous programming pattern.

```
void BeginPlatformStartup()
    {
        AsyncCallback callback = new AsyncCallback(EndPlatformStartup);
        this._collabPlatform.BeginStartup(callback, _collabPlatform);
    }

void EndPlatformStartup(IAsyncResult result)
{
    if (!result.IsCompleted)
        return;
    CollaborationPlatform platform = result.AsyncState
        as CollaborationPlatform;

    if (platform == null)
        return;
try
{
    platform.EndStartup(result);
    _platformReady = true;
}
catch (Exception excep)
{
    //Process any errors here.
}
    // proceed to other tasks
    ...
}
```

This asynchronous pattern of calling *BeginOperation* and passing in the *EndOperation* callback parameter is repeated in almost all operations in the UCMA. It provides the scalability of throughput required by the UCMA server applications.

> **Note** This asynchronous programming pattern must be used when the application is deployed in production. However, this can also make debugging the application during development difficult. One tip to make debugging easier is to turn the asynchronous operations into synchronous operations using the following programming pattern.
>
> ```
> //Synchronously call the Startup method
> _collabPlatform.EndStartup(_collabPlatform.BeginStartup(null, null));
> ```
>
> Here the callback routine and its input parameters are not used and the two corresponding input parameters of *BeginStartup* are both set to *NULL*.

Endpoints

Endpoints represent users or applications that engage each other in conversations or conferences, publish their own presence, or receive others' presence. In UC, endpoints are identified by the Session Initiation Protocol (SIP) Uniform Resource Identifiers (URIs) and other supplemental identification information, such as endpoint ID or application Globally Routable User Agent URI (GRUU). You can think of a SIP URI as an identification of a security principle, whereas endpoint ID or GRUU helps to differentiate different instances of endpoints that are owned by the same security principle.

The UCMA defines an abstract class named *LocalEndpoint* to represent endpoints that are used to communicate with each other and with Office Communications Server. Applications use either *ApplicationEndpoint* or *UserEndpoint*, both of which inherit from *LocalEndpoint*, depending on the scenario and requirements. It is possible, however, for an application to create and use both types of endpoints or to create and use multiple endpoints of the same type.

Using *ApplicationEndpoint*

An application endpoint represents a running instance of a UCMA application that must be trusted by Office Communications Server. It is encapsulated by the *ApplicationEndpoint* class. This type of application connects to the server using the MTLS protocol. Both the server and the application provide each other with a certificate issued by a mutually trusted certificate authority (CA). Application endpoints are mostly used for server or middle-tier applications providing UC services. An example of this type of application is an IM Web robot that is used to broadcast alerts across a network.

A UCMA application that uses *ApplicationEndpoint* must register the underlying *ApplicationEndpoint* instance with Office Communications Server if it wants to support presence and UC session services. This also means that the application must be a server application with the *CollaborationPlatform* instance configured by a *ServerPlatformSettings* instance. For most UC scenarios, registration with Office Communications Server is required. To enable or disable the registration, the *ApplicationEndpoint* and *ApplicationEndpointSettings* classes expose the *UseRegistration* property. When this property is set to *True*, the endpoint is registered with the server. If it is set to *False*, the endpoint is not registered with the server.

An application that is based on *ApplicationEndpoint* is identified by a SIP URI assigned to a *Contact* object in Active Directory Domain Services. Defining and configuring this *Contact* object is part of the application provisioning process. The SIP URI of the *Contact* object is used to register the application with Office Communications Server. The *Contact* object also defines a display name and a telephone URI that become associated with the application.

Creating an instance of *ApplicationEndpoint* involves first creating an instance of the *ApplicationEndpointSettings* class that defines the attributes of the *ApplicationEndpoint*. These attributes include the following:

- The SIP URI to be used to register with Office Communications Server
- The fully qualified domain name (FQDN) of the server running Office Communications Server
- The port number used to connect to Office Communications Server
- The GRUU that is assigned to the application in the provisioning process

In addition, the *UseRegistration* property on the *ApplicationEndpointSettings* instance can be set to *True* if the application requires that the endpoint must register with Office Communications Server to use the presence and session services. You can use the *UseRegistration* property on the *ApplicationEndpoint* instance for this purpose as well.

A newly created *ApplicationEndpoint* instance must be connected to Office Communications Server before it can be functional. To establish this connection, the application must start the process by calling the *ApplicationEndpoint.BeginEstablish* method and finish the process by calling the *ApplicationEndpoint.EndEstablish* method after the operation succeeded.

Listing 7-2 illustrates the programming pattern to create an *ApplicationEndpoint* instance and to establish the connection between the endpoint and the underlying server.

LISTING 7-2 Programming Pattern for an *ApplicationEndpoint* Instance

```
//Create the ApplicationEndpointSettings to define the properties of the Endpoint.
//These properties are the SIP URI that identifies this endpoint, the FQDN of the
//Office Communications Server to connect to, the TCP port to use for the connection,
//and the GRUU that uniquely identifies this application to the OCS server.

    ApplicationEndpointSettings settings =
    new ApplicationEndpointSettings(
    ConfigurationManager.AppSettings["appUri"],
    ConfigurationManager.AppSettings["ocsServerFqdn"],
    Int32.Parse(ConfigurationManager.AppSettings["ocsServerTlsPort"]),
    ConfigurationManager.AppSettings["appGruu"],
    cert);

    //Enable registration
    settings.UseRegistration = true;

    //Create the ApplicationEndpoint
    //Binding it to the CollaborationPlatform previously created
    ApplicationEndpoint _appEndpoint =
    new ApplicationEndpoint(_collabPlatform, settings);

    //asynchronously call the Establish method
    _appEndpoint.BeginEstablish(EndEstablishEndpoint, _appEndpoint);
```

Note that in Listing 7-2, the configuration settings for the endpoint come from the configuration file associated with the application. The call to *ApplicationEndpoint.BeginEstablish* is required to initialize the endpoint and must complete successfully before the endpoint can be used. The status of the *BeginEstablish* method is returned in the *EndEstablishEndpoint* callback.

Using *UserEndpoint*

A user endpoint registers with Office Communications Server using the SIP URI that is assigned to the corresponding Active Directory *User* object that has been enabled for Office Communications Server. This *User* object could be a user account defined specifically for this application or associated with a real user. The *UserEndpoint* class allows UCMA applications to register with Office Communications Server and perform operations on behalf of the user.

Creating a *UserEndpoint* instance involves configuring the settings for *UserEndpoint*. You use the settings to specify the SIP URI of the user and the FQDN and port number of the underlying server. These settings are encapsulated by the *UserEndpointSettings* class. The *UserEndpoint* instance must also be bound to an initialized *CollaborationPlatform* instance. This binding is specified when *UserEndpoint* is instantiated.

To establish a *UserEndpoint* instance, you use similar semantics and follow a similar programming pattern as you do to establish an *ApplicationEndpoint* instance. The main difference is that the *UserEndpoint* instance uses the SIP URI of an Active Directory *User* object, whereas an *ApplicationEndpoint* object uses the SIP URI of an Active Directory *Contact* object.

Listing 7-3 illustrates the programming pattern that establishes a *UserEndpoint* instance.

LISTING 7-3 C# Programming Pattern to Establish a *UserEndpoint* Instance

```
UserEndpointSettings settings = new UserEndpointSettings("sip:user1@domain",
ConfigurationManager.AppSettings["OCSserverFqdn"],
ConfigurationManager.AppSettings["applicationServerPort"]);
UserEndpoint _userEndpoint = new UserEndpoint(_collabPlatform, settings);
_userEndpoint.BeginEstablish(EndpointEstablish, _userEndpoint);
```

In Listing 7-3, the SIP URI that is used to register the endpoint with Office Communications Server is sip:user1@domain. Note that the *sip:* prefix is required. This SIP URI must belong to an Active Directory *User* object that is enabled for UC. In this code example, it is assumed that the UCMA application is trusted by Office Communications Server and Transport Layer Security (TLS) is used as the transport between the application and the server. If the application is not trusted by the server or Transmission Control Protocol (TCP) is used as the transport, you must specify the user credentials on the *UserEndpointSettings.Credentials* property to establish the endpoint.

Conversation, Call, and Call Flow

After you have successfully established a local endpoint, you can use it to create a conversation in which the local participant makes a call to a remote participant. The local endpoint can also join a conversation by accepting an invitation to a call from a remote participant. In UCMA, a conversation encapsulates the participants and the calls between or among the participants. Each call consists of a flow of media. Thus, a call has a specific modality. The supported modalities include IM and A/V. Programmatically, these modality-specific calls are encapsulated by the *InstantMessagingCall* and *AudioVideoCall* classes. The corresponding media flows are encapsulated by the *InstantMessagingFlow* and *AudioVideoFlow* classes.

> **Note** While the class name in UCMA 2.0 is *AudioVideoCall*, the only media type supported in this release is audio.

Figure 7-1 illustrates these concepts.

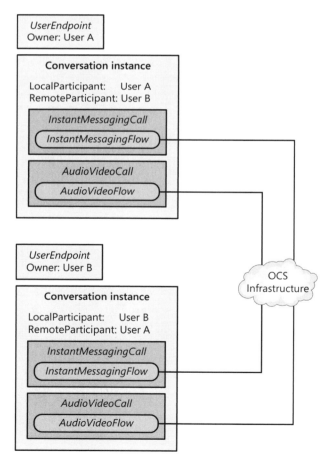

FIGURE 7-1 A two-party conversation involving IM and A/V calls.

A conversation involving more than two participants is referred to as a *conference*. In a two-party conversation (referred to as a *conversation* from here on) the participants make calls to each other directly. In a conference, the participants call each other with the help of MCUs that are configured for the supported Office Communications Server deployment. The media stream flows between a local participant's endpoint and the set of MCUs that are configured to support the conference.

Creating Calls

In UCMA, a call corresponds to a modality-specific channel in a communication session. The supported modalities include IM and audio. This means that a UCMA application can make and receive IM and audio calls, provided that an endpoint is established. The type of endpoint can be either an *ApplicationEndpoint* or *UserEndpoint*.

Calls are managed by sessions. In UCMA, a session is represented by an instance of the *Conversation* class. Programmatically speaking, a *Conversation* consists of participants and modality-appropriate calls.

The steps to establish calls between two parties are as follows:

1. Create a call.
2. Handle incoming calls.
3. Handle call flows.
4. Handle call state and messages.

The relationship between the classes involved in making calls was shown previously in Figure 7-1.

Classes are instantiated from the "outside in" using the asynchronous call pattern. Each class instance is not ready for operations until its *EndOperationCallback* method is called and returns a success status. For example, you must create a valid instance of the *Conversation* class before you can create an instance of the *Call* class. You need a valid instance of a *Call* class, such as *InstantMessagingCall* or *AudioVideoCall*, to create and use the *Flow* classes (that is, *InstantMessagingFlow* or *AudioVideoFlow*) to exchange data during the call.

The *Call* classes, *InstantMessagingCall* and *AudioVideoCall*, represent instances of sessions using the specified media type. The properties of the *Call* classes define things such as the participants, the state of the call (for example, established or ending) and also the instance of the *Flow* class that is associated with the call after it is established. The *Flow* class encapsulates the information related to the actual media (IM text or audio streams) that belong to the call. After a call has been established (that is, the call was offered by the caller and answered by the called party), the application uses the *Flow* class to send and receive the messages over the call.

Creating a Call

To simplify the code, the following example shows the steps to create and use a call using the synchronous design pattern. Note that you should use the asynchronous design pattern for production code. While this sample uses an IM call, the pattern is the same for audio calls. The steps for creating a call are as follows:

1. Use *_appEndpoint* to create a new instance of the *Conversation* class named *_currentConversation*.

2. Use *_currentConversation* to create a new instance of the *InstantMessagingCall* class named *_currentImCall*.

3. Use the *BeginEstablish* method of *_currentImCall* to set the parameters of the IM call, such as the target SIP URI.

4. Use the *Flow* class that is associated with *_currentImCall* when it is established to send a message saying "Hello World."

5. Use the *BeginTerminate* method of *_currentImCall* to signal that the call is being ended.

6. Use the *BeginTerminate* method of *_currentConversation* to delete and clean up the *Conversation* class.

The UCMA code to implement the logic in the previous list is as follows.

```
// Create the conversation
Conversation _currentConversation =
    new Conversation(_appEndpoint);
// Create the IM call
InstantMessagingCall _currentImCall =
    new InstantMessagingCall(_currentConversation);
// Establish the call synchronously
_currentImCall.EndEstablish(_currentImCall.BeginEstablish("SIP:user1@domain",
            null,         // Toast Message
            null,         // Callback routine is not used in the synchronous pattern
            null));       // Context object
// Send the message synchronously
_currentImCall.Flow.EndSendMessage(
    _currentImCall.Flow.BeginSendMessage("Hello World!", null, null));
// Terminate the call synchronously
_currentImCall.EndTerminate(_currentImCall.BeginTerminate(null, null));
// Terminate the conversation synchronously
_currentConversation.EndTerminate(_currentConversation.BeginTerminate(null, null));
```

Handling Incoming Calls

For an endpoint to accept incoming calls, it must register an event handler for the *RegisterForIncomingCall* event. This event handler is called whenever there is an incoming call request to this endpoint. For example:

```
_appEndpoint.RegisterForIncomingCall<InstantMessagingCall>(On_InstantMessagingCall_Received);
```

This code registers the callback method, *On_InstantMessagingCall_Received*. This callback method is called when an incoming IM call is received for the endpoint, *_appEndpoint*. Note that the *RegisterForIncomingCall* event is defined using a generic type. This allows the application to use the same event pattern for *AudioVideoCall* as well as for other custom modality types defined by the application. Custom modalities are defined by creating a new call type to pass as the specifier for the generic type. An example of a handler for an incoming *AudioVideoCall* is as follows.

```
_appendpoint.RegisterForIncomingCall<AudioVideoCall>(On_AudioVideoCall_Received);
```

The handler for this event is called when an incoming *AudioVideoCall* request is received. A typical handler can examine the details of the session request, such as the SIP URI of the sender, the subject of the call, the priority of the call, or any other call properties that are needed, to make the decision to either accept or reject the request.

> **Note** The other properties of the *Call* are defined in the UCMA documentation on the *Call* class.

The following is an example of an event handler for an IM call.

```
void On_InstantMessagingCall_Received(object sender,
CallReceivedEventArgs<InstantMessagingCall> e)
{
     _instantMessagingCall = e.Call;
     /* Register a handler for the Call.StateChanged event to receive the call state transitions. */
     _instantMessagingCall.StateChanged += new
EventHandler<CallStateChangedEventArgs>(_instantMessagingCall_StateChanged);
     // Remote Participant URI represents the caller.
     // Toast is the message set by the caller as the 'greet' message in the call.
     Console.WriteLine("Call Received! From: " + e.RemoteParticipant.Uri +
" Toast is: " + e.ToastMessage);
     // Now, accept the call.
     _instantMessagingCall.BeginAccept(EndAcceptCall, _instantMessagingCall);
}
```

Handling Call Flows

After the application accepts the call, the underlying UCMA code builds an instance of the *Flow* class that is used to handle the media, either IM or audio, that is a part of the call. Because this setup is handled automatically by the UCMA library, the application only needs to register for the event that has the modality of interest. In the following code example, the application registers for the *InstantMessagingFlowConfigurationRequested* event to track the progress of the creation of the *Flow* request. This registration can be done by using the *StateChanged* event handler for the call after the call state, which uses the *State* property of the call, is reported to be "Established." For an *AudioVideoCall*, the coresponding event is called *AudioVideoFlowConfigurationRequested*.

```
/* Subscribe for the flow event. When this event is received, the media flow associated
with the call is created and can be used to send and receive IM messages. */
_instantMessagingCall.InstantMessagingFlowConfigurationRequested +=
this.instantMessagingCall_FlowConfigurationRequested;
```

After the flow is available for the call, the media (that is, IM or audio) can be exchanged during the call. After the *InstantMessagingFlowConfigurationRequested* handler is called and indicates that the *Flow* is available, the application can register event handlers for *StateChanged* and *MessageReceived* events on the *Flow*. The *StateChanged* event signals a change in the *Flow* state of the call. This state is one of the values defined by the *MediaFlowState* enumeration (that is, *Idle*, *Active*, or *Terminated*). The *MessageReceived* event is raised when a message is received from the other participant in the call, as shown in the following code example.

```
/* Flow created indicates that there is a media flow class present that can be used for
media operations. */
public void instantMessagingCall_FlowConfigurationRequested(object sender,
InstantMessagingFlowConfigurationRequestedEventArgs e)
{
    _instantMessagingFlow = e.Flow;
    /* Bind to the event handlers to get notification of state changes and messages
received. */
    /* When the flow becomes active, (as indicated by the state changed event) the call is
ready to send IM messages. */
    _instantMessagingFlow.StateChanged += this.instantMessagingFlow_StateChanged;
    /* Message Received is the event used to indicate that a message has been recieved from
the remote participant. */
    _instantMessagingFlow.MessageReceived += this.instantMessagingFlow_MessageReceived;
}
```

Handling Call State and Incoming Message Events

After the *InstantMessagingFlow.StateChanged* event is signaled and the state of the call is *Active*, the messages can be sent. If the state is *Idle*, no action is needed. If the state is *Terminated*, the application can clean up any application resources associated with the call, if any, and end the call.

```
private void instantMessagingFlow_StateChanged(object sender,
MediaFlowStateChangedEventArgs e)
{
   // When flow is active, media operations (here, sending an IM) may begin.
   if (e.State == MediaFlowState.Active)
   {
       /* Send the message on the InstantMessagingFlow. _messageToSend is the text of the
message, EndSendMessage is the callback routine that is called to report the status of the
operation. */
       _instantMessagingFlow.BeginSendMessage(_messageToSend, EndSendMessage,
_instantMessagingFlow);
    }
}
```

Conferences

Conferences are calls that involve more than two participants. In many cases, applications start with a two-party call and then escalate the call to a conference when one or more participants are added to the two-party call. When the third participant is added, the media for the call is routed through a conferencing server, also referred to as an MCU. This server role is part of the Office Communications Server infrastructure. There are conferencing servers that support the various content types (such as IM, A/V, and Web conferencing media) and replicate the media generated by each participant to all of the participants in the conference. What appears to the participants to be one UC session may be composed of multiple calls, with each call being used to manage and transport one media type. The exact mechanism and behavior used to escalate a two-party session to a conference involving three or more parties is specific to the application.

Scheduling a Conference

The UCMA provides classes to schedule conference sessions for participants to connect to the conference. The main class that is used for scheduling is called *ConferenceScheduleInformation*. This class defines the media types that can be used during the conference, the participants, the roles assigned to participants, and the time and duration of the conference. An example of scheduling a conference is shown in Listing 7-4.

LISTING 7-4 Scheduling a Conference

```
// The base conference settings object, used to set the policies for the
//conference.
ConferenceScheduleInformation _conferenceScheduleInformation = new
ConferenceScheduleInformation();

// In an open meeting, any participant can join but authentication is required.
//No anonymous users are allowed. Other possible authentication values are: None,
//ClosedAuthenticated, and Anonymous.
_conferenceScheduleInformation.AdmissionPolicy =
ConferenceAdmissionPolicy.OpenAuthenticated;

// This flag determines whether the passcode is optional to join the conference.
//A value of "true" means that the passcode is optional.
_conferenceScheduleInformation.IsPasscodeOptional = true;

// The conference passcode.
_conferenceScheduleInformation.Passcode = "sample";

// The verbose description of the conference. _
conferenceScheduleInformation.Description = "Example Conference";

// This field indicates the date and time when the conference can be deleted.
_conferenceScheduleInformation.ExpiryTime = System.DateTime.Now.AddHours(5);
```

```
// Specify the set of modalities (here, only InstantMessage) to use during the
//conference. McuType is a UCMA defined enum type.
ConferenceMcuInformation _instantMessageMCU = new
ConferenceMcuInformation(McuType.InstantMessaging);
_conferenceScheduleInformation.Mcus.Add(_instantMessageMCU);

// Now that the ConferenceScheduleInformation class is specified, schedule the
//conference using the conference services of the Endpoint. The _callerEndpoint
//can be either an ApplicationEndpoint or a UserEndpoint.

// Note: the conference organizer is considered a leader of the conference
//by default.
_callerEndpoint.ConferenceServices.BeginScheduleConference
(_conferenceScheduleInformation,
EndScheduleConference, // Callback method
_callerEndpoint.ConferenceServices);
```

In Listing 7-4, the status of the scheduling operation is returned in the *EndScheduleConference* callback that is passed to the *BeginScheduleConference* call. After the *EndScheduleConference* callback is signaled, the conference is created and can be used by the participants. The conference's URI property uniquely identifies the conference. Participants use the conference URI to join the conference. The conference URI is available once the conference is scheduled and the value is retrieved by getting the *Conference.ConferenceUri* property. This URI is available only to the conference organizer and must be communicated to the other participants by using out-of-band methods (for example, e-mail), or an existing *Conversation* can be escalated to a conference using the *Conversation.BeginEscalateToConference* method.

Joining a Conference

To enable users to join a conference, the application creates a new *Conversation* class and uses the conference URI, as shown in the following code example.

```
// Now that the conference is scheduled, it's time to join it. As we
//already have a reference to the conference object populated from the
//EndScheduleConference call, we do not need to get the conference first.

//Initalize a conversation off of the endpoint, and join the conference
// from the URI provided above.
Conversation _callerConversation = new Conversation(_callerEndpoint);
ConferenceJoinInformation _confJoinInfo = new ConferenceJoinInformation(
new RealTimeAddress(_conference.ConferenceUri));

// Start the process of joining the conference using
//the ConferenceSession.BeginJoin call.
_callerConversation.ConferenceSession.EndJoin(
_callerConversation.ConferenceSession.BeginJoin(
_confJoinInfo,
null,
null));
```

```
// Since we are joining using the synchronous pattern, we can now create the calls
//for the conference media. Otherwise, the work to create the call would be done
//in the ConferenceSession.EndJoin callback handler by using the code below.
//Placing the calls on the conference-connected conversation connects to the
//respective MCUs. These calls may then be used to communicate with the MCUs.
InstantMessagingCall _instantMessagingCall =
    new InstantMessagingCall(_callerConversation);
```

If an audio call is included in the conference, the application creates an instance of *AudioVideoCall* to connect to that media stream.

Publish and Subscribe to Presence

Applications written using UCMA 2.0 can publish presence to Office Communications Server, as well as subscribe to presence notifications from the server. Both of these operations require the use of an endpoint that is registered with Office Communications Server.

When a user first signs in to Office Communications Server by using Microsoft Office Communicator, the client creates a set of containers for the user in Office Communications Server and publishes one or more instances of presence categories to those containers. Each container created is assigned access control entries (ACEs) to set permission levels. These permission levels define the access level contacts that are allowed. The default set of containers that are created are as follows:

- Blocked
- Public
- Company
- Team
- Personal

Office Communicator publishes the user's availability, calendar state, and many other presence categories to these containers. For more information about the containers, access control, and the categories, see the "Office Communicator 2007: Enhanced Presence Model White Paper" at *http://go.microsoft.com/fwlink/?linkid=143209*.

Note It is possible—in fact, it is common—for a single user to have more than one active endpoint registered with Office Communications Server at the same time. Each of those endpoints publishes one or more instances of the presence categories to one or more containers on behalf of the user. There is a server component on Office Communications Server, named the Aggregation Script, that takes these multiple instances of a given presence category that are published to the aggregation containers (that is, container 2 or 3) and aggregates them into one global value for each category. When user A subscribes to user B's presence information, the values of the categories returned are these aggregated values. A representation of this aggregation operation is shown in Figure 7-2.

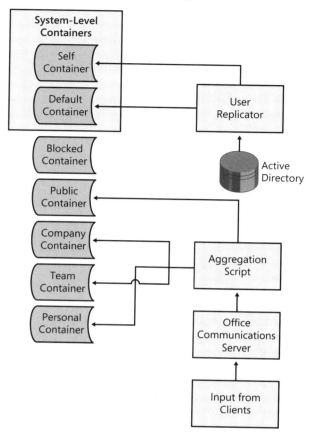

FIGURE 7-2 Containers and presence aggregation.

The publication of presence information is always performed in the context of a specific SIP URI and is published to one or more containers. When an application wants to subscribe to presence information, the subscription is always in the context of a specific SIP URI. A user has access to all of the presence categories that he or she published to Office Communications Server. This is referred to as "self presence." However, when a user subscribes to the presence information of other users, the information returned is limited to the set of presence information published to the container where the subscriber has been granted access.

In practice, what this means is that presence subscriptions for the self presence items might return multiple instances for each category that are published to different containers. Presence subscriptions for other users return a single instance of a presence category from a single container.

Publishing Presence

When an application uses UCMA 2.0 to publish presence, it first needs to create and initialize an instance of the *CollaborationPlatform* class and then create either a *UserEndpoint* or *ApplicationEndpoint* for the SIP URI that is used to publish presence information. Because the code for creating an endpoint was covered earlier in the chapter, this section assumes that these steps have been carried out and the resulting endpoint, *_userEndpoint*, has been created.

The steps to publish presence are as follows:

1. Create an instance of the *CustomPresenceCategory* class, *_customCat*, to hold the information that is published and supply the category name. This example uses *state* and the category data as a formatted Extensible Markup Language (XML) string.
2. Create an instance of the *List<CustomPresenceCategory>* class to hold the category instances to be published.
3. Add any *CustomPresenceCategory* entries that are to be published to the list.
4. Use the *BeginPublishPresence* method of the *LocalOwnerPresence* property of the endpoint to publish the presence value.

The following code provides an example of how to publish an instance of the *state* category. It is an Enhanced Presence category that is defined by Office Communicator. All the Enhanced Presence categories are declared as XML structures, and their formats are specified by the "Unified Communications Enhanced Presence Schemas for Microsoft Office Communications Server 2007" located at *http://go.microsoft.com/fwlink/?linkid=143305*. The *_stateXML* variable contains an XML string for the *state* category describing the availability of a user. Office Communicator uses an integer to indicate the availability. For example, setting 3500 as the availability value indicates that the user state is "Available"; a value of 6500 indicates that the user state is "Busy"; and 15500 indicates that the user state is "Away."

```
private static String _stateXml = "<state
xmlns=\"http://schemas.microsoft.com/2006/09/sip/state\"
xmlns:xsi=\"http://www.w3.org/2001/XMLSchema-instance\" manual=\"true\"
xsi:type=\"userState\"><availability>{0}</availability></state>";

// Create an instance of CustomPresenceCategory to hold the presence data
//to publish. In this case, we are publishing an instance of the "state"
//category with a value of busyAvailable, which corresponds to the value 6500.
//The format of the category is an XML document and the value is of type int.
CustomPresenceCategory _customCat = new CustomPresenceCategory("state",
String.Format(_stateXml, 6500));

List<PresenceCategory> _catPublish = new List<PresenceCategory>();
_catPublish.Add(_customCat);

// The LocalOwnerPresence class on the user endpoint means that we are
//publishing presence for the user that owns the endpoint. */
_userEndpoint.LocalOwnerPresence.BeginPublishPresence(_catPublish,
EndPublishPresenceCallback, _userEndpoint.LocalOwnerPresence);
```

Note that *BeginPublishPresence* takes a collection as an input parameter for the presence categories to be published. This allows this method to publish multiple different presence categories in a single operation.

The result of the presence publication operation is returned when the *EndPublishPresence-Callback* method is called. This callback method is registered during the *BeginPublishPresence* call. The following is an example of this callback.

```
private void EndPublishPresenceCallback(IAsyncResult result)
{
    /* Verify the result of the publication by calling EndPublishPresence. If there is no
    exception thrown, then the operation succeeded. */
    try
    {
        _userEndpoint.LocalOwnerPresence.EndPublishPresence(result);
    }
    catch (Exception excpt)
    {
        Trace.Writeline(excpt.Message +" in EndPublishPresence");
    }
}
```

Because this presence category instance was published using a *UserEndpoint*, which is bound to a user's SIP URI, it is published to the containers owned by that user. If the endpoint is an *ApplicationEndpoint*, the presence information is published to the containers owned by the *Contact* object that defines the SIP URI.

It is possible for the application to update any or all of the presence categories that it has published at any time.

Subscribing to Presence

When an application using UCMA subscribes to presence information, it must use an enabled endpoint to create the subscription. The application can subscribe to the signed-in user's presence information, called self presence, or it can subscribe to contacts' (that is, other users') presence information. When subscribing to self presence, the application has full access to the signed-in user's (that is, endpoint) categories published to the containers owned by the user. However, when subscribing to other users' presence information, the application has access only to the presence categories visible to the user represented by the signed-in endpoint.

Self Presence Applications generally subscribe to self presence to retrieve the user's information and update his or her presence information. An example of user information is the contact list, which is published as a set of instances of the *contactCard* category. To subscribe to the self presence information, the application creates a subscription from the *LocalOwnerPresence* property of the endpoint, as shown in the following code example. Note that the subscription to *LocalOwnerPresence* returns all of the presence categories that have been published for the user.

```
// Create a subscription to the "self presence" of the user represented by the endpoint.
_userEndpoint.LocalOwnerPresence.BeginSubscribe(SelfPresenceSubscribeCallback, null);
```

The callback handler, *SelfPresenceSubscribeCallback*, that is specified in the *BeginSubscribe* method receives notification of whether the subscription completed successfully. This callback method should then subscribe to the *CategoryNotificationReceived* event so that the application can be notified when instances of the presence categories are updated in Office Communications Server. This event is signaled the first time a category value is received and any time an updated value is received. The *SelfPresenceSubscribeCallback* method implementation is shown in the following code example.

```
private void SelfPresenceSubscribeCallback(IAsyncResult result)
{
    // Call EndSubscribe to check the status returned from the BeginSubscribe method.
    try
    {
        _userEndpoint.LocalOwnerPresence.EndSubscribe(result);
        /* Subscribe to the CategoryNotificationReceived events to be notified when category data is received. */
        _userEndpoint.LocalOwnerPresence.CategoryNotificationReceived +=
            new EventHandler<CategoryNotificationEventArgs>(Self_CategoryNotificationReceived);
    }
    catch (Exception excpt)
    {
        //There was an error in the subscription.
        Trace.WriteLine(excpt.ToString() + "in EndSubscribe");
    }
}
```

To receive notifications of event changes to the user's categories, the event handler, *Self_CategoryNotificationReceived*, that is registered for *CategoryNotificationReceived* events must be implemented. The following code example illustrates this implementation.

```
void Self_CategoryNotificationReceived(object sender, CategoryNotificationEventArgs e)
{
    //e.CategoryList is the list of category instances that were returned from the server
    foreach (PresenceCategoryWithMetaData cat in e.CategoryList)
    {
        // handle the categories of interest in the switch statement
        switch (cat.Category.CategoryName)   //Switch on the name of the category
        {
            case "state":
                //process the user state data.
                break;
```

```
                case "contactCard":
                    //process a contact card instance. We get one of these per contact in the list.
                    break;
                default:
                    break;
            }
        }
    }
}
```

Presence from Other Users   Subscribing to presence information from other users follows a similar pattern as the self-subscription process. The endpoint registered to the user is used to create the subscription. In this case, use the *RemotePresence* class to manage the subscription and subscribe to the *PresenceNotificationReceived* events of that class, as shown in the following code example. These events are used to receive notifications of presence changes from Office Communications Server. The *BeginAddTargets* method is used to create the subscriptions to one or more users.

```
//Create an instance of the RemotePresence
RemotePresence _remotePresence = _userEndpoint.RemotePresence;

//Create an event handler for the PresenceNotificationReceived events.
_remotePresence.PresenceNotificationReceived += new
EventHandler<RemotePresenceNotificationEventArgs>(
remotePresence_PresenceNotificationReceived);

//Create a list of the remote presentities to subscribe to.
RemotePresentitySubscriptionTarget _target1 = new
RemotePresentitySubscriptionTarget("sip:user1@domain", String.Empty;
RemotePresentitySubscriptionTarget _target2 = new
RemotePresentitySubscriptionTarget("sip:user2@domain", String.Empty);
RemotePresentitySubscriptionTarget _target3 = new
RemotePresentitySubscriptionTarget("sip:user3@domain", String.Empty);

// Once the target SIP URIs to subscribe to are defined, add them to a List class.
List<RemotePresentitySubscriptionTarget> _targetList = new
List<RemotePresentitySubscriptionTarget>();
_targetList.Add(_target1);
_targetList.Add(_target2);
_targetList.Add(_target3);

// First create the string array of category names to query.
string [] _cats = {"state", "contactCard"};

//Create the query passing the targets, the category list, the event
//handler that will receive the category data and null for the callback
//and the status object.
_remotePresence.EndPresenceQuery(_remotePresence.BeginPresenceQuery(_targetList, _cats,
remotePresence_PresenceNotificationReceived, null, null);
```

The *PresenceNotificationReceived* event handler is called whenever presence updates are received from the subscription. Presence information for multiple targets can be delivered in a single event. The data for the targets is delivered in a collection in the *Notifications* property of the event arguments. Each entry in the collection applies to one presentity whose SIP URI is indicated by the *URI* property on the notification. Each notification can contain data for multiple presence categories, as shown in the following code example.

```
void remotePresence_PresenceNotificationReceived(object sender,
RemotePresenceNotificationEventArgs e)
{
    foreach (RemotePresentityNotificationData notification in e.Notifications)
    {
        string targetUri = notification.Uri; //Presentity URI for this presence data.
        //Each Notification can contain multiple categories

        foreach (PresenceCategoryWithMetaData category in notification.Categories)
        {
            switch (category.Category.CategoryName) //Switch based on the category name.
            {
                case "state":
                //process state.
                break;

                default:
                break;
            }
        }
    }
}
```

Summary

The UCMA 2.0 Software Development Kit (SDK), available at *http://go.microsoft.com/fwlink/?linkid=143314*, contains significant new functionality compared to the 1.0 release. This new functionality provides the application programmer with the ability to do the following:

- Create and control multimodal calls that include both IM and audio.
- Create, schedule, and participate in multiparty conferences.
- Publish and subscribe to presence information.

The UCMA SDK is designed to operate using an asynchronous programming model that provides for maximum throughput for applications that need scalability. In addition, the design of the SDK enables you to extend the programming model to add new media types.

Additional Resources

- "Office Communicator 2007: Enhanced Presence Model White Paper" (*http://go.microsoft.com/fwlink/?linkid=143209*)

- Microsoft Unified Communications Managed API 2.0 SDK (32 bit) (*http://go.microsoft.com/fwlink/?linkid=143314*)

- Microsoft Unified Communications Managed API 2.0 SDK (64 bit) (*http://go.microsoft.com/fwlink/?LinkID=139195*)

- Unified Communications Enhanced Presence Schemas for Microsoft Office Communications Server 2007 (*http://go.microsoft.com/fwlink/?linkid=143305*)

Chapter 8
Publishing Custom Presence with UCMA

This chapter will help you to:

- Understand how to define custom presence categories in Microsoft Office Communications Server 2007 R2.
- Understand how to publish instances of custom presence categories for users using the Unified Communications Managed API (UCMA)

Creating Custom Presence Categories

Office Communications Server 2007 R2 supports a flexible presence data model and provides a generic framework for applications to publish, subscribe, and query for presence information for users of the system. The presence information is defined by the Enhanced Presence categories and is published to one or more presence containers on the server. Each container has one or more access control entries (ACEs) that defines the access to the instances of presence categories published to the container. For details about the presence model implemented in Microsoft Office Communicator 2007, see the "Enhanced Presence Model White Paper" at *http://go.microsoft.com/fwlink/?linkid=143209*.

For an application to subscribe to or publish presence information to Office Communications Server, the application must instantiate an endpoint and register that endpoint with Office Communications Server. While any registered endpoint can subscribe to or query for presence information, an endpoint can only publish presence data for the user whose identity is used to register the endpoint with Office Communications Server.

In the Unified Communications (UC) platform, applications are responsible for publication, querying, and subscription of presence information. This responsibility includes defining categories as Enhanced Presence data types and defining ACEs on containers to block or permit access to the published presence data. The content of Enhanced Presence data is opaque to the server. An application can define its own presence categories and implement the application-specific semantics for the data in the categories. It can also use the presence categories defined by other applications and follow the rules of publication and subscription defined by other applications.

For example, Office Communicator defines a rich set of Enhanced Presence categories, some of which are used to represent the user state. Office Communicator publishes them to indicate whether the user is available, on the phone, away, and so on. Other categories are used to represent the user's Free/Busy information from the user's calendar data. Office Communicator reads this information from Microsoft Office Outlook and Microsoft Exchange Server and makes it available to the user's contacts. Other UC applications can use the presence information defined and published by Office Communicator. They can also define and publish custom presence information. This chapter describes how to publish application-specific presence information using custom presence categories.

Common Custom Presence Application Scenario

A common custom presence application scenario is publishing presence information for users based on sources of information and applications other than Office Communicator. Before the release of Office Communications Server 2007 R2, there were two main issues with this scenario. The first issue was that to register an endpoint with the server running Office Communications Server for a user, the application required the user's domain credentials to validate the registration with Office Communications Server. This can be an issue with the security and operation of the application. The second issue was that none of the APIs that supported presence operations (Unified Communications Client API and Communicator Web Access Asynchronous JavaScript and Extensible Markup Language [XML] API) provided the behavior and scale required to support large numbers of simultaneous users. The UCMA API 2.0 provides solutions to both of these issues and makes a custom presence scenario easy to implement.

Choice of Technology

UCMA 2.0 provides the ability to create a middle-tier application that registers endpoints and publishes instances of presence categories on behalf of some users. Because an application written with UCMA can be registered with Office Communications Server as a trusted application, the application is not required to supply the user's domain credentials to register an endpoint for the user.

This chapter provides an example of how UCMA can be used to create an application that registers many user endpoints with Office Communications Server and publishes instances of a presence category on their behalf. For this example, we are publishing a custom presence category named GPSLocation, representing the Global Positioning System (GPS) location of the user, but the application can publish instances of any of the system-defined categories that are defined for users as well. However, when publishing the system-defined categories, there can be issues of coexistence with running instances of Office Communicator and multiple simultaneous endpoints for the user, which complicate the scenario. These potential issues are not addressed in this chapter; the application is created on the assumption that it is the only endpoint that is publishing GPS location information for users.

Overall Code Structure

The sample application is designed to support publication of an XML document containing the GPS coordinates of a user to Office Communications Server. It assumes that the GPS data to be published is generated in some other application or system and supplied to the sample application to be published as presence information for the specified user. This data can be passed into the publishing application, or the application can query the system that supplies the GPS information for the location at some interval.

The sample creates and registers endpoints for users of the *UserEndpoint* type at startup. After the application receives the location information, it uses the registered *UserEndpoint* instances to publish the new location value in a custom presence category instance.

The basic structure and flow of control in the sample is shown in Figure 8-1.

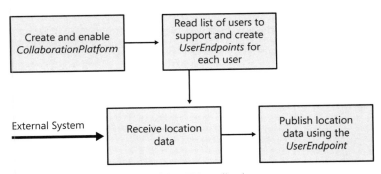

FIGURE 8-1 Overall structure of the GPS application.

Test Environment

To run and test this application, the custom category that is used to store the location information for users must be defined in Office Communications Server. This is accomplished by running a stored procedure on the Microsoft SQL Server database used by Office Communications Server to define the new category.

Various tools can be used to access the database, such as the command-line SQLCMD tool, but these details are not covered in this book. After the database is accessed, the stored procedure is invoked with the name of the new category as the input parameter. For example, the commands to the SQLCMD tool are as follows.

```
use rtc
exec RtcRegisterCategoryDef N'GPSLocation'
```

where *rtc* is the name of the database, *RtcRegisterCategoryDef* is the name of the stored procedure, and *GPSLocation* is the name of the new category to be defined.

After this category is defined, any application can publish instances of it to Office Communications Server or subscribe to it from Office Communications Server. Note that nothing in the process defines either the syntax or the semantics of the new category because the category is opaque to Office Communications Server. The only requirement that Office Communications Server imposes on instances of the category is that they be written in well-formed XML. It is up to the application that uses the category to define the structure of the instances, to ensure that they are well-formed when published to the server, and to parse them when they are delivered from the server. This is typically achieved by defining the structure of the category data by creating an XML Schema Definition (XSD) file that defines the schema for the custom category.

In Listing 8-1, the schema for the GPS location that is published to the custom category is defined.

LISTING 8-1 Schema for the *GPSLocation* Custom Presence Category

```xml
<GPSLocation xsi:type="gpslocation"
    xmlns:xsi="http://www.w3.org/2001/XMLSchema-instance"
    xmlns="http://schemas.microsoft.com/2006/09/sip/GPSLocation">
  <locationstring></locationstring>
</GPSLocation>
```

The only element that is supplied in the publication is the *<locationstring>* value, which is a string.

Detailed Code

For this sample, we are going to create a console application in C#. The application requires that the UCMA 2.0 Software Development Kit (SDK) is installed, along with any of its prerequisites.

1. Start by running Microsoft Visual Studio 2008 and creating a new Microsoft Windows project for a console application using C#. Name the project PublishPresence.

2. At the top of the Program.cs file, add the following lines to reference the class libraries that the application uses.

   ```csharp
   using System.Configuration;
   using System.Security.Principal;
   using System.Security.Cryptography.X509Certificates;
   using System.Net;
   using System.Diagnostics;
   ```

```
using Microsoft.Rtc.Signaling;
using Microsoft.Rtc.Collaboration;
using Microsoft.Rtc.Collaboration.Presence;
```

3. On the Solution Explorer tab, right-click the References node, and add references to System.Security, System.Configuration, and Microsoft.Rtc.Collaboration, as shown in Figure 8-2.

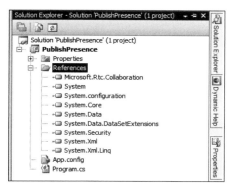

FIGURE 8-2 Adding assembly references.

4. Inside the Program class, insert the following definitions.

```
private static CollaborationPlatform _collabPlatform;
private static Dictionary<string, UserEndpoint> _userDictionary;
private static string OCS_SERVER_FQDN = "contosoocs2.uc.contoso.com";
private static int APPLICATION_PORT = 9000;
private static string APPLICATION_GRUU = "sip:contosolabts.uc.contoso.
com@uc.contoso.com;gruu;opaque=srvr:contosolabts.9000:NXnkPWAgokGhv62DhnHAIgAA" ";
    private static string GPSLOCATION_SCHEMA = "<GPSLocation xsi:type=\
"gpslocation\" xmlns:xsi=\"http://www.w3.org/2001/XMLSchema-instance\" xmlns=\
"http://schemas.microsoft.com/2006/09/sip/GPSLocation\"><locationstring>{0}
</locationstring></GPSLocation>"
```

These lines of code define the following:

- The *CollaborationPlatform* for connecting to the server running Office Communications Server
- The *Dictionary* used to store the *UserEndpoints* for the users the program supports, indexed by the Session Initiation Protocol (SIP) Uniform Resource Identifier (URI)
- The fully qualified domain name (FQDN) of the Office Communications Server to which to connect
- The Transmission Control Protocol (TCP) port that the application uses
- The Globally Routable User Agent URI (GRUU) defined for the application provisioned
- The XML syntax to publish the GPS location information

These settings are defined when the application is provisioned. The values for the FQDN of the Office Communications Server, the TCP port, and the GRUU used by the application will need to be modified based on the environment where the application is run.

5. Navigate to the *Main* method in Program.cs and add the following lines to the body of the procedure.

```
static void Main(string[] args)
{
    string inputLine;
    char sep = new char();
    sep = Convert.ToChar(",");
    _userDictionary = new Dictionary<string, UserEndpoint>();
    //Prepare the CollaborationPlatform for use.

    PreparePlatform();
        Console.Write("Enter sipuri,Gpslocation to publish user GPS location or exit to quit");
        inputLine = Console.ReadLine();
        while (!inputLine.ToLowerInvariant().StartsWith("exit"))
        {
            if (!string.IsNullOrEmpty(inputLine))
            {
                string[] input = inputLine.Split(sep);
                //input[0] is the SIP URI for the user and input[1] is the GPS string.
    PublishUserLocation(input[0], input[1]);
                inputLine = Console.ReadLine();
            }
            else
            {
                Console.WriteLine("Enter sipuri,Gpslocation to publish user GPS location or exit to quit");
                inputLine = Console.ReadLine();
            }
        }
        ShutdownAndCleanup();
}
```

Here, we are initializing the Dictionary and then calling the *PreparePlatform()* method to instantiate the *CollaborationPlatform*. The code then prints a message to the console and waits for user input. The input is either the word **exit** to exit the program or a SIP URI followed by a comma and the GPS string that defines the user's location. If the user enters **exit**, then the *ShutdownAndCleanup()* method is called to clean up and exit the program. Note that this sample assumes that the data entered for the GPS string is formatted as a valid XML string. Production code should validate the format.

6. Below the *Main* procedure, create the *PreparePlatform* method by entering the following.

```
private static void PreparePlatform()
{
    //Get the certificate that is used for making the TLS connection to OCS
    X509Certificate2 cert = GetLocalCertificate();
```

```
//Create the ServerPlatformSettings to define the connection to the server.
ServerPlatformSettings settings = new ServerPlatformSettings(
    "GPSPublisher",
    Dns.GetHostEntry("localhost").HostName,
    APPLICATION_PORT,
    APPLICATION_GRUU,
    cert);

    //Create the platform and call BeginStartup to make it usable.
    _collabPlatform = new CollaborationPlatform(settings);
    _collabPlatform.BeginStartup(BeginPlatformStartupCallback, _collabPlatform);
}
```

The *CollaborationPlatform* will be ready to use after the *BeginPlatformStartupCallback* method is called.

7. Just below the *PreparePlatform* method, add the following code to implement the *BeginPlatformStartupCallback* method.

```
private static void BeginPlatformStartupCallback(IAsyncResult result)
{
    try
    {
        //Call EndStartup to finish initializing the platform and then create the
        //set of UserEndpoints that are registered with Office Communications Server
        //to publish the user's location.
        _collabPlatform.EndStartup(result);
        CreateUserEndpoints();
    }
    catch (ConnectionFailureException connFailEx)
    {
        // ConnectionFailureException will be thrown when the platform cannot connect.
        Trace.WriteLine(connFailEx.Message + " in " + connFailEx.TargetSite);
    }
}
```

After the call to *CollaborationPlatform.EndStartup()* is called, the *CollaborationPlatform* is ready to use. Next, the callback method calls the method, *CreateUserEndpoints()*, to create the endpoints of users to publish their location.

8. Below the definition of *BeginPlatformStartupCallback,* insert the following code to implement the *CreateUserEndpoints* method.

```
private static void CreateUserEndpoints()
{
    List<string>_users = new List<string>();
    //Populate the list of users to publish.
    //In this case we just hardcoded this list for simplicity.
    _users.Add("sip:adamb@uc.contoso.com");
    _users.Add("sip:shannonb@uc.contoso.com");
    //For each user, create and establish a UserEndpoint.
    foreach (string userUri in _users)
    {
        UserEndpoint _endPoint;
        UserEndpointSettings settings =
```

```
            new UserEndpointSettings(userUri, OCS_SERVER_FQDN);
    _endPoint = new UserEndpoint(_collabPlatform, settings);
    _endPoint.BeginEstablish(EndPointBeginEstablishCallback, _endPoint);
}
```

In this method, the code creates a *UserEndpoint* for each user that it is configured to support. Although we use a hard-coded list here, in production code it would have been populated dynamically from a database, Active Directory Domain Services, or another source. The code in the *EndPointBeginEstablishCallback* stores the prepared endpoints for later use. This is done in the next step.

> **Note** Depending on the usage pattern for the endpoints, it might be a better design to defer creating the endpoints until there is data that needs to be published and to terminate the endpoints after the publication has finished. For example, if the presence data to be published changes infrequently, it might be a better choice to tear down the endpoints after the publication operation. This reduces the memory consumption of the application. If the data to be published changes frequently, then keeping the endpoints active is more efficient than the overhead of creating them again and again at a high rate. However, for simplicity in this sample, we create them once at startup and terminate them only when the application shuts down.

9. Just below the code for *CreateUserEndpoints*, add the following code to implement the *EndPointBeginEstablishCallback* method. This method is called for each endpoint that is created by the code in *CreateUserEndpoints*.

```
private static void EndPointBeginEstablishCallback(IAsyncResult result)
{
    UserEndpoint userEndpoint = result.AsyncState as UserEndpoint;
    try
    {
        userEndpoint.EndEstablish(result);
        //When the endpoint is ready to use, we will add it to the dictionary
        //that stores the endpoints for later use in publishing presence. The entries
        //in the dictionary are indexed by the SIP URI of the user and the value is a
        //reference to the endpoint itself.

        SipUriParser parsed = new SipUriParser(userEndpoint.EndpointUri);
        //Use just the "name@domain" portion of the URI.
      _userDictionary.Add(parsed.UserAtHost, userEndpoint);

    }
    catch (Exception except)
    {
        Trace.WriteLine(except.Message + " in " + except.TargetSite);
    }
}
```

At this point, we have a valid *UserEndpoint* for each user. We will use this endpoint to publish the location information of that user. The logic to publish the user's location information is encapsulated in the *PublishUserLocation* method.

10. Below the code for *EndPointBeginEstablishCallback,* add the following code to implement the *PublishUserLocation* method.

```
public void PublishUserLocation(string userUri, string location)
{
    //Check to make sure that we have been passed a user SIP URI that is in our list.
    //If so, then publish the location for that user.
    UserEndpoint userEndpoint;
    if ((!string.IsNullOrEmpty(userUri) && (_userDictionary.TryGetValue(userUri, out userEndpoint))))
    {
        PublishLocation(userEndpoint, location);
    }
    else
    {
        Trace.WriteLine("Entered URI: " + userUri + " was not found");
        Console.WriteLine("Entered URI: " + userUri + " was not found");
    }
}
```

This method is called to publish the location information for the specified user, which in this example is called from *Main*. In practice, it is called whenever the user's location changes or at regular intervals. It validates the user and calls the *PublishLocation* method to publish the user's location.

11. Below the *PublishUserLocation* method, add the following code to implement the *PublishLocation* method.

```
private static void PublishLocation(UserEndpoint userEndpoint, string location)

    Try

    {

        //Insert the location string into the XML blob.
        string catToPublish = string.Format(GPSLOCATION_SCHEMA, location);

        //Create an instance of the GPSLocation category and specify that it is
        //to be published in the 200 container which only makes it available to
        //contacts in the "Team" level.  See the Enhanced Presence Model whitepaper in the
        //"Additional Resources" section of this chapter for details.

        PresenceCategoryWithMetaData loc =

            new PresenceCategoryWithMetaData(1,

                200,

                new CustomPresenceCategory("GPSlocation", catToPublish));
```

```
            loc.ExpiryPolicy = ExpiryPolicy.Persistent;   //Keep this data until overwritten.

            //Add it to the list of items to be published.
            List<PresenceCategoryWithMetaData> itemList = new
List<PresenceCategoryWithMetaData>();

            itemList.Add(loc);

            //Publish the category instance to the user's endpoint.

            userEndpoint.LocalOwnerPresence.BeginPublishPresence(itemList,
PublishPresenceCallback, userEndpoint.LocalOwnerPresence);

        }

        catch (Exception excpt)

        {

            Trace.WriteLine(excpt.Message + " in " + excpt.TargetSite);

        }

    }
```

The status of the publication is returned in the *PublishPresenceCallback* method.

12. Just below the *PublishLocation* code, add the following code to implement the *PublishPresenceCallback* method.

```
private static void PublishPresenceCallback(IAsyncResult result)
{
    try
    {
        //Get the instance of LocalOwnerPresence from the result and end.
        LocalOwnerPresence userPresence = result.AsyncState as LocalOwnerPresence;
        userPresence.EndPublishPresence(result);

        SipUriParser parsed = new SipUriParser(userPresence.SubscriberEndpoint.
EndpointUri);

        Console.WriteLine("Presence published for: " + parsed.UserAtHost);

    }
    catch (Exception excpt)
    {
        Trace.WriteLine(excpt.Message + " in " + excpt.TargetSite);
    }
```

13. Finally, add the three utilty methods, *GetLocalCertificate, ShutdownAndCleanup,* and *BeginTerminateCallback. GetLocalCertificate* is used to get the certificate that the application uses to connect to Office Communications Server using Transport Layer Security (TLS). This certificate must be stored in the local computer's certificate store and have

a *SubjectName* that matches the FQDN of the computer where the application is running. The *ShutdownAndCleanup* method is called when the application exits. It terminates the cached *UserEndpoints* that the application created and then shuts down the *CollaborationPlatform* instance.

Just below the code for *PublishPresenceCallback*, add the following code to implement these three methods.

```
private static X509Certificate2 GetLocalCertificate()
{
    //Get a handle to the local machine's certificate store
    X509Store store = new X509Store(StoreLocation.LocalMachine);
    //Open the store
    store.Open(OpenFlags.ReadOnly);
    //Get a handle to a collection of the certificates installed on the machine
    X509Certificate2Collection certificates = store.Certificates;
    //Loop through the certificates looking for one
    //where the SubjectName matches the FQDN of the
    //Local Machine and the private key is available.
    foreach (X509Certificate2 certificate in certificates)
    {
        if (certificate.SubjectName.Name.Contains
            (Dns.GetHostEntry("localhost").HostName)
            && certificate.HasPrivateKey)
        {
            //Return the certificate that matches
            return certificate;
        }
    }
    //If not certificates match, return null
    return null;
}

internal static void ShutdownAndCleanup()
{
    foreach (UserEndpoint endpoint in _userDictionary.Values)
    {
        IAsyncResult result = endpoint.BeginTerminate(BeginTerminateCallback,
endpoint);
        result.AsyncWaitHandle.WaitOne();
    }

    _collabPlatform.BeginShutdown(
                            new AsyncCallback(
                                delegate(IAsyncResult aResult)
                                {
                                    _collabPlatform.EndShutdown(aResult);
                                }
                            ),
                            null);

}
```

```
private static void BeginTerminateCallback(IAsyncResult result)
{
    UserEndpoint userEndpoint = result.AsyncState as UserEndpoint;
    userEndpoint.EndTerminate(result);
}
```

Summary

This sample shows how to use UCMA 2.0 to build an application that publishes custom presence items on behalf of a set of users. Because the Office Communications Server infrastructure is configured to treat the application as a trusted application, the application does not need to supply domain credentials to Office Communications Server for each user when it publishes their presence.

Additional Resources

- Microsoft Unified Communications Managed API 2.0 SDK (32 bit) (*http://go.microsoft.com/fwlink/?LinkID=140790*)
- Microsoft Unified Communications Managed API 2.0 SDK (64 bit) (*http://go.microsoft.com/fwlink/?LinkID=139195*)
- "Unified Communications Managed API 2.0 Core SDK Documentation" (*http://go.microsoft.com/fwlink/?linkid=126312*)
- "Office Communicator 2007: Enhanced Presence Model White Paper" (*http://go.microsoft.com/fwlink/?linkid=143209*)

Part V
Debugging, Tuning, and Deploying Unified Communications Applications

Microsoft Office Communications Server is a complex product, and deploying your Unified Communications (UC) application can be a weighty process that requires proper configurations of the application development components and the underlying infrastructure components. As a developer, you may not be proficient enough with deploying and managing Office Communications Server to know what needs to be done to ensure that your application is properly deployed and successfully executed. Chapter 9, "Preparing the UC Development Environment," covers the details for properly configuring Office Communications Server and application development components to run your application.

In any application development, bugs are inevitable and debugging is necessary. Because UC applications are inherently asynchronous, bugs may seem mysterious and debugging may be frustrating. Bugs, however, are not an option. They are an expected part of any development effort. The key is to release your product with as few bugs as possible, within the time and resource constraints available. Chapter 10, "Debugging a Unified Communications Application," provides guidance on how to debug your application and troubleshoot the system when problems arise.

Chapter 9
Preparing the UC Development Environment

This chapter will help you to:

- Understand essential components of a Microsoft Unified Communications (UC) application development environment.

- Understand the basic requirements of the infrastructure and application development components of a UC application development environment.

- Prescribe how to configure basic infrastructure components for developing UC applications in a single-domain forest.

- Prescribe how to configure application development components for developing UC applications by using Microsoft Office Communicator Automation API, Unified Communications Managed API (UCMA) Core, and UCMA Workflow.

UC Application Development Environment Components

As an application developer, you can think of the UC environment as consisting of infrastructure components and application development components. The infrastructure components are configured and managed by system administrators, and the application development components can be configured and administered by application developers who may have to obtain appropriate permissions from a system administrator.

The infrastructure components include all of the Office Communications Server roles, the dependent Active Directory Domain Services (AD DS), and supporting services such as Domain Name System (DNS), public key infrastructure (PKI), Structured Query Language (SQL), reverse proxy, firewalls, and hardware load balancers for an Enterprise pool. For Office Communications Server Standard Edition, the supported server roles include Front End Server, Web Conferencing Server, Audio/Video (A/V) Conferencing Server, Web Components Server, and Application Sharing Server. The Web Conferencing Server and A/V Conferencing Server roles are not required if the Office Communications Server Standard Edition deployment is intended to support only instant messaging and presence. For more information about server roles in Office Communications Server, see the "Microsoft Office Communications Server 2007 R2" documentation at *http://go.microsoft.com/fwlink/?LinkID=133608* and the Microsoft Office Communications Server 2007 R2 Resource Kit at *http://go.microsoft.com/fwlink/?LinkId=141203*.

The server roles in Office Communications Server 2007 R2 must be deployed on computers that are running Windows Server 2003 Service Pack 2 (SP2) or later or Windows Server 2008. You must have appropriate permissions to install, activate, and administer the server components. For example, to install, activate, and administer an infrastructure component, you must be a member of an appropriate Real-Time Communications (RTC) Universal security group that is created and configured by AD DS preparation before Office Communications Server is installed. Most administrative tasks require that you belong to the RTCUniversalServerAdmins group. Some tasks may also require that you have local or domain administrator permissions. For example, to install and activate a server running Office Communications Server 2007 R2 Standard Edition, you must be a domain administrator who belongs to the RTCUniversalServerAdmins security group. To configure a server running Office Communications Server 2007 R2 Standard Edition, you must be a member of the RTCUniversalServerAdmins group. In general, if you are a domain administrator and you belong to the RTCUniversalServerAdmins security group, you can perform all three administrative functions. For more information about the administrative credentials that are required for deploying Office Communications Server, see the "Accounts and Permissions Requirements" topic (*http://technet.microsoft.com/en-us/library/dd425321(office.13).aspx*) in the "Microsoft Office Communications Server 2007 R2" documentation at *http://go.microsoft.com/fwlink/?LinkID=133608*.

The application development components can be grouped by the types of UC applications. These application groups include client-side applications that log on to Office Communications Server, middle-tier applications that mediate between the client and Office Communications Server, and server-side applications that extend the functionality of Office Communications Server. Some of these application development components include Office Communicator, Office Communicator Automation application programming interface (API), Unified Communications Client API, Unified Communications Managed API (UCMA) Core, UCMA Workflow, and Office Communications Server API. This book mainly discusses developing applications by using the Office Communicator Automation API, UCMA Core, and UCMA Workflow.

You can build UC applications by using one or more application development components. As an application developer, you are responsible for configuring and using the application development components while relying on system administrators in your organization to configure and manage the infrastructure components.

For testing purposes, you might want to isolate your development environment from the production deployment of the UC infrastructure. For example, you can do this when you reconfigure infrastructure components to enable custom features or extensions. In these cases, bugs or other unwanted actions can cause UC to operate abnormally in your organization. It will be helpful to create an isolated Office Communications Server environment to use for your application development.

This chapter explains how to create an isolated Office Communications Server environment for developing and testing your UC applications in an autonomous manner. Prescriptive step-by-step instructions explain how to deploy Office Communications Server Standard Edition for internal users in a single-domain forest topology. This is the most basic environment required to run Office Communications Server.

Before explaining how to set up and configure the UC application development environment, this chapter reviews the three central components of the environment: AD DS for managing a network, Office Communications Server roles, and the UC APIs.

AD DS for Managing a Network

Office Communications Server uses AD DS to store global settings and the user information that it needs. Global settings include information about the deployment configurations, routing rules, security groups, and specific access permissions for the Office Communications Server infrastructure. User information is used to authenticate the user when an endpoint tries to sign in to Office Communications Server. It is also used to assign user rights to the user's endpoints.

In AD DS, users enabled for UC are represented by Active Directory *User* objects. Applications or services that are trusted by Office Communications Server are represented by Active Directory *Contact* objects. Two AD DS server roles support these features: AD DS and Active Directory Certificate Services.

When Office Communications Server is deployed, the AD DS schema must be extended. These extensions involve adding new classes and attributes into the AD DS schema and configuring these settings. This process is known as *AD DS preparation* and must be performed before Office Communications Server can be deployed in an AD DS forest.

To start AD DS preparation, you must select an existing AD DS forest or create a new one. This chapter is based on the assumption that you intend to set up a new UC network for development. This means that you must install AD DS on a separate network. The resulting AD DS forest can consist of one or more domains. Make sure that the following requirements are met:

- All domain controllers and global catalog servers must run Windows Server 2003 SP1, Windows Server 2003 R2, or Windows Server 2008 in either the 32-bit or 64-bit edition.
- All domains must have a domain functional level of Windows Server 2003 or Windows Server 2008. Domain functional levels of Microsoft Windows 2000 Mixed, Windows 2000 Native, or Windows Server 2003 interim domain are not supported.
- The AD DS forest must have a forest functional level of Windows Server 2003 or Windows Server 2008. Forest levels of Windows 2000 Mixed, Windows 2000 Native, or Windows Server 2003 interim forest are not supported.

AD DS preparation for Office Communications Server consists of three major tasks:

- Prep Schema
- Prep Forest
- Prep Domain

Prep Schema extends the AD DS schema to support new classes and attributes that are specific to Office Communications Server. To review the schema extensions that the Office Communications Server created, see the schema file Schema.ldf., which is located in the installation root directory of the Office Communications Server. To perform Prep Schema, you must be an administrator in the Schema Admins security group, and you must perform this operation on a schema master server. Successful AD DS schema preparation creates AD DS classes and attributes that have names that contain the prefix *ms-RTC-SIP-*. After Prep Schema completes successfully, you can verify the *ms-RTC-SIP* prefix by using ADSI Edit, as shown in Figure 9-1. (The instructions to navigate in ADSI Edit are given in the "Verifying Extended AD DS Schemas" section later in this chapter.)

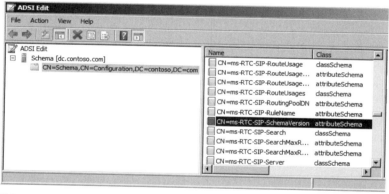

FIGURE 9-1 AD DS schema preparation.

Prep Forest creates global settings that are used by every server running Office Communications Server that is installed in the AD DS forest. To perform this task, you must log on as a member of the Enterprise Admins or Domain Admins group in the AD DS forest root domain. Active Directory Prep Forest does the following:

- Creates universal service and administration groups for Office Communications Server
- Adds default access control entries (ACEs) to the newly created security groups

Figure 9-2 shows these groups as they appear in Active Directory Users And Computers.

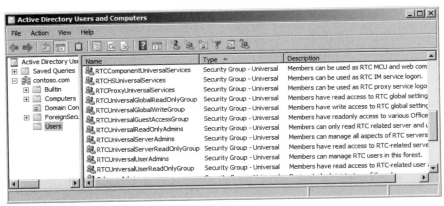

FIGURE 9-2 Security groups created by AD DS forest preparation.

Prep Domain assigns appropriate permissions to objects and attributes that are enabled for Office Communications Server in each AD DS domain in the chosen forest. To perform this task, you must be logged on as a member of the Enterprise Admins or Domain Admins group for the domain from which you plan to run this task. You must perform Prep Domain for all of the domains in which you want to deploy Office Communications Server and any domains where your Office Communications Server users reside. Preparing an AD DS domain does the following:

- Adds the necessary ACEs to the universal groups that are used to grant permissions to domain controllers and to manage users in the domain
- Adds ACEs to the domain root and the following built-in containers: Users, Computers, and Domain Controllers

Office Communications Server Roles

Office Communications Server supports several server roles and is available in two editions: Standard Edition and Enterprise Edition. Standard Edition is intended for small-scale deployment. You can deploy the Standard Edition server roles on a single host. Enterprise Edition is used for mid- and large-scale deployments that involve multihomed server roles and may require a load balancer. This chapter uses Standard Edition to demonstrate how to set up a UC application development environment in an AD DS forest topology that consists of a single domain.

A successful deployment of Office Communications Server 2007 R2 Standard Edition results in the automatic installation of the following server roles, in addition to Microsoft SQL Server 2005 Express Edition SP2 and four applications (Conferencing Attendant, Conferencing Announcement Service, Response Group Service, Outside Voice Control):

- Front End Server
- Web Conferencing Server
- Audio/Video (A/V) Conferencing Server

- Web Components Server
- Application Sharing Server

To make sure that the deployment of these server roles is successful, verify that the following minimum operating system and hardware requirements are met:

- **Hardware requirements** Use only a 64-bit computer that is running a 64-bit edition of Windows Server for all Office Communications Server 2007 R2 Standard Edition server roles. Other technical specifications include the following:
    - CPU   Quad-core 2.0 gigahertz (GHz) or greater dual processor or dual-core 2.0 GHz 4-way processor
    - RAM   8 gigabytes (GB) of memory
    - Hard drive   2 x 72 GB hard drives of 15 kilobytes (K) or 10K revolutions per minute (RPM), RAID 0 (striped), or equivalent
    - Network adapter   2 x 1 gigabit-per-second (Gbps) network adapter (2 refers to dual port)
- **Operating system requirements** Use only the 64-bit edition of Windows Server 2003 SP2, Windows Server 2003 R2 SP2, or Windows Server 2008.
- **Database requirements** Use SQL Server 2005 Express Edition SP2 (32-bit version). This is included with Office Communications Server 2007 R2.
- **Administrative tools requirements** Install Office Communications Server Administrative Tools manually, which includes the Office Communications Server Microsoft Management Console (MMC) snap-in, the Active Directory Users And Computers snap-in, the Computer Management console snap-in extension, the Communicator Web Access snap-in, and the Response Group Service snap-in. The administrative tools can be installed independent of the Office Communications Server deployment on a computer that is running the 32-bit or 64-bit edition of Windows Server 2003 SP2, Windows Server 2003 R2 SP2, Windows Server 2008, Windows Vista Business, or Windows Vista Enterprise with SP1.

UC APIs

UC APIs are major application development components of a UC application development environment. This chapter discusses the basic requirements for using Office Communicator Automation API, UCMA Core, and UCMA Workflow.

Office Communicator Automation API

Office Communicator Automation API is a Component Object Model (COM)–based Automation API that enables an application to manipulate a locally running instance of Office Communicator. It can be used by applications in C/C++, Visual Basic, Visual Basic

Scripting Edition (VBScript), and other scripting languages, and any .NET-based programming languages, such as C#. For security reasons, some API calls are disabled for scripting languages. To use this API, download the Office Communicator 2007 Software Development Kit (SDK) from MSDN located at *http://go.microsoft.com/fwlink/?LinkId=141225*.

Running an Office Communicator Automation API application requires that Office Communicator be installed on the application's local computer. To build and run the Win32 C/C++ application by using the Office Communicator Automation API, you must include the COM interface IDs definition file in the application's project. This file is not included automatically in the Office Communicator 2007 Automation API SDK. However, you can generate it easily from the type library by using the Microsoft Interface Definition Language (MIDL) compiler.

To build and run the Microsoft .NET Framework application by using the Office Communicator Automation API through the COM interop service, you must add to your application's project a reference to the primary interop assembly (PIA) that is appropriate to your application's target platform. For more information about how to do this, see the "Configuring the Office Communicator Automation API" section later in this chapter.

UCMA Core

UCMA Core is a .NET-based API that supports the full suite of UC functionality: instant messaging, telephony, A/V conferencing, and presence. It is optimized for use by middle-tier applications mediating between a UC client and Office Communications Server. The API is supported in the run-time environment of either the latest service pack of Windows Server 2003 R2 (64-bit edition) or the latest service pack of Windows Server 2008. It also requires .NET Framework 3.5 SP1 and Visual C++ 2008 Redistributable Package. For more information about the deployment requirements, see "Unified Communications Managed API 2.0 Core SDK Documentation" at *http://go.microsoft.com/fwlink/?LinkID=133571*. To use the API, download the Microsoft UCMA 2.0 SDK from the Download Center at *http://go.microsoft.com/fwlink/?LinkID=139195*.

In most application scenarios, a UCMA application must be trusted by Office Communications Server. In these cases, you must have an Active Directory *Contact* object created and configured for the application. To do this yourself, you can use the ApplicationProvisioner sample application provided with the UCMA Core API SDK. For more information, see the "Configuring UCMA Core" section later in this chapter.

UCMA Workflow

The UCMA Workflow extends the Windows Workflow Foundation activities to support instant messaging activities, speech activities, and presence activities that are used in UC. As with the UCMA Core, this workflow API requires you to install .NET Framework 3.5 SP1, which includes Windows Workflow Foundation 3.5. It also requires Microsoft Visual Studio 2008 and at least one Microsoft Speech Server Language Pack. To use this API, you must install the Microsoft UCMA 2.0 SDK at *http://go.microsoft.com/fwlink/?LinkID=139195*. In addition, you must install at least one of the Speech Server 2007 Language Packs that are distributed as part of Office Communications Server. For more information, see the "Configuring UCMA Workflow" section later in this chapter.

Deploying Office Communications Server Standard Edition

Deploying Office Communications Server Standard Edition on a private network involves the following tasks:

1. Build an AD DS forest.
2. Prepare AD DS for Office Communications Server.
3. Configure DNS for automatic sign-in by UC clients.
4. Set up the Office Communications Server host computer.
5. Install Office Communications Server Standard Edition.
6. Configure UC user accounts.
7. Validate Office Communications Server functionality.
8. Install Office Communicator and other clients to verify that users can log on.

The following sections explain how to perform each of these tasks.

Building an AD DS Forest

To build a separate AD DS forest on a private network for development purposes, the minimal topology is a single-domain forest topology.

> **Note** Skip this task if you plan to use an existing AD DS forest to set up your development environment.

The AD DS forest fully qualified domain name (FQDN) for the example in this chapter is "contoso.com." A single computer running Windows Server 2008 hosts all of the network services. This includes the domain controller and DNS server. The FQDN of this computer is "dc.contoso.com."

After the AD DS forest is configured, another computer running Windows Server 2008, set up to host Office Communications Server Standard Edition and one or more application hosting computers, is added to the network. For a client application, the operating system on the host computer can be Microsoft Windows XP, Windows Vista, or Windows Server. For a server or middle-tier application, the host computer may have to be a server running Windows Server. In this example, the host computer running Office Communications Server to be added has the FQDN "ocs.contoso.com". The application computer names follow the naming convention of "app*N*.contoso.com," where *N*=0, 1, 2, and so on.

To build an AD DS forest, you must make one computer that is running Windows Server the root domain controller. You must also create and configure the DNS server if you do not intend to use existing domain name services. Finally, you must install an Enterprise certificate authority (CA) that issues certificates to users, computers, and applications within the network. Having a private CA is especially useful for a private network that consists of a set of virtual servers and clients. In such a network, it is recommended that Dynamic Host Configuration Protocol (DHCP) is disabled and that static Internet Protocol (IP) addresses be assigned to each computer manually.

Assigning Static IP Addresses for Domain Controller and Other Computers

For the domain controller, you must perform these steps before you run the DCPromo tool to make the server into a domain controller. In this chapter, the IP address of 192.168.100.1 is assigned to the domain controller of the private network.

> **Note** Skip this task if you rely on a DCHP server in your network to assign IP addresses to the computers in your network.

1. On Windows Taskbar, click Start, and then click Control Panel.
2. In Control Panel, double-click Network And Sharing Center.
3. In Network And Sharing Center, under the Tasks pane, click Manage Network Connections.
4. In Network Connection, right-click Local Area Connection, and then click Properties.
5. Select the Internet Protocol version 4 (TCP/IPv4) options, and then click Properties.
6. In the IPv4 Properties window, as shown here, select Use The Following IP Address, assign an IP address as the static IP address for the computer, and type **255.255.255.0** in the Subnet Mask field. Select Use The Following DNS Server Addresses, type the IP address of the domain controller in the Preferred DNS Server field, and then click OK.

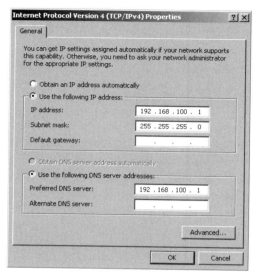

For a private network, the IP address should use the format *192.168*.x.y, where *x* and *y* are integers between 0 and 255, inclusive. In the Default Gateway field, use the IP address of the gateway, if there is one. Otherwise, leave this field blank. The specification of the preferred DNS server is based on the assumption that your DNS server is hosted on the domain controller as well.

For other computers, repeat the previous steps and change the host IP address values so that each computer in the network is identified by a unique IP address. You must do this before the computer joins the domain.

Promoting a Computer Running Windows Server to a Domain Controller

To promote a computer running Windows Server to an AD DS domain controller, run DCPromo.exe on the computer as explained in the following steps. In the example in the following procedure, the FQDN of the domain is contoso.com.

1. Move the computer running Windows Server into a workgroup if it is currently joined to an AD DS domain. To do this on a computer running Windows Server 2008, click Start, right-click Computer, and then click Properties.

2. In the System pane, under Computer Name, Domain, and Workgroup Settings, click Change Settings, as shown here.

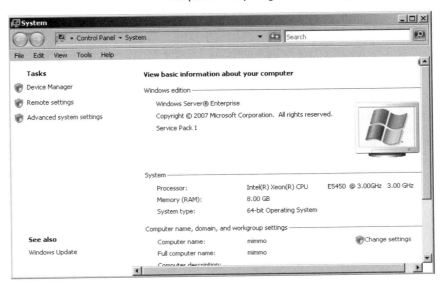

3. On the System Properties page, click Change...

4. On the Computer Name/Domain Changes page, select Workgroup, if it is not already selected, and then click OK, as shown here.

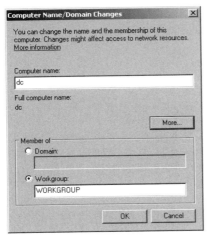

5. When you are prompted to restart the computer, click OK. In this example, "dc" is the name of the domain controller that is hosted on a computer running Windows Server 2008.

6. To promote the server to a domain controller, from an elevated command prompt, run DCPromo. This starts the Active Directory Domain Services Installation Wizard.

7. On the Welcome To The Active Directory Domain Services Installation Wizard page, shown here, click Next.

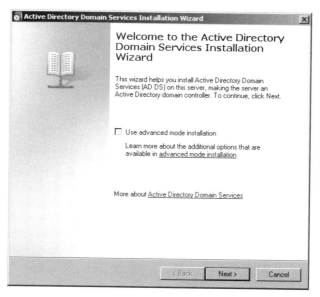

8. On the Operating System Compatibility page, click Next.

9. On the Choose A Deployment Configuration page, select Create A New Domain In A New Forest, as shown here, and then click Next.

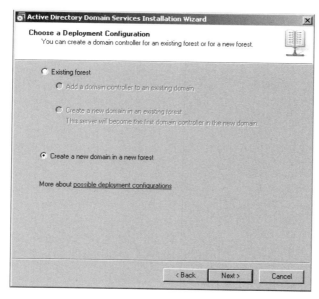

10. On the Name The Forest Root Domain page, in FQDN Of The Forest Root Domain, enter the FQDN of the forest root, as shown here, and then click Next. In this example, the FQDN is contoso.com.

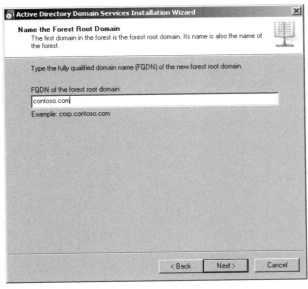

11. On the Set Forest Functional Level page, as shown here, from the Forest Functional Level drop-down menu, select Windows Server 2008, and then click Next.

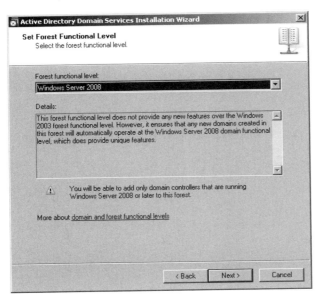

12. On the Additional Domain Controller Options page, select DNS Server (if it is not already selected), as shown in the following screen, and then click Next.

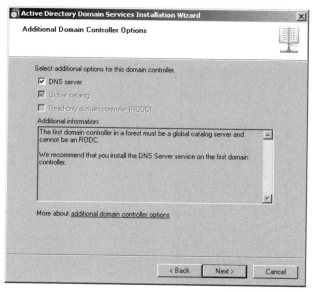

13. If your network uses a DHCP server to assign IP addresses, and you receive a prompt that asks you whether you want to continue without assigning IP addresses, select No, I Will Assign Static IP Addresses To All Physical Network Adapters, as shown here.

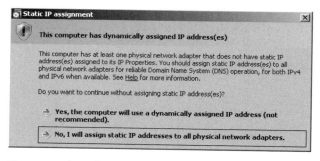

You must then figure out which network adapter has a DHCP-assigned IP address. If the IPv6 address is not static, you can either disable IPv6 or assign a static address to it. After this is done, go back to the Additional Domain Controller Options page, shown in step 12, and click Next.

14. On the following page, click Yes to continue without creating a delegation for the new DNS server.

15. On the Location For Database, Log Files And SYSVOL page, shown in the following screen, leave the default values, and then click Next.

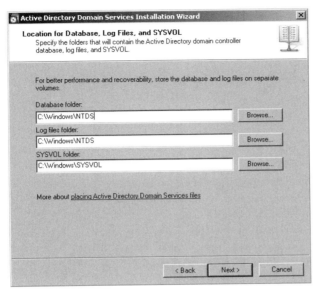

16. On the Directory Services Restore Mode Administrator Password page, enter and confirm the password of the restore mode administrator, and then click Next.

17. On the Summary page, review your selections. If your selections are correct, click Next to finish the installation of AD DS.

18. On the Completing The Active Directory Domain Services Installation Wizard page, click Finish to exit the installation. This step takes awhile to complete.

19. Restart your computer for the settings to take effect.

Verifying that the DNS Server Role Is Running

After the domain controller is configured, it is a good practice to verify that the newly installed DNS server role is running. To verify the DNS server role, perform the following steps:

1. Log on to the domain controller as a domain Administrator.

2. Click Start, click Administrative Tools, and then click Server Manager.

3. In Server Manager, expand the Roles node in the tree view on the left pane.

4. Select DNS Server.

5. Verify that the Status value of the DNS Server row in System Services is Running.

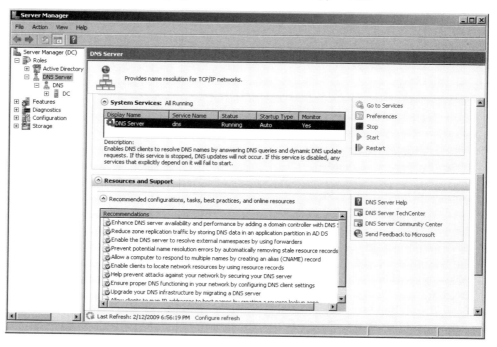

Installing a Domain CA

Office Communications Server requires certificates to be issued to servers by a trusted CA. In a private development environment, the most convenient way to issue certificates is to install a private CA in the domain. You can install this CA by adding the Active Directory Certificate Services role to the domain controller. (For more information, see the "Active Directory Certificate Services Step-by-Step Guide" on TechNet at *http://technet.microsoft.com/en-us/library/cc772393.aspx*.)

> **Note** If you choose other certificate services such as Certification Authority Web Enrollment or Online Responder, you are prompted to add other supporting roles, such as the Application Server and Web Server (IIS) roles, that have not been enabled. In such cases, make sure that you enable the IIS 6 Management Compatibility service of each Web Server (IIS) role during the installation.

To install a CA for the domain, perform the following steps:

1. Log on to a computer running Windows Server 2008 that is a domain controller.
2. In Server Manager, right-click Roles, and then click Add Roles.
3. On the Before You Begin page, click Next.
4. On the Select Server Roles page, shown here, select Active Directory Certificate Services, and then click Next.

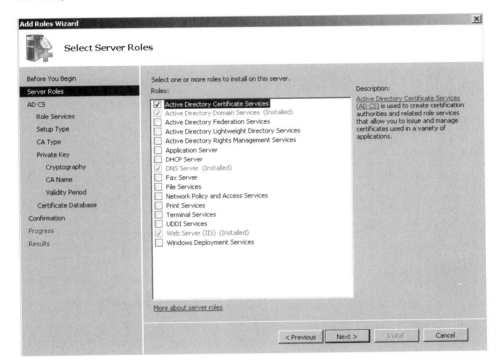

5. On the Introduction page, click Next.
6. On the Select Role Services page, shown here, select Certification Authority, and, if you want, select Certification Authority Web Enrollment or Online Responder. Then click Next.

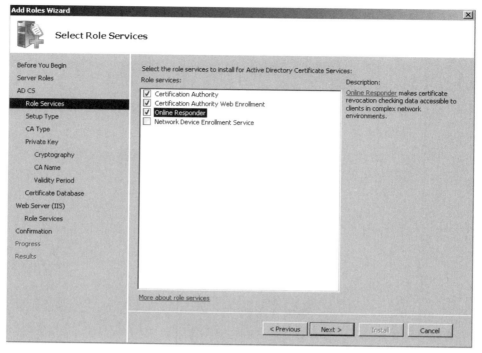

7. On the Specify Setup Type page, select Enterprise, and then click Next.
8. On the Specify CA Type page, select Root CA, and then click Next.
9. On the Set Up Private Key page, select Create A New Private Key, and then click Next.
10. On the Configure Cryptography For CA page, use the default setting, and then click Next.
11. On the Configure CA Name page, use the default setting, and then click Next.

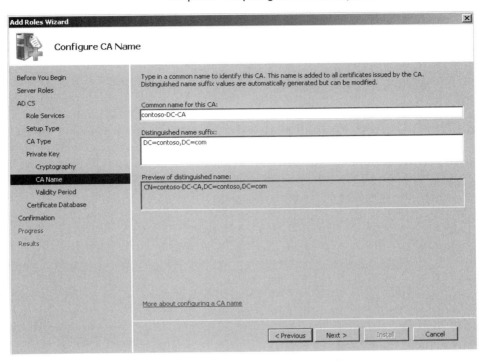

12. On the Set Validity Period page, use the default setting, and then click Next.

13. On the Configure Certificate Database page, use the default setting, and then click Next.

14. On the Confirm Installation Selections page, confirm the settings that you selected, and then click Install if the settings are correct.

15. On the Installation Results page, verify the installation status, and then click Close.

Preparing AD DS for UC

To prepare AD DS for UC, you must perform the following tasks:

1. Extend and verify AD DS schemas for Unified Communications.
2. Configure and verify the Active Directory Forest for Unified Communications.
3. Configure and verify the Active Directory Domain for Unified Communications.

The following sections explain the steps to perform each of these tasks.

Extending AD DS Schemas

To extend AD DS schemas, you can use the Microsoft Office Communications Server 2007 R2 Deployment Wizard to complete the following steps:

1. Log on to the domain controller by using a domain account that is a member of the Schema Admins security group on the schema master.

2. Double-click SetupSE.exe to start the Microsoft Office Communications Server 2007 R2 Deployment Wizard. If you are prompted to install the Microsoft Visual C++ 2008 Redistributable package, click Yes. If you are prompted to install .NET Framework 3.5 SP1, click Yes. After the added packages are installed, restart the computer and repeat the procedure from the beginning.

3. On the starting page of the Office Communications Server 2007 R2 Standard Edition Deployment Wizard, click Prepare Active Directory.

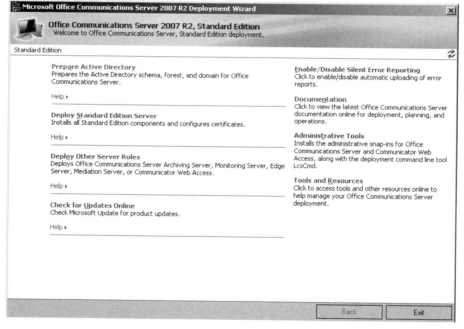

4. On the Prepare Active Directory For Office Communications Server page, shown here, under Step 1: Prep Schema, click Run.

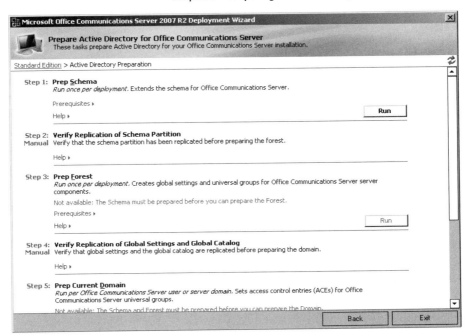

5. On the Welcome To The Schema Preparation Wizard page, click Next.

6. On the Directory Location Of Schema Files page, either use the default option or explicitly specify the path of a directory in which the Office Communications Server extensions of the AD DS schemas are located, and then click Next. The default schema file name is Schema.ldf, which is located in the same folder as SetupSE.exe.

7. On the Ready To Prepare Schema page, click Next to start the schema preparation.

8. On the Schema Preparation Wizard Has Completed Successfully page, click Finish.

Installing Office Communications Server Administrative Tools

To perform the tasks prescribed from here on, you need to use the Office Communications Server 2007 R2 Administrative Tools. To install the administrative tools on the AD DS domain controller, perform the following steps:

1. Log on as a member of the DomainAdmins group to the AD DS domain controller.

2. Start the Microsoft Office Communications Server 2007 R2 Deployment Wizard (SetupSE.Exe).

3. On the Microsoft Office Communications Server 2007 R2 Standard Edition introduction page, click Administrative Tools, as shown here.

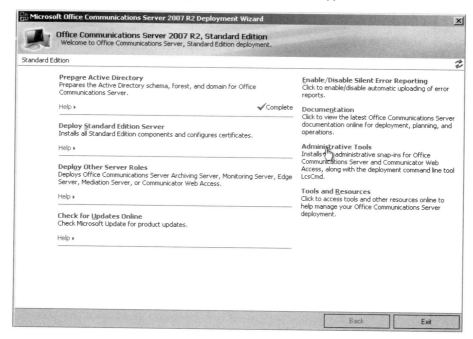

4. Follow the steps of the wizard to finish the installation.

Verifying Extended AD DS Schemas

After the Prep Schema step completes successfully, verify that the schema extensions are replicated throughout the system. You may have to wait for this to happen, especially with a topology that involves multiple domains. However, with a topology that has a single domain in a single forest, replication is almost instantaneous and verification can start immediately.

> **Note** Performing the procedure described in this section satisfies the requirement set forth by the Step 2: Verify Replication Of Schema Partition In The Prepare Active Directory For Office Communications Server page of the Microsoft Office Communications Server 2007 R2 Deployment Wizard.

To verify that the AD DS schemas are extended properly, perform the following steps:

1. Log on to the domain controller where the Office Communications Server Administrative Tools are installed.

2. On the Windows Tasksbar, click Start, Administrative Tools, and then ADSI Edit. This assumes that you have already installed the Office Communications Server Administrative Tools that are available from the Office Communications Server Installation Wizard.

Chapter 9 Preparing the UC Development Environment

3. In ADSI Edit, click Action, and then click Connect To.

4. In the Connection Settings box, shown here, select the Select A Well Known Naming Context option, select Schema on the drop-down menu, and then click OK.

In the ADSI Edit console, expand the Schema node in the left pane, and select CN=Schema. In the right pane, scroll down until you find CN=*ms-RTC-SIP-SchemaVersion*, as shown here.

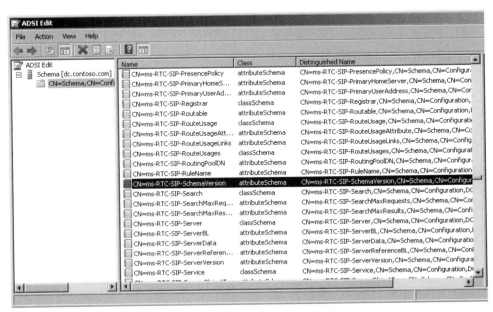

If the schema preparation is replicated, you see results that are similar to those shown here. If the schema extensions are not shown, wait for the schema changes to replicate in the system, and then refresh the view.

5. Double-click the *ms-RTC-SIP-SchemaVersion* attribute to open the Properties dialog box. Scroll down to locate the *rangeUpper* property value, and verify that it is 1008, as shown here. This value confirms that the schema for Office Communications Server 2007 R2 has been prepared successfully. Once the schemas are extended successfully, click OK, and then close the ADSI Edit console.

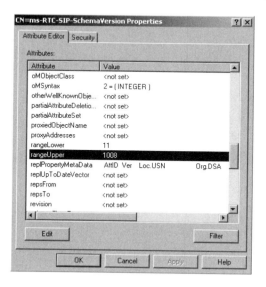

Preparing the AD DS Forest

After you verify that the schema preparation was successful, follow these steps to perform AD DS forest preparation:

1. In the Microsoft Office Communications Server 2007 R2 Deployment Wizard (SetupSE.exe), on the Prepare Active Directory For Office Communications Server page (shown here), under Step 3: Prep Forest, click Run.

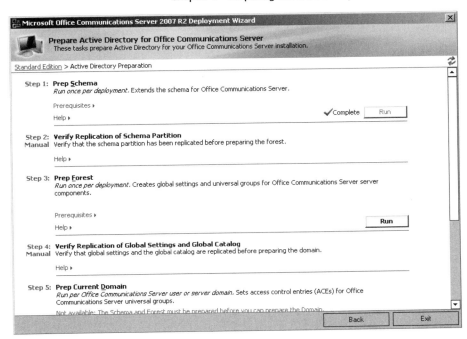

2. On the Welcome To The Forest Preparation Wizard page, click Next.

3. On the Select Location To Store Global Settings page, shown here, use the default settings, and then click Next.

4. On the Location Of Universal Groups page, select the domain that you want, and then click Next. In this example, there is only one domain from which to choose: the contoso.com domain.

5. On the SIP Domain Used For Default Routing page, select the Session Initiation Protocol (SIP) domain that you want, and then click Next. Again, in this example, there is only one domain from which to choose.

6. On the Ready to Prepare Forest page, review the settings, and then click Next to start the forest preparation.

7. On the Forest Preparation Wizard Has Completed Successfully page, select View The Log When You Click Finish, and then click Finish.

8. Verify that the log was created. The log should look similar to the screenshot shown here. Expand Execute Action, and then confirm that the value next to Forest Settings is *Ready*. This indicates that the forest preparation process has completed successfully.

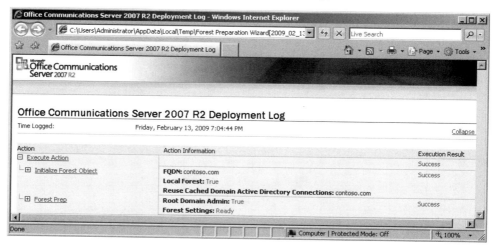

Verifying Prepared Forest Settings

If the value next to Forest Settings indicates that the forest is not ready, wait for the system to replicate the changes, and then type the following command at a command prompt to verify the forest settings:

LcsCmd /forest /action:CheckForestPrepState /PDCRequired:FALSE

The results that are obtained while executing this command are written to a Hypertext Markup Language (HTML) file. By default, the file path is %Userprofile%\AppData\Local\Temp\2\Forest_CheckForestPrepState[<date>][<time>].html.

Preparing the AD DS Domain

After you verify that AD DS forest preparation was successful, you are ready to prepare the AD DS domain by performing the following steps:

1. In the Microsoft Office Communications Server 2007 R2 Deployment Wizard (SetupSE.exe), on the Prepare Active Directory For Office Communications Server page, under Step 5: Prep Current Domain, click Run.
2. On the Welcome To The Domain Preparation Wizard page, click Next.
3. On the Domain Preparation Information page, click Next.
4. On the Ready To Prepare Domain page, click Next to start domain preparation.
5. On the Domain Preparation Wizard Has Completed Successfully page, click Finish. Select View The Log When You Click Finish, and then click Finish. The wizard creates a log file and displays the results in a browser, as shown here.

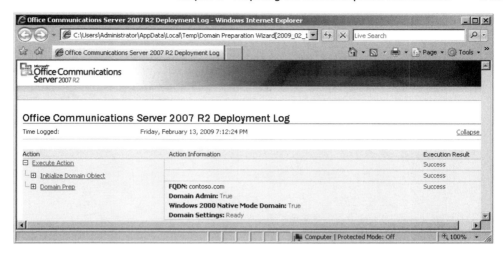

6. In the log file, expand Execute Actions. If the value next to Domain Settings is *Ready*, the domain preparation succeeded. If not, wait for the domain changes to replicate, and then perform the steps in the following section.

Verifying Prepared Domain Settings

To verify domain settings, type the following command at a command prompt:

LcsCmd /domain /action:CheckDomainPrepState /PDCRequired:FALSE

This command creates an HTML file that you can view in a browser. The default location of this file is %Userprofile%\AppData\Local\Temp\2\Domain_CheckDomainPrepState[<*date*>] [<*time*>].html. In the Action pane, expand Execute Action. If domain preparation succeeds, the value next to Domain Settings is *Ready,* as shown in the screenshot under step 5 in the section entitled "Preparing the AD DS Domain," earlier in this chapter.

When the domain settings are ready, you can return to the Microsoft Office Communications Server 2007 R2 Deployment Wizard, and under Step 7: Delegate Setup And Administration, shown in the screenshot under step 1 in the section entitled "Preparing the AD DS Domain," click Run. You need to perform this step only if you want users who are not members of an authorized security group to be able to set up and manage Office Communications Server roles.

Configuring DNS for Automatic Sign-In

Before you install Office Communications Server Standard Edition, you can configure the DNS records that are used to resolve the server's FQDN to its IP address. Doing this supports automatic sign-in by clients.

Two kinds of DNS records are used. One record type is SRV, and the other is A. The A record is required for Office Communications Server Enterprise Edition. Office Communications Server Standard Edition installation automatically creates the A record. An SRV record of _sipinternaltls._tcp.<*domain*> over port 5061 (for Transport Layer Security [TLS]) or _sipinternal._tcp.<*domain*> over port 5060 (for Transmission Control Protocol [TCP]) resolves to the FQDN of the server running Office Communications Server Standard Edition. Office Communicator and other UC clients use this SRV record to locate an Office Communications Server before they can sign in.

Creating a DNS SRV Record

To create a DNS SRV record for a server running Office Communications Server Standard Edition, perform the following steps:

1. On the DNS server (domain controller), click Start, Administrative Tools, and then DNS.

2. In DNS Manager, in the left pane, select DC, and then expand Forward Lookup Zones, right-click the domain in which you want to install the Office Communications Server (that is, contoso.com), and then click Other New Records.

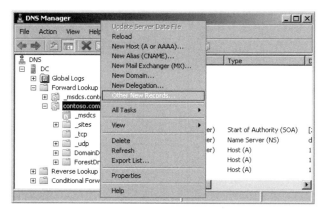

3. In the Resource Record Type dialog box, under Select A Resource Record Type, select Service Location (SRV), as shown here, and then click Create Record.

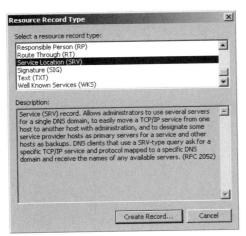

4. In the New Resource Record dialog box, in the Service box, type **_sipinternaltls** for TLS or **_sipinternal** for TCP.

5. In the Protocol box, type **_tcp**. This applies to both TLS and TCP.

6. In the Port Number box, type **5061** for TLS or **5060** for TCP.

7. In the Host Offering This Service box, type the FQDN of the server running Standard Edition, as shown here.

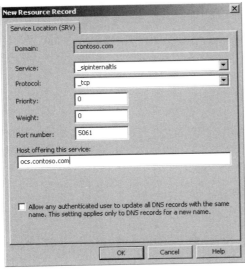

8. Click OK, and then click Done when the operation is finished.

Verifying DNS Records Creation

To verify the DNS records that you have created, perform the following steps:

1. Log on to any computer in the domain by using an account that is a member of the Administrators group or an account that has equivalent permissions.
2. Click Start, and then click Run.
3. In the Open box, type **cmd**, and then click OK.
4. At the command prompt, type **nslookup**, and then press Enter.
5. Type **set type=srv**, and then press Enter.
6. Type **_sipinternaltls._tcp.contoso.com** for the TLS record, and then press Enter. The output is shown here.

7. When you are finished, at the command prompt, type **exit**, and then press Enter.

Setting Up the Office Communications Server Host Computer

Before you install Office Communications Server, make sure that the server has been properly prepared. The following recommendations assume that the server is running Windows Server 2008:

- Apply operating system updates for Windows Server 2008 by following the instructions in Microsoft Knowledge Base article 953582, "You may be unable to install a program that tries to register extensions under the IQueryForm registry entry in Windows Server 2008 or Windows Vista," at *http://go.microsoft.com/fwlink/?LinkId=131392*. You must install this update before you install Office Communications Server 2007 R2 Administrative Tools.
- Also, apply the operating system update described in the Microsoft Knowledge Base article, "AV at mscorwks!SetAsyncResultProperties," located at *http://go.microsoft.com/fwlink/?linkid=143318*.

- Start Windows Firewall if you plan to use it and if it is not already enabled. If Windows Firewall is running when Office Communications Server is installed, activating the server automatically adds exceptions that are needed for Office Communications Server. Otherwise, you have to set the exceptions manually. For more information, see the "Prepare Windows for Setup" topic in the Microsoft Office Communications Server 2007 R2 documentation at *http://go.microsoft.com/fwlink/?linkid=143299*.

- Disable Windows services that are not required by any of the Office Communications Server roles that are installed on the server. The required Windows services include Hypertext Transfer Protocol (HTTP) Secure Sockets Layer (SSL), Windows Management Instrumentation (WMI), WMI Driver Extensions, and Message Queuing (MQ). HTTP SSL is a collective reference of HTTP, Internet Information Services (IIS) Admin Service, Remote Procedure Call (RPC), and Security Account Manager (SAM). WMI can include the Event Log and RPC. MQ is for archiving and can include MQ access control, NT LAN Manager (NTLM) Security Support Provider (NTLMSSP) service, RRPC, RMCAST [Pgm] Protocol Driver, TCP/IP Protocol Driver, Internet protocol security (IPsec) driver, and SAM.

For more information, see the "Microsoft Office Communications Server 2007 R2" documentation at *http://go.microsoft.com/fwlink/?LinkID=133608*.

Joining the Server Running Office Communications Server to the Domain

The host computer must be joined to the properly prepared AD DS domain of your choice before you install or activate Office Communications Server. Before you join the computer to the domain, make sure the computer uses your network's DNS server as the preferred DNS server. You can verify and change these settings on the Internet Protocol Version 4 (TCP/IPv4) Properties console, as shown in Figure 9-3.

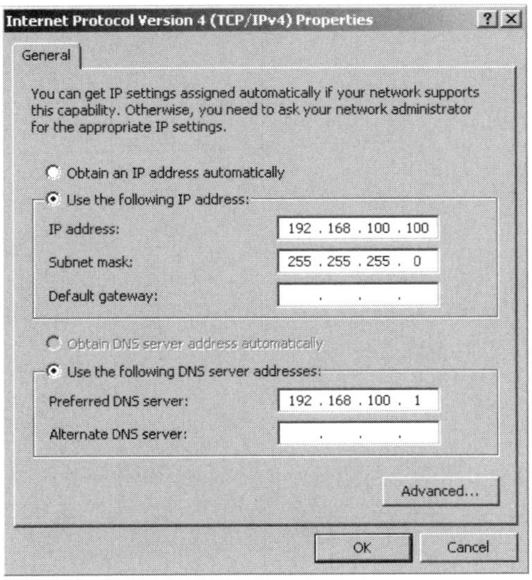

FIGURE 9-3 Specifying the DNS server for your host computer running Office Communications Server.

250   Part V   Debugging, Tuning, and Deploying Unified Communications Applications

To join a server running Windows Server 2008 to your domain, perform the following steps:

1. On the computer running Windows Server 2008 that you want to host Office Communications Server, in the Initial Configuration Tasks dialog box, shown here, click Provide Computer Name And Domain.

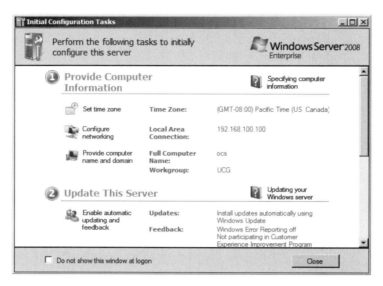

2. In System Properties, on the Computer Name tab, shown here, click Change.

3. In Computer Name/Domain Changes, select the Domain option, type the name of the domain that you want this computer to join, as shown here, and then click OK.

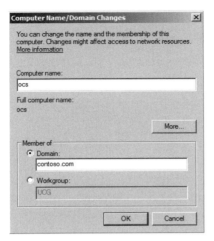

4. When prompted, enter the credentials of a network account that has the permissions to perform the operation.

5. Restart the computer.

Verifying Office Communications Server Host Name Resolution

To verify Office Communications Server host name resolution, perform the following steps:

1. Log on to a computer in the domain.

2. Click Start, and then click Run.

3. In the Open box, type **cmd**, and then click OK.

4. At the command prompt, type **ping ocs.contoso.com**. This is the FQDN of the server running Office Communications Server Standard Edition that you specified earlier in this chapter. Press Enter.

5. Verify that the IP address that is returned is correct for the server specified earlier.

Installing and Configuring Office Communications Server Standard Edition

To install and configure Office Communications Server Standard Edition, perform the following steps:

1. Log on to the computer running Windows Server 2008 that has been set up for Office Communications Server by using an account that is a member of the Domain Admins and RTCUniversalServerAdmins security groups in the domain.

2. Install the Web Server (IIS) server role on the computer where Office Communications Server will be installed.

3. Install and activate Office Communications Server Standard Edition.
4. Configure Office Communications Server Standard Edition.
5. Install a server certificate on the local computer.
6. Install a Web Server certificate for the Web Components Server role.
7. Configure A/V and Web conferencing.

The following sections describe how to perform these tasks.

Becoming a Member of the RTCUniversalServerAdmins Security Group

To deploy Office Communications Server, you must be a member of the Domain Admins and RTCUniversalServerAdmins security groups. If you are not already a member of these security groups, perform the following steps to add yourself to these groups.

1. Log on as a domain administrator to the domain controller or a computer that is joined to the domain in which the Active Directory Administrative Tools are installed.
2. Click Start, Administrative Tools, and then Active Directory Users And Computers.
3. In the left pane, under the name of the domain that you want to use, click Users, right-click a user in the right pane, and then click Properties. In this example, the user is Administrator.
4. Click the Member Of tab to view the group memberships that are available. If the user is already a member of required groups, click Cancel. Otherwise, click Add to add the user to a group.
5. To add the user to the RTCUniversalServerAdmins group, type the group name in the Enter The Object Names To Select box, as shown here. Click Check Names, and then click OK.

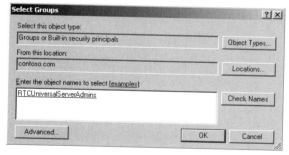

6. Click OK to close the User Properties page.

Installing the Web Server (IIS) Server Role

Office Communications Server requires that the Web Server (IIS) server role be installed on the host computer. To install Web Server (IIS), perform the following steps:

1. Click Start, Administrative Tools, and then Server Manager.
2. Click Action, and then click Add Roles to start the Add Roles Wizard, as shown here.

3. On the Before You Begin page, click Next.
4. On the Select Server Roles page, select Web Server (IIS).
5. On the Add Features Required For Web Server (IIS) page, click Add Required Features.
6. On the Select Server Roles page, click Next.
7. On the Web Server (IIS) page, click Next.
8. On the Select Role Service page, shown here, select the following role services:
    - Under Application Development, select ASP.NET.
    - Under Security, Select Windows Authentication.
    - Under Management Tools, select IIS 6 Management Compatibility.

    Once you have selected the services, click Next.

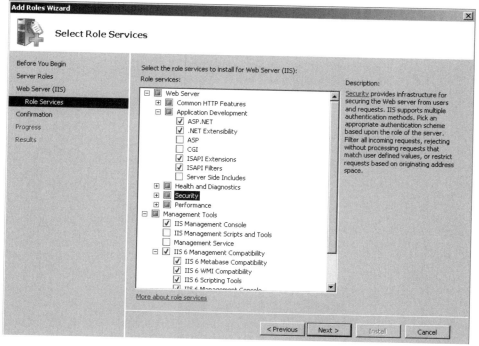

9. On the Confirm Installation Selections page, click Install.
10. Click Close after the installation finishes.

Installing Office Communications Server Standard Edition

To install Office Communications Server Standard Edition, perform the following steps:

1. Log on as a member of the DomainAdmins and RTCUniversalServerAdmins groups to the computer on which you want to install Office Communications Server Standard Edition.

2. Insert the Microsoft Office Communications Server 2007 R2 CD, and then click Standard Edition to start the Microsoft Office Communications Server 2007 R2 Deployment Wizard. Alternatively, browse to the \Setup\Amd64\ folder on a network share and double-click SetupSE.exe. When asked to install Microsoft Visual C++ 2008 Redistributable, click Yes. When asked to install Microsoft .NET Framework 3.5, click Yes. You may need to restart the computer and then repeat steps 1 and 2 of this procedure.

3. In the wizard, click Deploy Standard Edition Server.

Chapter 9 Preparing the UC Development Environment

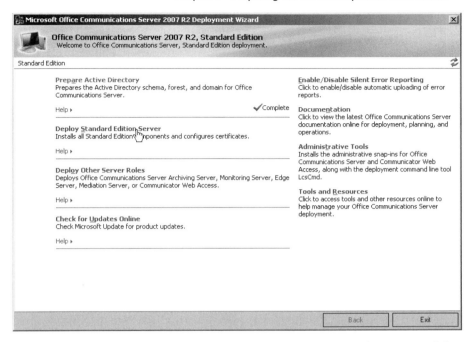

4. On the Deploy Standard Edition Server page, under Step 1: Deploy Server, click Run, as shown here.

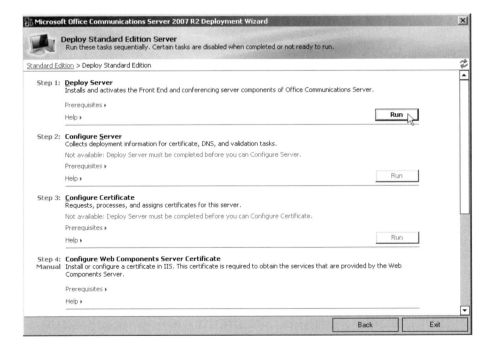

5. On the Welcome To The Deploy Server Wizard page, click Next.

6. Review the license agreement, click I Accept The Terms In The License Agreement, and then click Next.

7. On the Location For Server Files page, use or modify the default location for server files, and then click Next.

8. On the Application Configuration page, use the defaults or modify the selections that you want to install, and then click Next.

9. On the Main Service Account For Standard Edition Server page, type the name of a new or existing service account that will run the core Office Communications Server service on this computer. Also, type the password for the account, as shown here, and then click Next.

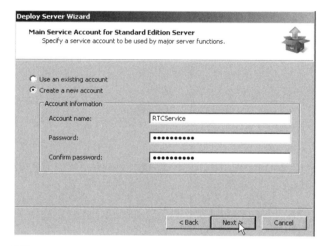

When you create a new account, activation might fail until the account has been replicated in AD DS. If activation fails, wait until the account has been replicated, and then try again.

10. On the Component Service Account For This Standard Edition Server page, type the name of a new or existing service account that will run the A/V Conferencing Server and Web Conferencing Server components on this computer. Type and then confirm the password for the account, as shown here, and then click Next. The default account is RTCComponentService.

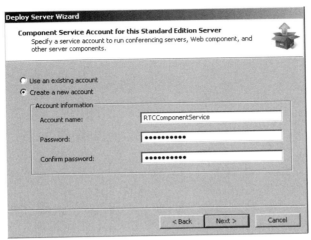

11. On the Web Farm FQDNs page, verify that Internal Web Farm FQDN displays the FQDN of your server, as shown here, and then click Next. This FQDN is used by internal users for client downloading of Web conferencing content, distribution group expansion, and Address Book information.

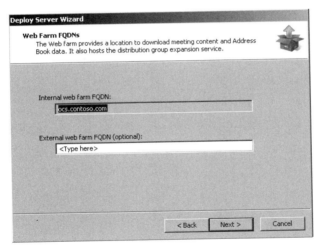

If you plan to enable external user access to Web conferences, under External Web Farm FQDN, type the FQDN that resolves to the external IP address of your reverse proxy server. This FQDN is used by external users for client downloading of Web conferencing content, distribution group expansion, and Address Book information. It is also used by anonymous and federated users to download Web conference content. For more information, see "Deploying Edge Servers for External User Access" at *http://go.microsoft.com/fwlink/?LinkID=143690*.

12. On the Location For Database Files page, accept or modify the default directories for user database and transaction log files, and then click Next.

13. On the Ready To Deploy Server page, shown here, review the settings that you specified, and then click Next.

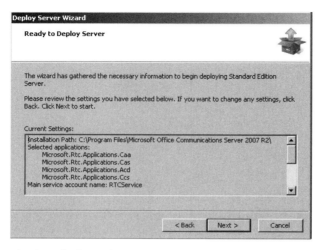

14. The installation takes some time to complete. After the files are installed and the wizard has completed, select the View The Log When You Click Finish check box, and then click Finish.

15. In the log file, verify that Success appears under the Execution Result column at the end of each task, as shown here, and then close the log window.

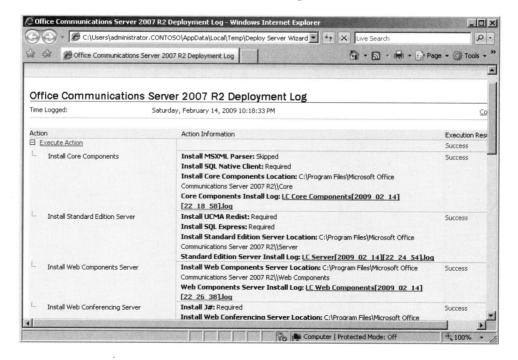

Configuring Office Communications Server Standard Edition

The server running Office Communications Server Standard Edition that you installed must be configured by using the Configure Server Wizard. To configure the server, perform the following steps:

1. On the Deploy Standard Edition Server page of the Microsoft Office Communications Server 2007 R2 Deployment Wizard, under Step 2: Configure Server, click Run, as shown here.

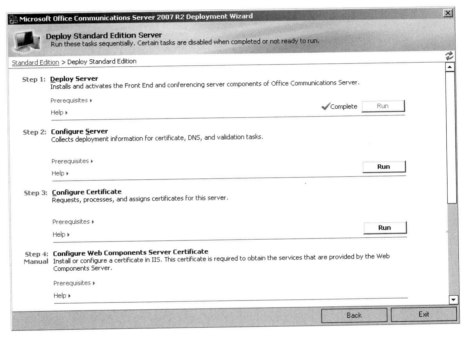

2. On the Welcome To The Configure Pool/Server Wizard page, click Next.

3. On the Server Or Pool To Configure page, shown here, select the server that you just installed from the list, and then click Next.

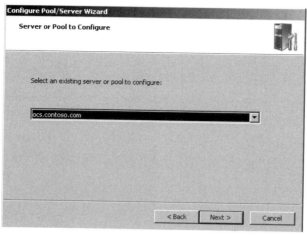

4. On the SIP Domains page, shown here, verify that your SIP domain appears in the list. If it does not, type your SIP domain in the SIP Domains In Your Environment box, and then click Add. Repeat these steps for all SIP domains that the server running Standard Edition Server supports. When you are finished, click Next.

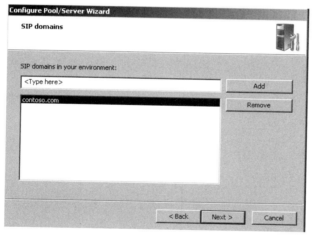

5. On the Client Logon Settings page, shown here, select Some Or All Clients Will Use DNS SRV Records For Automatic Logon, and then select Use This Server Or Pool To Authenticate And Redirect Automatic Client Log Requests. The above choices require that you have followed the steps given in the "Configuring DNS for Automatic Sign-In" section earlier in this chapter. If you did not configure DNS for automatic sign-in, you must select the Client Will Manually Configure For Logon option, and then click Next.

Chapter 9 Preparing the UC Development Environment   261

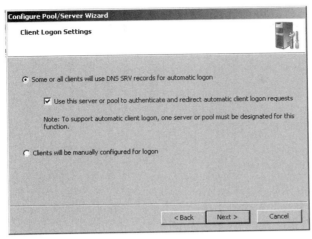

6. On the next page, select the domains that the server supports for automatic sign-in, and then click Next.

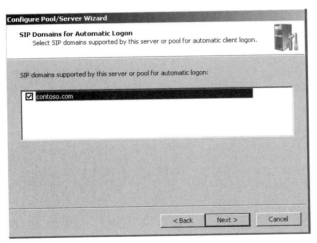

7. Unless you plan to test the external user access scenario, on the External User Access Configuration page, select Do Not Configure For External User Access Now, as shown here, and then click Next.

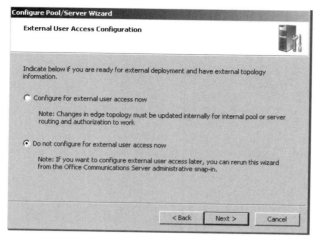

8. On the Ready To Configure Server Or Pool page, review the settings that you specified, and then click Next.

9. When the files are installed and the wizard is complete, select the View The Log When You Click Finish check box, and then click Finish.

10. In the log file, under Action, expand Execution Action. In the Execution Result column, verify that Success appears next to each task, as shown here. When you are done, close the log window.

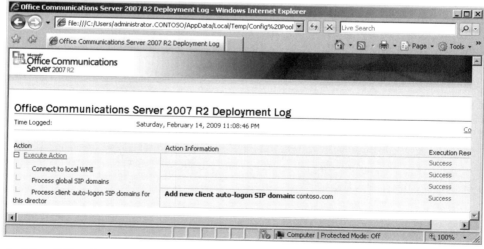

As part of the Office Communications Server 2007 R2 installation, the Address Book Server is configured automatically. In addition to the configuration steps described earlier, you can configure additional Office Communications Server roles. For more information about server roles, see the "Microsoft Office Communications Server 2007 R2" documentation at *http://go.microsoft.com/fwlink/?LinkID=133608*.

Installing and Configuring TLS/MTLS Certificates for Servers Running Standard Edition

Office Communications Server requires that each host computer be configured with a trusted server certificate to communicate with each other. For a trusted application, the application hosting computer must also have a certificate installed. The certificates are required for establishing Mutual Transport Layer Security (MTLS) connections among servers, between a trusted application hosting computer and the server running Standard Edition, and for establishing a TLS connection between a client and the server. These certificates are also used by the Web Component Server role to support SSL over Hypertext Transfer Protocol Secure (HTTPS), which is described in detail in the "Configuring IIS Certificates for the Web Components Server Role" section later in this chapter.

To install the certificates that are required for TLS or MTLS on the computer that is running Office Communications Server, perform the following steps:

1. Log on to the server on which you want to install the certificate by using an account that has permissions to request a certificate from your CA and to install it on the local computer.

2. Insert the Microsoft Office Communications Server 2007 R2 CD, and then click Standard Edition; if you are installing from a network share, browse to the \Setup\Amd64\ folder on the network share, and then double-click SetupSE.exe.

3. On the Deploy Standard Edition Server page, under Step 3: Configure Certificate, click Run.

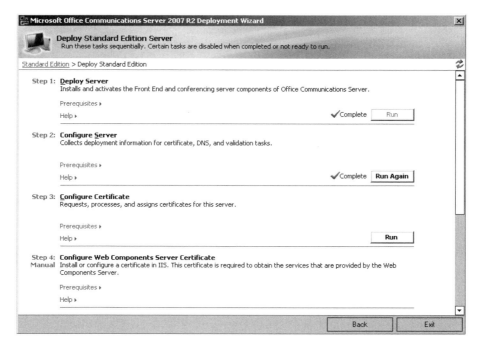

264 Part V Debugging, Tuning, and Deploying Unified Communications Applications

4. On the Welcome To The Certificate Wizard page, click Next.
5. On the Available Certificate Tasks page, select the Create A New Certificate option, as shown here, and then click Next.

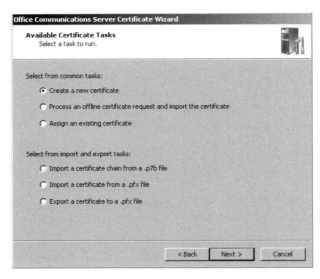

6. On the Delayed Or Immediate Request page, select Send The Request Immediately To An Online Certification Authority, and then click Next.
7. On the Name And Security Settings page, in the Name box, type the name of the certificate that you want to use. In the Bit Length box, select the bit length to be used for encryption. Use the default settings for the other options. Click Next.

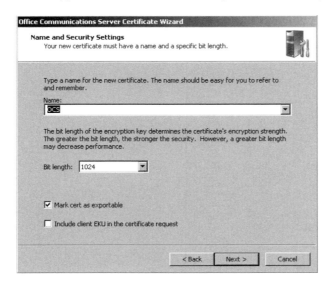

8. On the Organization Information page, shown here, type the names of your organization and organizational unit, and then click Next.

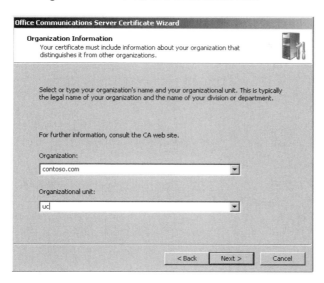

9. On the Your Server's Subject Name page, shown here, in the Subject Name box, type the FQDN or pool name of your server. Leave the default value in the Subject Alternative Name box, and then click Next. For the server running Standard Edition, the subject name should be the FQDN of the server and the Subject Alternative Name (SAN) is not required.

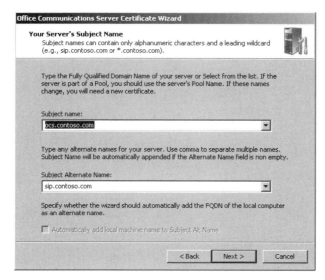

10. On the Geographical Information page, fill in the Country/Region, State/Province, and City/Locality boxes, and then click Next.

11. On the Choose A Certification Authority page, use the default option, and then click Next.

12. On the Request Summary page, verify that the settings are correct, and then click Next.

13. On the Assign Certificate Task page, select Assign Certificate Immediately, and then click Next.

14. On the Configure The Certificates Of Your Server page, click Next.

15. On the Certificate Wizard Completed Successfully page, click Finish.

Configuring IIS Certificates for the Web Components Server Role

The MTLS-required certificate for a server that is running Office Communications Server Standard Edition is also used for the Web Components Server role. The certificate is required for establishing SSL connections over HTTPS. However, you must use the Internet Information Services (IIS) Manager to assign the certificate to the Web Components Server role. The following steps explain how to configure the Web Components Server certificates by using IIS 7 on Windows Server 2008:

1. Log on to the server that is running the Web Components Server as a member of the Administrators group. In the example in this chapter, this would be the computer running Office Communications Server Standard Edition.

2. Click Start, Administrative Tools, and then Computer Management.

3. Expand the Services And Applications node, as shown here, and then expand the Internet Information Services (IIS) Manager node.

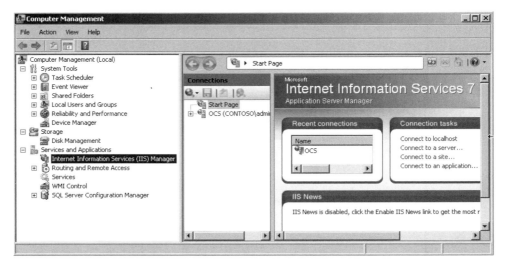

4. In the Connections pane, expand Web Components Server.

5. Expand Sites, and then click Default Web Site.

6. In the Default Web Site Home pane, under IIS, click Authentication, as shown here.

7. In the Actions pane, shown here, click Bindings.

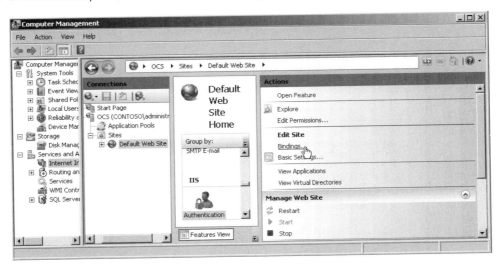

8. In the Site Bindings dialog box, click Add.

9. In the Add Site Binding box, from the Type drop-down menu, click Https. Verify that IP Address is set to All Unassigned. Verify that Port is set to 443. Select the certificate for the Web Components Server in the SSL Certificate drop-down menu, as shown here, and then click OK.

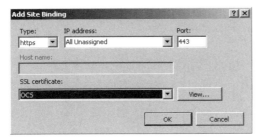

10. On the Site Bindings page, click Close to finish configuring the Web Components Server certificate.

Starting Office Communications Server Services

Before starting the services, make sure that the changes made to AD DS have been replicated and Windows Firewall is running. As part of starting the services, Office Communications Server opens the required ports in the firewalls. If the firewalls are not running, Office Communications Server does not open the required ports. For Office Communications Server to open the required ports, you must add the necessary exceptions to the firewall. To add the required exceptions automatically, complete the following steps while the firewalls are running:

1. On the Deploy Standard Edition Server page, under Step 6: Start Services, click Run, as shown here.

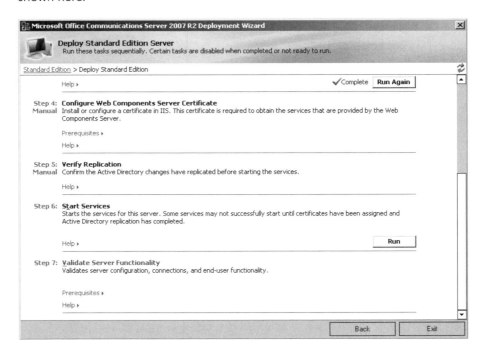

2. On the Welcome To The Start Services Wizard page, click Next.

3. On the Start Office Communications Server 2007 R2 Services page, click Next.

4. On the last page of the Start Services Wizard page, select View The Log When You Click Finish, and then click Finish. If the services do not start successfully, examine the log file, fix the problem, and then restart the Office Communications Server services. Some services might fail to start because they have not been installed or enabled; these can be ignored safely. For example, the Monitoring Agent service fails to start when a Monitoring Server role is not installed. You can also use Windows Event Log to examine the cause of the failure.

Configuring Audio/Video and Web Conferencing

To configure or modify the configuration of A/V and Web conferencing, perform the following procedure.

> **Note** You do not have to complete the steps in this section if you plan to use only instant messaging and presence.
>
> For detailed instructions, see Microsoft Office Communications Server 2007 R2 documentation.

1. Log on to a server running Office Communications Server as a member of the RTCUniversalServerAdmins group.

2. Click Start, Control Panel, Administrative Tools, and then Office Communications Server 2007 R2.

3. Right-click the forest node, point to Properties, and then select Global Properties, as shown here.

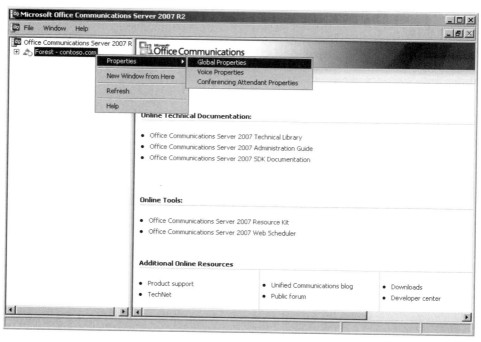

4. To add or modify a Meeting policy, click the Meetings tab, and, from the Anonymous Participants drop-down list, select one of the available options, as shown here.

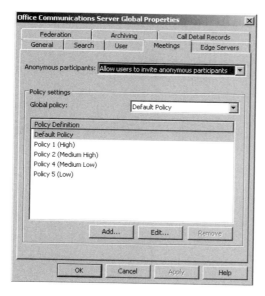

The Anonymous Participants options are as follows:

- **Allow Users To Invite Anonymous Participants** Allows all users to be able to organize Web conferences that allow anonymous participants to join

- **Disallow Users From Inviting Anonymous Participants** Prevents all users from organizing Web conferences that allow anonymous participants to join

- **Enforce Per User** Allows only some users to be able to organize Web conferences that allow anonymous participants to join

Note By default, all users can create Web conferences that allow anonymous participants to join unless you deny this privilege to organizers on an individual basis. For more information about how to deny this privilege, see the "Configuring UC User Accounts" section later in this chapter.

5. To add or modify a policy, do one or all of the following under Policy Settings:

    - To modify a policy, select an entry from the Policy Definition list (for example, Default Policy), and then click Edit to modify the policy definition.

    - To add a new policy, click Add to specify the new policy definition.

6. In the Add Policy or Edit Policy dialog box, perform the following steps:

    a. For a new policy, type a name in the Policy Name text box. For an existing policy, skip this step.

    b. To set or change the maximum number of meeting participants, set an appropriate integer value in the Maximum Meeting Size text box.

    c. To enable or disable Web conferencing, select or clear the Enable Web Conferencing check box, and make appropriate selections for other Web conferencing settings.

    d. To enable or disable audio or video, select or clear the Enable IP Audio check box and make appropriate selections for other A/V settings, as shown here.

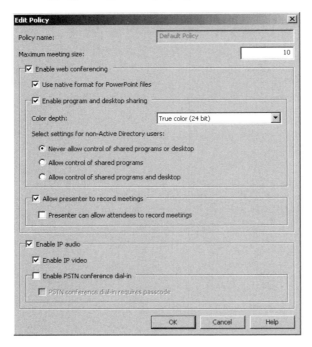

   e. Click OK.

7. On the Office Communications Server Global Properties page, click Apply.

8. After you have finished adding the new features or editing the enabled features by each policy, decide which policy to apply to Web conferences that are organized by users, and then select one of the following options:

   ❑ To apply the same policy to all users, click Global Policy, and then click the name of the policy that defines the features you want to enable for all users.

   ❑ To apply different policies to different users, click Global Policy, and then click Use Per User Policy.

9. Click OK.

Configuring UC User Accounts

Users created in AD DS are not enabled automatically for Office Communications Server. Enabling users for Office Communications Server requires that you complete the following tasks in sequence:

1. Create a user account in AD DS.

2. Enable the user account for Office Communications Server.

3. Configure the user account for UC.

To perform these tasks, you need to access the Active Directory Users And Computers management console, which has been enabled for Office Communications Server. You can meet these requirements by installing the Office Communications Server 2007 R2 Administrative Tools and then running the tasks on the domain controller.

The following sections explain in detail how to complete these steps.

Creating a User Account in AD DS

To create a user account in AD DS, perform the following steps:

1. Log on as a member of the DomainAdmins group to your AD DS domain controller that has the Office Communications Server Administrative Tools installed.
2. Click Start, and then click Run.
3. In the Open box, type **dsa.msc**, and then press Enter.
4. Right-click the Users container or another organizational unit in which you want to create users, click New, and then click User.
5. Complete the New Object – User Wizard, shown here, and click Next.

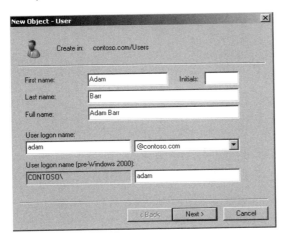

6. On the following page, click Finish.

Enabling a User for Office Communications Server

After a user account is created in AD DS, it must be enabled for Office Communications Server. To enable a user for Office Communications Server, perform the following steps:

1. Log on as a member of the RTCUniversalUserAdmins group on to your AD DS domain controller that has the Office Communications Server 2007 Administrative Tools installed.
2. Click Start, and then click Run.

3. In the Open box, type **dsa.msc**, and then click OK.

4. Navigate to the Users folder or to the organizational unit where your users are.

5. Select one or more users that you want to enable for Office Communications Server, right-click your selection, and then click Enable Users For Communications Server, as shown here.

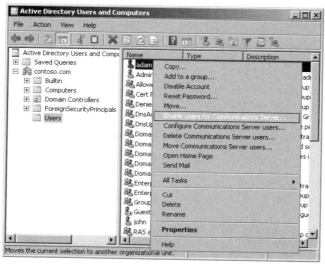

6. On the Welcome To The Enable Office Communications Server Users Wizard page, click Next.

7. On the Select Server Or Pool page, shown here, select the Office Communications Server that you want to place the users on from the drop-down list (in this example, the server is ocs.contoso.com), and then click Next.

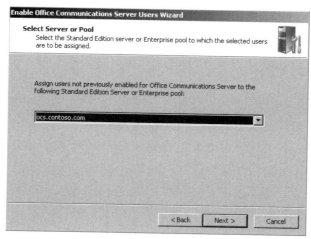

8. On the Specify Sign-In Name page, select one of the following options to specify how to generate the SIP address:

- ❑ To generate the SIP address from the user's e-mail address, click Use User's E-mail Address. Select this option only if you have configured an e-mail address for your users.
- ❑ To generate the SIP address from the user's principal name, click Use UserPrincipalName.
- ❑ To generate the SIP address by using the user's full name, click Use The Format: <first name>.<lastname>@, and then select the Office Communications Server domain.
- ❑ To generate the SIP address by using the user's SAM account, click Use The Format: <SAMAccountName>@, and then select the Office Communications Server domain.

> **Note** If you need to configure SIP addresses by using a different format from the options presented, you can enable users individually or build your own management tool to bulk-enable users with the SIP address format of your choice.

9. On the Ready To Enable Users page, click Next.
10. On the Enable Operation Status page, verify that the users were enabled successfully, and then click Finish.

Verifying Replication of Users Enabled for Office Communications Server

After you enable users for Office Communications Server, make sure that the settings for the newly enabled users replicate before you configure the users. To verify that user settings have replicated, perform the following steps:

1. Log on as a member of the RTCUniversalUserAdmins group to your AD DS domain controller that has the Office Communications Server 2007 R2 Administrative Tools installed.
2. Click Start, Control Panel, Administrative Tools, and then Office Communications Server 2007 R2.
3. Expand the forest node and the pool node, and then click Users. For the server running Standard Edition, the pool node corresponds to the server node.
4. Confirm that the users that you enabled for Office Communications Server are listed in the details pane on the right, as shown here.

Configuring Users for Office Communications Server

After users are enabled and their settings replicated, you can configure them to be enabled or disabled for federated access, remote access, public instant messaging, or Enhanced Presence. When applicable, you can also enable or disable the supported archiving capabilities for the users. To configure users for Office Communications Server, perform the following steps:

1. Log on as a member of the RTCUniversalUserAdmins group to your AD DS domain controller that has the Office Communications Server 2007 R2 Administrative Tools installed.

2. Click Start, Control Panel, Administrative Tools, and then Office Communications Server 2007 R2.

3. Expand the forest node, expand the Standard Edition Servers node, select your Office Communications Server computer, and then click Users.

4. From the list of users in the details pane on the right, select one or more users, right-click the selection, and then click Configure Communications Server Users, as shown here.

5. On the Welcome To The Configure Users Wizard page, click Next.

6. In the Configure Office Communications Server Users Wizard, on the Configure User Settings page, select the check box next to the settings that you want to configure, and then click Enable or Disable to configure those settings, as shown here. Then click Next.

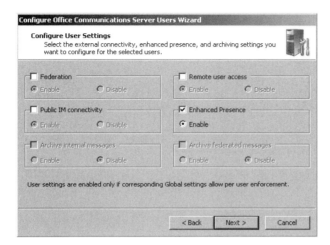

Note If you did not configure the global properties on the Meetings tab to enforce per-user settings, some of the options are not available because they are enforced by the global policies you configured.

7. On the Configure Meeting Settings page, select the Organize Meetings With Anonymous Participants check box, click Allow or Disallow to enable or disable this option, and then click Next. The options may or may not be enabled, depending on the existing global settings.

8. On the Configure User Settings page, under Meeting Policy (if it is enabled), select the Change Meeting Policy check box. From the Select A Meeting Policy For The Users list, select the name of the policy that you want to apply to the users, and then click Next.

9. On the Configure Enterprise Voice page, select the Change Enterprise Voice Settings check box. Then select the Enable Voice check box (if you want to enable Enterprise Voice and configure the Enterprise Voice policy that will be applied to the selected users), and then click Next.

10. On the Configure Enterprise Voice Settings And Location Profile page, under the Voice Policy list, select the name of the policy that you want to apply to the selected user or users, if one is enabled, select Location Profile for selected users if you want, and then click Next.

Note To configure a particular Enterprise Voice setting for a specific user, the corresponding setting under Voice Properties must be configured to allow enforcement on a per-user basis. For more information about Enterprise Voice, see the "Microsoft Office Communications Server 2007 R2" documentation.

11. On the Ready To Configure User page, verify the status of each user configuration, and then click Next.

12. On the Configure Operation Status page, verify that the operation is successful, and then click Finish to complete configuring the selected user or users.

Validating Server Functionality

After Office Communications Server is installed, configured, and activated and the user accounts are enabled and configured for UC, perform the following steps to validate the deployed server functionality:

1. Log on as a member of the RTCUniversalServerAdmins group to the computer on which Office Communications Server has been installed.

2. On the Deploy Standard Edition Server page of the Microsoft Office Communications Server 2007 R2 Deployment Wizard, click the Step 7: Validate Server Functionality link, as shown here.

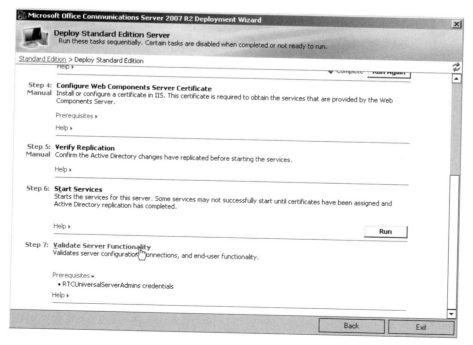

3. On the Validate Pool Or Server Functionality page, shown here, run Steps 1-6 and complete the rest of the steps in the wizard to verify that the required Office Communications Server roles are installed properly.

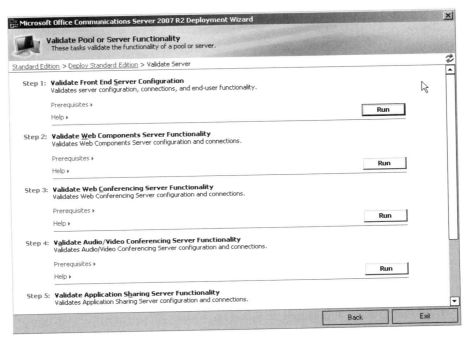

Configuring Application Development Components

Most requirements of an application development component in the UC environment are met if the corresponding SDK is installed correctly. This section focuses on the following APIs, which are the primary interest in this book, and highlights the configuration issues that are not addressed by the SDK installation:

- Office Communicator Automation API
- UCMA Core
- UCMA Workflow

Office Communicator Automation API is distributed in the Office Communicator Automation API SDK. The UCMA Core and UCMA Workflow are contained in the Microsoft UCMA 2.0 SDK. The SDKs for these APIs are available from the Download Center on MSDN.

Configuring the Office Communicator Automation API

To build an application by using the Office Communicator Automation API, you must have the Office Communicator Automation API SDK installed on your development computer. To run and debug the application, you must have Office Communicator installed on the computer where your application runs.

Installing Office Communicator

Office Communicator is distributed as part of Office Communications Server. It can be installed in various configurations and by using different methods. For more information about how to install Office Communicator 2007 R2, see the "Deploying Office Communications Server 2007 R2" topic in the "Microsoft Office Communications Server 2007 R2" documentation at *http://go.microsoft.com/fwlink/?LinkID=133608*.

Installing the Office Communicator Automation API SDK

To build applications using the Office Communicator Automation API, you must install the Office Communicator Automation API SDK (Ocsdk.msi) on your development computer. You can download the API SDK from the Download Center on MSDN at *http://go.microsoft.com/fwlink/?linkid=143206*. This SDK must be installed on computers running Windows 2000 SP4, Windows Server 2003, Windows Server 2008, Windows Vista, or Windows XP SP2.

Generating Class and Interface IDs for Win32/C++ Office Communicator Automation API Applications

When you build a Win32 C/C++ application by using the Office Communicator Automation API, you must include several C/C++ header files in your project. These include Msgrua.h and Msgrua_i.c. The latter contains the globally unique identifier (GUID) definition for the class ID (CLSID) and interface ID (IID). If you do not include these files, your project does not build because of unresolved link errors. In this case, you can create the required GUID definition file by using the following MIDL command from the Visual Studio 2008 SDK console in the directory where Msgrua.idl is located:

```
midl.exe /ms_ext /char unsigned /c_ext msgrua.idl /out c:\myOcAPIAppProj
```

Using a Visual Studio SDK console, you can invoke Midl.exe without specifying its full path. These details are based on the assumption that you have Visual Studio 2008 installed on your application development computer. If you still use Visual Studio 2005, you can open the Visual Studio 2005 SDK console to execute this command.

The previous MIDL command produces the interface and class ID file, Msgrua_i.c, in the specified output directory C:\MyOcAPIAppProj. You can then add the resultant Mgrua_i.c file to your Win32/C++ application project. In Visual Studio, perform the following steps:

1. In the Solution Explorer pane, right-click Header Files, as shown here, and then click Add.

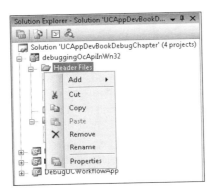

2. Select Existing Item.

3. Navigate to the folder where the Msgrua_i.c file is located, and then double-click the file to add it to the project.

Configuring UCMA Core

There are two requirements that must be met before a UCMA Core application can be developed, tested, and deployed. The first one is to enable TLS or MTLS connections between the application and Office Communications Server. The second requirement is to establish trust between the application and Office Communications Server. The steps that you must take to meet these requirements are explained in the following sections. The discussion also includes where to download and install the SDK.

Installing the UCMA SDK

You can install the UCMA SDK on computers that are running Windows 2000 SP4, Windows Server 2003, Windows Server 2008, or Windows Vista. Before installing the UCMA SDK, make sure that you have Visual Studio 2008 SP1 and .NET Framework 3.5 SP1 installed on your application development computer.

To install the Microsoft UCMA 2.0 SDK, visit the MSDN Download Center at *http://go.microsoft.com/fwlink/?LinkID=139195* and follow the instructions.

To summarize, the installation is a two-stage process, briefly described as follows:

1. Download and run the UcmaSdkWebDownload.msi file. This installs the component SDKs and other supporting resources on your local drive. By default, all of the files are installed in the C:\Microsoft Unified Communications Managed API 2.0 SDK Installer Package folder.

2. Go to the Installer Package folder on the local drive and navigate to the SetupUCMASdk.exe file. Double-click this file to start the SDK installation. In the last step of the installation, follow the Language Pack Download link to install at least one speech language package. The language pack is required by the UCMA Workflow application.

Enabling TLS or MTLS Connections

When a UCMA Core application communicates with Office Communications Server over TLS or MTLS, the application host computer and Office Communications Server host computer must have valid certificates installed. The certificates must be issued from a mutually trusted CA. With TLS, the application host presents its certificate to Office Communications Server before the connection can be established. The connection succeeds only if Office Communications Server can validate the certificate that is presented by the application. With MTLS, Office Communications Server also presents its certificate for the application to validate. The connection succeeds only if both Office Communications Server and the application can validate the certificates from one another. MTLS is required if the application is trusted by Office Communications Server. TLS can be used when the application does not have this kind of trust relationship. A UCMA Core client application can use either TLS or MTLS; however, a UCMA Core server application must use MTLS.

On the server running Office Communications Server, the certificate can be created and installed by using the Certificate Wizard in Office Communications Server. For more information, see "Configuring Internal and External Interfaces and Certificates for Edge Servers" at *http://go.microsoft.com/fwlink/?linkid=143222*.

On the computer that hosts the application, you must create or import a certificate and install it in the Console Root\Certificates (Local Computer)\Personal\Certificates folder. In addition, a root certificate must be installed in the Console Root\Certificates (Local Computer)\Trusted Root Certificate Authorities\Certificates folder on the server running Office Communications Server. For more information about how to configure the certificates, in the Microsoft Office Communicator 2007 R2 documentation, see the "Set Up Certificates for the Internal Interface" topic at *http://go.microsoft.com/fwlink/?linkid=143691* and the "Set Up Certificates for the External Interface" topic at *http://go.microsoft.com/fwlink/?linkid=143692*.

Verifying That the Certificate Is Installed on the Application Hosting Computer

A UCMA application using the server platform or TLS or MTLS must be trusted by Office Communications Server. To enable this trust relationship, the application must present to the server the hosting computer's certificate, which must be issued by a CA common to both the application hosting computer and the Office Communications Server hosting computer. To ensure that the UCMA application runs successfully, make sure that the proper certificate is installed. If the required certificate is not installed, you must then install the required certificate.

To verify that the certificate is installed on the computer that hosts the UCMA application, perform the following steps:

1. Log on as a local administrator to the computer where the UCMA Core application is to be installed and run.

2. Open an MMC window, shown here, by typing **mmc** at a command prompt or by clicking Start, Run, and then typing **mmc**.

3. On the MMC Console Root, click File, and then click Add/Remove Snap-In to add the Certificates Snap-In, as shown here.

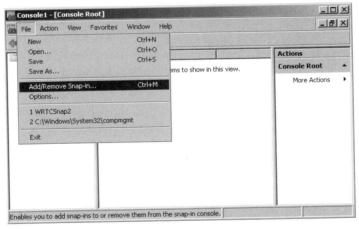

4. In the Add Or Remove Snap-Ins dialog box, under Available Snap-Ins, double-click Certificates.

5. On the Certificates Snap-In page, under This Snap-In Will Always Manage Certificates For, select Computer Account, as shown here, and then click Next.

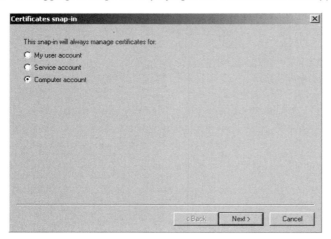

6. On the Select Computer page, select Local Computer, as shown here, and then click Finish. This adds the local computer to the selected snap-in list.

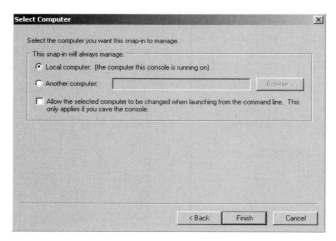

7. In the Add/Remove Snap-Ins window, click OK to display the certificates that are installed on the computer.

8. In the Console Root, in the left pane, expand Personal, and then click Certificates to display available computer certificates, as shown here.

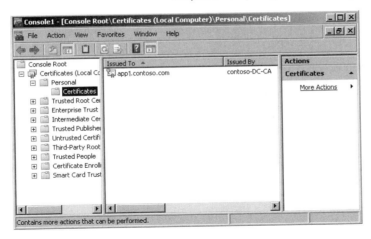

Installing a Computer Certificate on a UCMA Application Host

If the computer certificate has not been installed, you can request one from your CA or import one that you have requested previously. To request a new certificate and install it on an application computer, perform the following steps:

1. In the Console Root, in the left pane, expand Personal, right-click Certificates, click All Tasks, and then click Request New Certificate, as shown here.

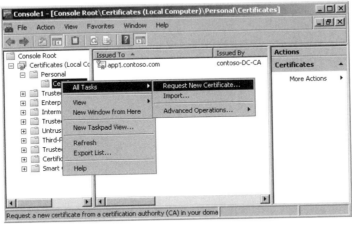

2. In the Certificate Enrollment Wizard, on the Before You Begin page, click Next.
3. On the Request Certificates page, select Computer, and then click Enroll.

Enabling a Trusted Application

As a server application or middle-tier application, a UCMA Core application must be trusted by Office Communications Server. This includes applications that use the *ApplicationEndpoint* object. A trusted application has a mutual trust relationship between the application and

Office Communications Server. This trust relationship is enabled when a trusted service entry (TSE) is created in AD DS for the application. This trusted application is represented by an Active Directory *Contact* object and is identifiable by its SIP Uniform Resource Identifier (URI) and the application Globally Routable User Agent URI (GRUU).

In general, you may need to be a member of the Domain Admins security group to create a TSE for your application. This may not be practical if you are working in a large enterprise network. Fortunately, the UCMA Core SDK provides an application provisioning tool (ApplicationProvisioner.exe) that you can use to enable an application to be trusted by Office Communications Server. However, you still have to belong to an RTCUniversalServerAdmins group to perform this operation.

The application provisioning tool is installed as part of the installation of the UCMA Core SDK. By default, this application project is located in the C:\Program Files\Microsoft Office Communications Server 2007 R2\UCMA SDK 2.0\UCMACore\Sample Applications\Collaboration\ApplicationProvisioner directory. Before you can use the application provisioning tool to make your applications trusted by Office Communications Server, perform the following steps to build the application provisioning tool executable:

1. Make sure that your domain user account is a member of the RTCUniversalServerAdmins group before you run ApplicationProvisioner.Exe.

2. Click Start, All Programs, Microsoft Visual Studio 2008, Visual Studio Tools, and then Visual Studio 2008 Command Prompt.

3. At the command prompt, change the directory to the ApplicationProvisioner solution directory where the ApplicationProvisioner.sln file is located, and then run MSBuild against the solution. This builds the ApplicationProvisioner.exe image.

   For example, the following commands are based on the assumption that the application solution folder is in the default installation path and the application executable is output to the Bin\Debug subdirectory.

   ```
   cd c:\program files\Office Communications Server 2007 R2\UCMA SDk 2.0\UCMACore\Sample Applications\Collaboration\ApplicationProvisioner
   MSBuild
   ```

After the Application Provisioner executable (ApplicationProvisioner.exe) is built, perform the following steps to enable a UCMA application that is trusted by Office Communications Server:

1. Run the ApplicationProvisioner.exe executable as a member of the RtcUniversalServerAdmins group to create a *Contact* object for the application (in this case, the application is named myUcmaApp). In the Application Provisioner dialog box, shown here, in the Application Name box, type **myUcmaApp**, and then click Find Or Create.

Chapter 9 Preparing the UC Development Environment 287

2. If the application name myUcmaApp does not exist, in the Create Application Pool dialog box, shown here, select or specify the following, and then click OK:

   ❑ From the OCS Pool Fqdn drop-down list, select the FQDN of your Office Communications Server host.

   ❑ In the Application Server Fqdn text box, specify the FQDN of the computer hosting this application that is to be trusted by the Office Communications Server host.

   ❑ In the Listening Port input field, specify a port number for the application.

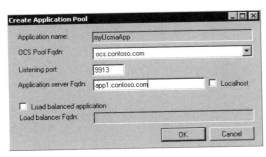

3. In the Application Provisioner dialog box, shown here, under Contacts, click Add to open the Create Contact dialog box.

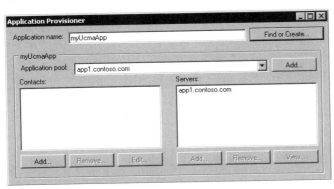

4. In the Create Contact dialog box, enter the contact's SIP URI, display name, and TEL URI for the application. These values are entered in the Contact Uri, Display Name, and Phone Uri boxes, respectively, as shown here. If specified, the value entered for the TEL URI must be unique. Using a TEL URI that is already assigned to another contact causes an exception to be thrown.

5. Click OK to have the newly specified *Contact* object created and added to AD DS and to return to the Application Provisioner dialog box.

6. In the Application Provisioner dialog box, the specified application should now appear under Contacts, as shown here.

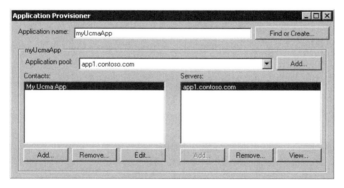

7. To view the application's GRUU, in the Application Provisioner dialog box, under Servers, double-click an application server that is listed there. Then select and copy the GrUU value that you want to view.

In this example, the complete value of the application's GRUU is *sip:app1.contoso.com@ contoso.com;gruu;opaque=srvr:myUcmaApp:GtUqdL83qkGvVYLhCt2AKAAA*. In addition to the application type, a trusted application can be identified by a combination of the application's host computer name and the port number. Therefore, the following string is also a valid GRUU for this application: *sip:app1.contoso.com@contoso. com;gruu;opaque=srvr:* **app1**.*9901:GtUqdL83qkGvVYLhCt2AKAAA*. The difference is shown by the substring that is in bold type.

8. To verify that the specified *Contact* object was created in AD DS, open Active Directory Users And Computers, and then click the Search button on the toolbar, as shown here.

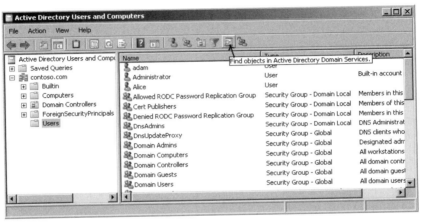

9. In the Find Users, Contacts, And Groups dialog box, in the In text box, specify Entire Directory as the search scope. In the Name box, enter the display name of the *Contact* object that you just created ("My Ucma App"), as shown here, and then click Find Now.

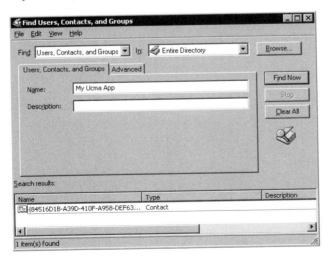

10. Verify that the *Contact* object shown in the Search Results pane is correct by double-clicking the result entry.

11. Verify that the *Contact* object is a member of the RTCUniversalReadOnlyGroup and RTCUniversalWriteGroup groups that have Read and Write permissions, respectively. If not, click the Member Of tab, as shown here, and then click Add to add the contact to these two groups.

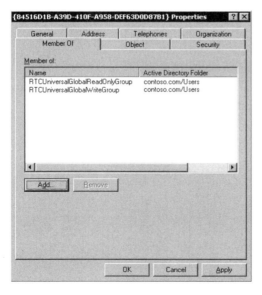

Configuring UCMA Workflow

UCMA Workflow extends the Windows Workflow Foundation to UCMA. The core requirements are the same as those of the UCMA Core, the details of which are presented in the "Configuring UCMA Core" section earlier in this chapter.

In addition, the UCMA Workflow activities require that .NET Framework 3.5 is installed on the application development computer. Visual Studio 2008 automatically installs this component.

UCMA Workflow is installed as part of the UCMA SDK installation. There is no separate download.

Installing a Language Pack

UCMA Workflow supports speech-enabled activities, such as speech recognition and text-to-speech. This requires a language-specific engine and other related resources that are available in the Speech Server Language Pack for the language that you plan to use. The language packs are distributed as part of the Speech Server. You must install at least one language pack before you use the API. You can also download language packs individually from MSDN by following the link provided in the last step of the UCMA SDK installer.

Summary

This chapter divided the UC development environment into two categories of components: infrastructure components and application development components. Configuring, activating, and administering infrastructure components is the responsibility of a system administrator unless it is delegated to someone else. Setting up application development components is the responsibility of the application developer, who may not be a system administrator.

To develop a UC application by using a UC API, you must have the corresponding SDK installed on the development computer. To download and install the SDKs, see the links provided in the next section.

To run an Office Communicator Automation API application, Office Communicator 2007 R2 must be installed locally. To run a UCMA Core or UCMA Workflow application as an Office Communications Server trusted service, you can use the ApplicationProvisioner tool provided as a sample application in the UCMA Core SDK.

If you want to set up an autonomous development environment, this chapter discussed what is required to set up and administer the most basic infrastructure components based on Office Communications Server 2007 R2 Standard Edition.

Additional Resources

- Microsoft Office Communicator 2007 SDK (*http://go.microsoft.com/fwlink/?LinkID=141199*)
- Microsoft Unified Communications Managed API SDK (*http://go.microsoft.com/fwlink/?LinkID=127908*)
- "Unified Communications Managed API 2.0 Core SDK Documentation" (*http://go.microsoft.com/fwlink/?LinkID=133571*)
- Microsoft Office Communications Server 2007 R2 Resource Kit (*http://go.microsoft.com/fwlink/?LinkId=141203*)
- "Microsoft Office Communications Server 2007 R2" documentation (*http://go.microsoft.com/fwlink/?LinkID=133608*)
- "Configuring Internal and External Interfaces and Certificates for Edge Servers" (*http://go.microsoft.com/fwlink/?linkid=143222*)
- "Prepare Windows for Setup" (*http://go.microsoft.com/fwlink/?linkid=143299*)
- "Deploying Edge Servers for External User Access" (*http://go.microsoft.com/fwlink/?LinkID=143690*)
- Microsoft Knowledge Base article 953990: "AV at mscorwks!SetAsyncResultProperties" (*http://go.microsoft.com/fwlink/?linkid=143318*)
- Microsoft Knowledge Base article 953582: "You may be unable to install a program that tries to register extensions under the IQueryForm registry entry in Windows Server 2008 or in Windows Vista" (*http://go.microsoft.com/fwlink/?LinkId=131392*)
- IQueryForm Registry Entry (*http://go.microsoft.com/fwlink/?LinkId=131392*)

Chapter 10
Debugging a Unified Communications Application

This chapter will help you to:

- Understand the general framework for debugging a Unified Communications (UC) application.
- Understand how to debug or troubleshoot a Microsoft Office Communicator Automation API application.
- Understand how to debug or troubleshoot a Unified Communications Managed API (UCMA) Core application.
- Understand how to debug or troubleshoot a UCMA workflow application.

Debugging in the UC Platform

Just as with the development of any application, errors can occur when you develop UC applications. This includes the errors discussed in this book. Errors can be caused by invalid user input, incorrect API calls, or server and network issues. To help debug programming errors and troubleshoot failure conditions, all UC APIs support a mechanism for reporting errors and failures. This reporting occurs in the form of exceptions that are thrown during run time, or trace statements that are logged to log files. The UC platform also provides tools for parsing and viewing the logs.

In an Office Communicator Automation API, errors are reported as *HRESULT* values or *COMException* instances. Trace statements are written to one or more log files or logged to the Windows Event Log. In the UCMA Core and the UCMA Workflow, errors are reported as *Exception*, together with its derived classes that are supported by the Microsoft .NET Framework. In addition, the Microsoft Office Communications Server 2007 R2 Logging Tool (OCSLogger.exe) can be used to enable tracing on the server and to analyze the network traffic generated by applications that use these APIs to communicate with the server.

Sources of Errors and Failures

Errors and failures can generally be classified as compile-time errors or run-time exceptions. Common compile-time errors occur when, for example, a compiler detects a type-checking error in the source code. These errors can be caused by typographical errors in the programming statements, use of undefined or mismatched types, or incorrect calls to members of a type;

for example, setting a value on a read-only property. With modern compilers, such errors are easily caught and making the corrections is usually straightforward.

Run-time exceptions occur during the execution of an application. The source of these errors has a wider scope and the resolution is frequently more involved. Providing incorrect user account information that prevents a user from logging on successfully is one example of a run-time error. Another example, although less common, is linking to a reference or dependent assembly that is formatted incorrectly. This prevents the assembly from loading and causes the application to end prematurely when a call is made to the library. For example, a 64-bit application must link and load the 64-bit Office Communicator Automation API library assembly. Otherwise, a run-time exception of the *BadImageFormatException* type is thrown at the first call to the API.

Usually, run-time exceptions are caused by defects in application logic, misunderstandings of the behaviors of the underlying API, or failures to follow expected programming patterns. For example, in any UC application, an endpoint connection must be established to Office Communications Server before it can function. Calling properties or methods on an endpoint that is not enabled throws an exception.

Run-time exceptions can also be traced to unexpected network disruptions caused by the server being unavailable or heavy network traffic. All of this can make debugging run-time errors a more challenging task. In addition, the asynchronous nature of the UC applications adds more complexity to the application interaction.

Error Codes and Exception Classes

In a native Win32/COM API such as Office Communicator Automation API and Unified Communications Client API, errors and failures are described by error codes. Some of these APIs may provide a descriptive message in addition to the error code. In Component Object Model (COM), an error code is represented by an *HRESULT* value, which is a 32-bit integer defined by a COM component or API. Each number identifies a specific error or warning and is associated with a named constant that communicates more user-readable information. As an example, Table 10-1 shows some *HRESULT* values with the corresponding named constants that are defined in Office Communicator Automation API.

TABLE 10-1 Examples of Error Codes Defined in Office Communicator Automation API

HRESULT	Named Constant	Error Condition
0x80004005	MSGR_E_FAIL	Unspecified failure
0x81000301	MSGR_E_CONNECT	Failed to connect
0x81000302	MSGR_E_INVALID_SERVER_NAME	Failure due to invalid server name
0x81000303	MSGR_E_INVALID_PASSWORD	Failure due to invalid password

For a complete listing of Office Communicator Automation API error codes and the corresponding named constants, see the "Microsoft Office Communicator Automation API Error Codes" documentation at *http://go.microsoft.com/fwlink/?linkid=143321*.

When a Win32 or COM API is called from a .NET Framework application, the COM error is wrapped in a *COMException* class by the COM Interop service. The *ErrorCode* property of the *COMException* class represents the original *HRESULT* value. The *Message* property holds the error description, if any, that is supplied by the API. For more information about the *COMException* class, see the section titled "Additional Resources" later in this chapter.

The Office Communicator Automation API is built on top of Microsoft Office Communicator, which in turn is built on top of the Unified Communications Client API (UCC API). When troubleshooting an Office Communicator Automation API application, see the "Unified Communications Client API Error Codes" documentation at *http://go.microsoft.com/fwlink/ ?linkid=143308.aspx* and the "Microsoft Office Communicator Automation API Error Codes" documentation at *http://go.microsoft.com/fwlink/?linkid=143321* for an *HRESULT* value.

UC also supports managed APIs, such as Unified Communications Managed API (UCMA). UCMA consists of three parts: the UCMA Core, the UCMA Workflow, and UCMA Speech. For these APIs, errors and failures are described by .NET Framework exception classes. A managed application can use *a try/catch* code block to handle the run-time exceptions by catching and handling one or more exceptions thrown by the managed API and the underlying framework components. Exception classes provide more debugging information than what is returned through the *HRESULT* type of error code. In addition to the error message, an exception class provides a call stack that leads to the origin of the exception. The call stack is exposed as the *StackTrace* property on the exception class.

In the .NET Framework, all exception classes are derived from the *System.Exception* type. In the UCMA, the object model of the exception classes starts from the *RealTimeException* class. This is the base exception class for other exceptions specific to this API. More specific exception classes cover operation-specific error and failure conditions. For example, a *ConnectionFailureException* instance indicates that an application did not connect to the server. The *RegisterException* class encapsulates an error that occurred in an attempt to register or unregister a user. For a complete list of the exception models, see the "Unified Communications Managed API 2.0 Core SDK Documentation" at *http://go.microsoft .com/fwlink/?linkid=126312*.

Session Initiation Protocol Error Codes

Another source of run-time errors reflects the exceptions that are encountered in network operations. Such exceptions are recorded and reported, as part of the signaling, by using the error codes, also known as the response codes, that are defined by the Session Initiation Protocol (SIP). For example, an attempt to invite an offline contact to a conversation results in a failure in the SIP dialogue. The following is a summary of the different SIP response code levels:

- **1xx** These provisional or informational responses indicate that a request has been received and is being processed. For more information, see *http://www.voip-info .org/wiki/view/SIP+response+class1*.

- **2xx** These successful responses indicate that a request is received, accepted, and processed successfully. For more information, see *http://www.voip-info.org/wiki/view/SIP+Response+class2*.

- **3xx** This class of responses indicates that a request is redirected because further action is needed to complete the request. For more information, see *http://www.voip-info.org/wiki/view/SIP+Response+class3*.

- **4xx** This class of responses indicates client errors that the request contains syntactic errors or cannot be processed by the server. For more information, see *http://www.voip-info.org/wiki/view/SIP+Response+class4*.

- **5xx** This class of responses indicates server errors that a server is unable to process an apparently valid request. For more information, see *http://www.voip-info.org/wiki/view/SIP+Response+class5*.

- **6xx** This class of responses indicates global failures that no server is able to process the request. For more information, see *http://www.voip-info.org/wiki/view/SIP+Response+class6*.

For more information about the SIP error codes, see RFC 3261, "SIP: Session Initiation Protocol," at *http://go.microsoft.com/fwlink/?LinkID=143940* and "[MS-SIP]: Session Initiation Protocol Extensions" at *http://go.microsoft.com/fwlink/?linkid=143292*.

Tracing

In addition to error codes and exceptions, the UC platform supports the ability to log the execution state of an application. The records may contain errors, warnings, events, or general information. They can be logged in one or more text files that can be examined by using a text editor or the Office Communications Server 2007 protocol analysis tool (Snooper.exe). Snooper is a tool that enables you to examine log files more conveniently. You can also enable Event Tracing for Windows (ETW) for a UC application. ETW tracing is logged into one or more .etl files. These files can then be used by Microsoft Technical Support for troubleshooting.

Office Communicator can be used to enable tracing that is logged to text files or .etl files. You can use the Office Communications Server Logging Tool (OCSLogger.exe) to enable tracing and to analyze the logs for applications that use the UCMA Core and the UCMA Workflow. For more information about how to enable tracing for various UC applications, see the sections titled "Debugging Office Communicator Automation API Applications" and "Debugging UCMA Core Applications" later in this chapter.

While error codes and exceptions can help you to determine errors on the application call stack, tracing is especially useful for uncovering failures that originate from the protocol stack. For example, an asynchronous application may not execute as expected if it mishandles an event or misses handling it altogether. In this case, the application may not break; instead, it may appear nonresponsive because the protocol requires that the event be handled properly. Examination of trace logs can help uncover such problems even when no error is reported or no exception is thrown.

Debugging Tools for UC Applications

The following sections focus on three tools that you can use to debug or troubleshoot UC applications:

- **Visual Studio Debugger**   This is part of the Microsoft Visual Studio development environment and can be used to trace the call stack of an application. You can use this tool to follow the execution of an application step by step from the beginning of the application or from a preset breakpoint. The execution can also jump to the next breakpoint, followed by more step-by-step examination. This procedure continues until the application exits normally, an error is reported, or an exception is thrown.

- **Snooper**   This tool (Snooper.exe) can be installed from the UCMA Software Development Kit (SDK) or the Office Communications Server Resource Kit. It can be used to parse and view tracing log files. This is especially useful when there is a need to follow the SIP protocol stack. The information displayed in Snooper can help you understand the underlying network topology and error or failure conditions that are detected by the server or caused by nonfunctioning remote partners.

- **OCSLogger**   The Office Communications Server Logging Tool (OCSLogger.exe) is available from the UCMA SDK. This tool is required to enable tracing for a UCMA application and to analyze the resulting logs. For more information about how to use this tool, see the section titled "Debugging UCMA Core Applications" later in this chapter.

Using Visual Studio Debugger

Visual Studio Debugger is a powerful debugging tool that can be used to debug any UC application. This tutorial is provided for those who are new to Visual Studio. If you are an experienced Visual Studio user, you may want to skip it.

Using Visual Studio Debugger to debug an application involves performing the following steps:

1. Set a breakpoint.

   To set a breakpoint, select a code statement for which you want to set the breakpoint, and then press the F9 key on the keyboard. (Note that the F9 key is a toggle switch; repeatedly pressing it turns a breakpoint on or off for the currently selected statement, as shown by the highlighted code statement in the following screenshot).

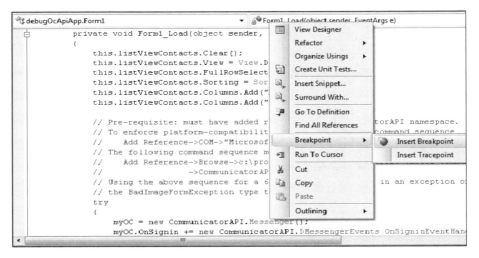

Normal execution of the application stops at the beginning of the statement that is marked by a breakpoint. Alternatively, you can set a breakpoint by right-clicking the target statement, clicking Breakpoint, and then clicking Insert Breakpoint, as shown here.

2. Run the application to a breakpoint.

To run the application to the next breakpoint by using Visual Studio Debugger, press the F5 key on the keyboard. Alternatively, you can click Debug on the Visual Studio toolbar, and then click Start Debugging, as shown here.

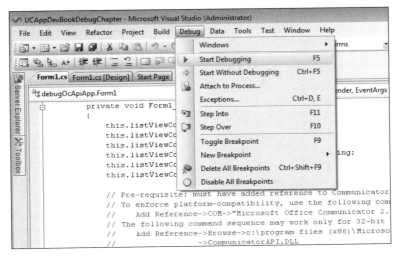

3. **Step through the code execution.**

   From any breakpoint, you can step through application execution one statement at a time by pressing either the F10 or F11 key. When the statement involves a call to a method or property that is defined in the running application, pressing the F10 key causes the application to step over the member call, and pressing the F11 key causes the application to step into the member call. You can also complete these actions by using the Debug menu, as shown in step 2.

4. **Examine execution states at each step.**

   As you step through the code, the execution stops and the execution context, including variables and constants, can be examined. Rest the mouse pointer on a variable of interest to open a tooltip that displays the value of the variable at the current execution step, as shown here.

   In this example, the tooltip shows the value of the variable *collabPlatform*, which is an instance of the *CollaborationPlatform* type of the UCMA Core. Expanding the tooltip window reveals the member values of the platform object.

   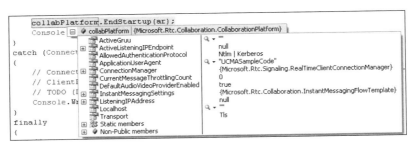

More Info Visual Studio Debugger provides other ways for you to examine the execution states as well. For more information, see Visual Studio Help.

Using Snooper

Snooper is used to parse and display logs. Logging can be enabled by using Office Communicator or OCSLogger. Figure 10-1 displays the output produced by Snooper as it traces a SIP dialogue in a failed invitation from one user to another. Four entries in the log file are relevant here, which are shown at the bottom of the display. They record the sequence of a call that one user sent to another at the times between 12:01:12.601 and 12:01:12:901.

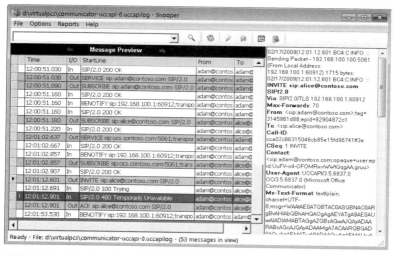

FIGURE 10-1 Display of a log using Snooper.

In the example shown in Figure 10-1, at 12 hours, 1 minute, and 12.602 seconds, a *SIP INVITE* is sent on behalf of a user (sip:adam@contoso.com) to another user (sip:alice@contoso.com). The SIP message detail is displayed in the right pane when the *INVITE* entry is selected in the Message Preview pane.

The next record entry displays a SIP response of *SIP/2.0 100 Trying* from the server. It is a provisional response indicating that the server has received the *INVITE* request and that the request is being processed. The response details are displayed in the right pane when the entry is selected. The *Call-ID* value in the details pane confirms that this response is part of the SIP dialogue initiated in the previous entry.

Continuing to the next entry, Snooper shows another response of *SIP/2.0 480 Temporarily Unavailable* from the server that arrived at 12 hours, 1 minute, and 12.901 seconds. The

480 response indicates a client error. This response was returned because the invited user was offline when the request was made. The entry is colored red because it indicates a failure condition of the *INVITE* request.

The last entry related to this dialogue was recorded at the same time (12:01:12.901) as when the previous 480 response was received. It shows that the inviting client closed the dialogue by sending an acknowledgement to the server.

Best Practice for Debugging or Troubleshooting a UC Application

Most robust UC applications offer multiple communication capabilities, including instant messaging, audio/video calling, conferencing, and application sharing. Some UC applications also take advantage of Enhanced Presence information to integrate the communications features with the user's availability, capabilities, and willingness. The asynchronous nature of the UC APIs can add even more complexities to developing such applications.

When you debug or troubleshoot a complex application, a prudent approach is to isolate problems in separate areas and then troubleshoot or debug each area. For a UC application, one way to divide the problem space is by feature or operation, as shown in the following groupings:

- **Initialization** In this phase, possible error sources include incorrect application configuration or a mismatched assembly format of a referenced or dependent library. Errors and failures are shown by exceptions and the premature termination of the application.

- **Endpoint sign-in** Possible sources of errors in this phase include invalid user account information that might include user credentials and signaling settings. Other possible failures may be caused by an unresponsive server and server policy.

- **Communication sessions** During instant messaging, audio/video calling, conferencing, and data sharing, possible failure sources include insufficient or unsupported media capabilities, blocked media traversal across firewalls, or unrecognized application protocol.

- **Presence publication, subscription, and query** Possible error or failure sources include invalid presence schemas; loss of interoperability between applications, for example, name collisions or semantic conflicts with existing presence schemas; and use of incorrect programming patterns.

- **Integration with other applications** Possible failure sources include the mishandling of messages or application protocols.

Each group can be divided further into subgroups if necessary. Some examples of applying this problem-solving strategy are explained later in this chapter.

Debugging Office Communicator Automation API Applications

Debugging Office Communicator Automation API applications also involves enabling tracing and examining logs. This section explains how to perform each task and then shows you some examples of debugging or troubleshooting an Office Communications Automation API application.

Enabling Tracing

An Office Communicator Automation API application can use Office Communicator to turn tracing on or off. Tracing can be logged to the Windows Event Log or a log file that uses plain text. To enable tracing to log files, the Windows Event Log, or both, perform these steps in Office Communicator:

1. In Office Communicator, click the presence-setting button, and then click Options, as shown here.

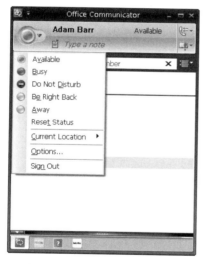

2. Under Logging, select Turn On Logging In Communicator to log tracing to log files that are generated by Office Communicator. Select the Turn On Windows Event Logging For Communicator to log tracing to the Windows Event Log, as shown here. Click OK.

Chapter 10 Debugging a Unified Communications Application

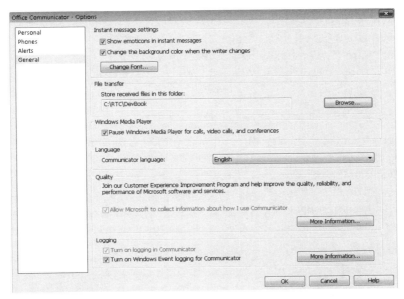

3. An optional step is to specify the location of the log files. To specify the location of the log files, open Registry Editor. Under HKEY_CURRENT_USER\Software\Microsoft\Tracing\uccapi\Communicator, edit the *FileDirectory* key, as shown here.

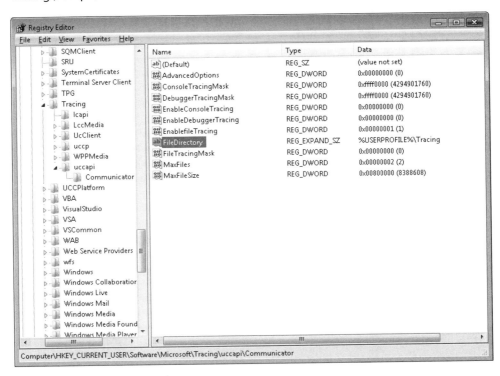

Examining the Windows Event Log

When the Windows Event Log is enabled for logging on Office Communicator, any failure that originates from an operation by Office Communicator is logged and can be viewed in the application logs by using the Windows Event Viewer. To view these logs, perform the following steps:

1. Click Start, right-click Computer, and then click Manage.

2. In Server Manager (on a server) or Computer Management (on a client) in the left pane, expand Event Viewer, expand Windows Logs, and then click Application to display a list of events in the center pane. When you select an event in the Event Viewer, as shown here, information about the selected event is displayed in the pane underneath. The *Warning* event from Office Communicator indicates that a SIP request made by Office Communicator failed because the invited party is not available to answer the call.

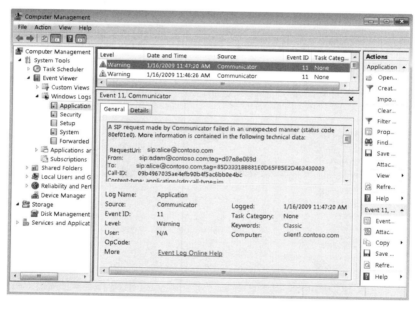

Examining Log Files

When file tracing is enabled, Office Communicator creates a detailed log to one or more log files. The maximum number of files is set by the MaxFiles registry key for Office Communicator. By default, this number is set to 2. The name of the first log file is communicator-uccapi0.uccapilog. When the log file reaches its maximum size as set in the *MaxFileSize* registry key value, Office Communicator continues logging to the next log file and increments the file name by 1. If MaxFiles is set to 2, the name of the next file becomes communicator-uccapi1.uccapilog. After the last file is full, Office Communicator returns to the first log file and overwrites the existing information in that log file.

The Office Communicator–generated log files are text-based files that contain the following four types of information:

- Run-time exceptions (ERROR)
- SIP traffic and information about the application's execution context (INFO)
- Application-generated tracing (TRACE) for Office Communicator and the Office Communicator Automation API
- Cautionary information (WARN)

Figure 10-2 shows an excerpt of a log file displayed in Notepad. For more information about the format and meaning of the different types of log entries, see the Office Communications Server 2007 R2 Resource Kit.

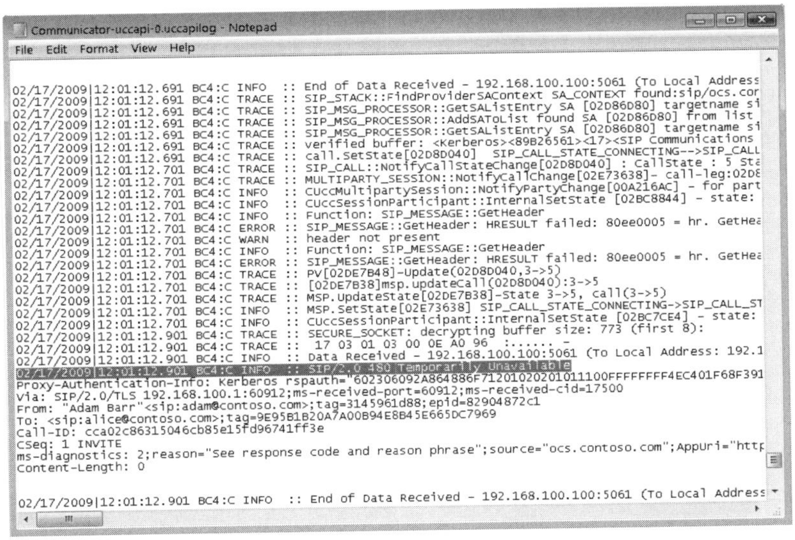

FIGURE 10-2 Plain text log file generated by Office Communicator.

When you compare Figure 10-2 and a corresponding Windows Event Viewer display, you see that a log file contains more information than what is displayed in Snooper. However, examining log files by using a text editor can be cumbersome and overwhelming when the file is large. It is easier to review the log files by using Snooper. For more information, see the section titled "Using Snooper" earlier in this chapter.

Handling Exceptions Using *HRESULT* Error Codes

In Office Communicator Automation API applications, run-time errors that originate from the API are COM errors. They can be detected in Office Communicator, the Automation API, the underlying UCC API, and the COM library. These errors are reported as *HRESULT* values.

The way they are handled in an Office Communicator Automation API application depends on how the application is written. If the application is of a native Win32/COM type, every call to an interface member or COM library function returns an *HRESULT* value. The application must check the return value of every method call for possible errors. Only when the *HRESULT* value does not indicate any error or failure should the application go to the next call. Listing 10-1 is a C++ code example that illustrates this point.

LISTING 10-1 C++ Code Example Illustrating the Use of the *HRESULT* Value to Gauge the Call Status of C++ Statements

```cpp
// Initialize the COM library
HRESULT hr = CoInitializeEx(NULL, COINIT_APARTMENTTHREADED);
if (FAILED(hr))
    return hr;

// Create Messenger co-class and retrieve IMessenger interface pointer
// and connect to the running Communicator instance.
    hr = CoCreateInstance(CLSID_Messenger,NULL,CLSCTX_LOCAL_SERVER,
            (IID_IMessenger), (LPVOID *)&m_pIMessenger);

if (FAILED(hr))
    return hr;

// The IMessenger interface is now ready for use. Gets the caller's display name
BSTR myName;
hr = m_pIMessenger->get_MyFriendlyName(&myName);
if (FAILED(hr))
     return hr;
CoUninitialize();
```

Listing 10-1 performs the task of creating a *Messenger* co-class instance to retrieve the *IMessenger* interface pointer before it obtains the display name of the caller. Notice that every statement is followed by a check of the returned *HRESULT* value. An *HRESULT* value of *0* (zero) indicates that the operation succeeded. Any nonzero *HRESULT* value suggests an error or warning, and, in this case, the *FAILED* macro returns *True*.

If the *HRESULT* value indicates a failure, the example returns the error code to the caller. In production code, the *HRESULT* value can be checked against a known error code and the situation handled accordingly. An Office Communicator Automation API application may have to check against the *HRESULT* values that are defined in the Automation API and in the underlying UCC API. For more information about error codes, see "Unified Communications Client API Error Codes" at *http://go.microsoft.com/fwlink/?linkid=143308*.

In a .NET Framework application that uses the COM API, COM errors are encapsulated by the *COMException* class that is defined in the *System.Runtime.InteropServices* namespace. The *HRESULT* value in the original COM error is represented by the *ErrorCode* property of the *COMException* type, and the COM error message is represented by the *COMException. Message* parameter. The call stack is exposed in the *COMException.StackTrace* property.

You can catch the COM exception by using a *try/catch* block, as shown in the following C# code example.

```csharp
CommunicatorAPI.Messenger myOC;
try
{
        myOC = new CommunicatorAPI.Messenger();
        myOC.OnSignin += new CommunicatorAPI.DMessengerEvents_OnSigninEventHandler
(myOC_OnSignin);
        myOC.OnSignout += new CommunicatorAPI.DMessengerEvents_OnSignoutEventHandler
(myOC_OnSignout);
        myOC.AutoSignIn();
}
catch (COMException cex)
{
        // Exception handling
        int hr = cex.ErrorCode;
        string msg = cex.Message;
}
```

Troubleshooting Office Communicator Automation API Applications

To understand the practical aspects of debugging and troubleshooting a UC application that is built by using the Office Communicator Automation API, the following case studies are given. Each case study in this section explores the cause of detected bugs.

Troubleshooting Automation API Assembly Initialization Failures

The first step in building an Office Communicator Automation API application is to create an instance of the *CommunicatorAPI.Messenger* class. Then you must register for notifications of events to be raised by the newly instantiated *Messenger* object. In a .NET Framework application, the referenced Automation API library is loaded into the application process just in time before the constructor is called on the *Messenger* class. Depending on how the assembly reference is added to the project, an exception of the *BadImageFormatException* type may be raised when an attempt is made to create an instance of the *Messenger* class. The C# code example in Listing 10-2 illustrates this process.

LISTING 10-2 C# Code Example in Which a *BadImageFormatException* Type Will Be Raised if the Referenced Office Communicator Automation API Assembly Is Targeted for a Platform Different from the Target Platform of the Application

```csharp
public partial class Form1 : Form
{
    CommunicatorAPI.Messenger myOC;
    string userName;
    public Form1()
    {
        InitializeComponent();
```

```
            myOC = new CommunicatorAPI.Messenger();
            myOC.OnSignin +=
                    new CommunicatorAPI.DMessengerEvents_OnSigninEventHandler
(myOC_OnSignin);
            myOC.OnSignout +=
                    new CommunicatorAPI.DMessengerEvents_OnSignoutEventHandler
(myOC_OnSignout);
        }

        void myOC_OnSignout()
        {
        }

        void myOC_OnSignin(int hr)
        {
        }
    }
}
```

Even though the code in Listing 10-2 should compile without errors, you might encounter a run-time exception of the *BadImageFormatException* type when you execute the code. This is illustrated in Figure 10-3, which shows the Visual Studio Debugger.

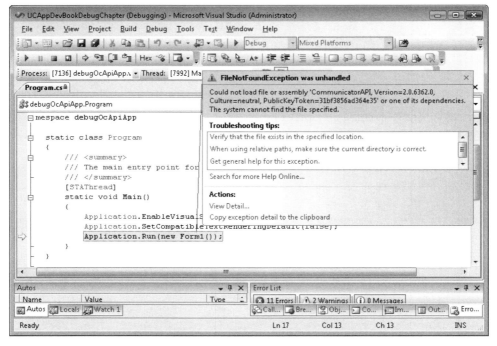

FIGURE 10-3 Unhandled *BadImageFormatException* thrown at run time.

The message associated with this run-time exception indicates that the required *CommunicatorAPI.DLL* assembly cannot be loaded correctly. This message helps point out that in the *Form1* class constructor, an attempt is made to create an instance of the *Messenger* class, which involves loading the assembly. The exception shows up here because the error was not handled in the code for the *Form1* class constructor.

The cause of this exception is most likely that a 32-bit primary interop assembly (PIA) was used when adding a reference for a .NET Framework application that was targeted for the 64-bit platform. The Office Communicator Automation API SDK contains the 32-bit version of the *CommunicatorAPI.DLL* assembly as a PIA for the API. By default, it is installed in the \Program Files (x86)\Microsoft Office Communicator\SDK directory (in a 64-bit operating system). You can use this assembly only when your application is targeted for the 32-bit (x86) platform. If you choose this assembly to add references to a 64-bit project, you receive the run-time exception.

The run-time exception situation in this code example can become trickier if you delay calling the *Messenger* class constructor until after the *Form1* class constructor. For example, if you create an instance of the *Messenger* class while the form is being loaded (*Form1_Load*), you might not be notified of the exception when the application stops functioning correctly.

To ensure the platform-appropriate *CommunicatorAPI* assembly is referenced in an Automation API application, use the *Communicator.EXE* assembly to add a reference to an application's project. This causes Visual Studio to create the platform-appropriate PIA (*Interop.CommunicatorAPI.DLL*). To do so, perform these steps in Visual Studio:

1. In Visual Studio Solution Explorer, right-click your application project, and then click Add Reference.

2. On the COM tab, scroll down and select Microsoft Office Communicator 2007 API Type Library, as shown here. You also have the option to select Microsoft Office Communicator 2007 API Private Type Library, if you plan to call the interfaces exposed in this assembly. Then click OK.

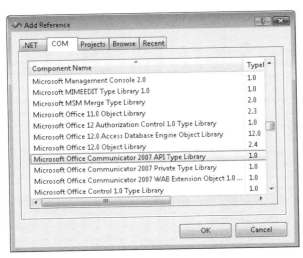

Troubleshooting COM Interop Errors

As with any COM Automation API, the *VARIANT* type is used extensively in the Office Communicator Automation API. There are many advantages to using *VARIANT* in COM. For example, the use of *VARIANT* as an input parameter type gives you the flexibility to take on different types of input values, such as a string, integer, object, or array. Through COM Interop, the *VARIANT* type in COM is translated into an object type in a .NET Framework application. However, the usual typecasting in the .NET Framework cannot be applied to such a translated object type.

To better understand how this works, consider a situation in which you want to start an instant messaging call. To do this, you have to add code that calls the *IMessengerAdvanced.StartConversation* method and starts an instant messaging session that has a list of selected contacts. The signature of this method, as specified in the Interface Definition Language file (ocapi.idl), is documented in the Office Communicator Automation API 2007 documentation. The second input parameter is of the *VARIANT* type and can take on an array of *BSTR* strings or an array of *IMessengerContact* objects. Because of COM Interop, this parameter becomes an object type in a .NET Framework application. The following C# code example calls the method by passing in an array of *IMessengerContact* instances as an array of the .NET Framework object type (*List<object>.ToArray*). This implementation runs without exceptions.

```
IMessenger myOC =...; // Assuming that this Messenger object has been created and enabled.
IMessengerAdvanced msgAdv = myOC as IMessengerAdvanced;

void StartImCall(IMessengerContact[] contacts)
{
    List<object> list = new List<object>();
    foreach (IMessengerContact c in contacts)
        list.Add(c);
    objct contactsAsObj = list.ToArray();  // casting object[] into object.
    object obj = msgAdv.StartConversation(
                    CONVERSATION_TYPE.CONVERSATION_TYPE_IM,
                    contactsAsObj,
                    null,
                    null,
                    null,
                    null);
}
```

However, if an array of *IMessengerContact* instances is passed in instead, as shown in the following C# code example, the implementation throws an exception that has the *HRESULT* value *0x80010105 (RPC_E_SERVERFAULT)*.

```
void StartImCall(IMessengerContact[] contacts)
{
    // the following statement throws the exception:
    //          HRESULT: 0x80010105 (RPC_E_SERVERFAULT)
    object obj = msgAdv.StartConversation(
                    CONVERSATION_TYPE.CONVERSATION_TYPE_IM,
```

```
        contacts as object, // This will cause run-time error
        null,
        "Hello",
        null,
        null);
}
```

Troubleshooting Common Operational Failures

In any UC application, operational failures can occur when, for example, the server is unresponsive, a client does not have the required media capabilities, a message has an incorrect format, the server refuses a request because of a violation of existing policy, or the client does not catch an event notification from the server. In these situations, the Visual Studio Debugger may be insufficient. However, tracing and examining log files provide more useful information about the source of failures. The following examples show the kind of debugging information that is available in the logs.

Common sign-in failures   The following excerpt from a log file generated by Office Communicator shows that a sign-in request failed because the strict Domain Name System (DNS) name checking failed for _sipinternaltls._tcp.contosohost.com.

```
03/20/2008|14:37:34.018 10DC:163C WARN :: GetDnsResults - failure in do the DNS query.

03/20/2008|14:37:34.018 10DC:163C ERROR :: QueryDNSSrv GetDnsResults query:
_sipinternaltls._tcp.contosohost.com failed 1a9ee37

03/20/2008|14:37:34.018 10DC:163C WARN :: DNS_RESOLUTION_WORKITEM::ProcessWorkItem query DNS
SRV records failed, Error 8007232b

03/20/2008|14:37:34.018 10DC:163C ERROR :: DNS_RESOLUTION_WORKITEM::ProcessWorkItem
ResolveHostName failed 8007232b

03/20/2008|14:37:34.019 10DC:1CA4 TRACE :: ASYNC_WORKITEM::OnWorkItemComplete(15D30498)
Enter
```

Sign-in operations can fail because the server is unavailable   A sign-in request can fail because the client does not have the correct certificate. In the following log file excerpt, *connection failed error 80072746* and *failed to send security negotiation token* ERROR entries (shown in bold) may be logged if tracing is enabled on Office Communicator.

```
16:56:01.797 1F10:360 INFO :: Outgoing 0231BC60-<sip:john@uc.contoso.com>,
local=(null)

16:56:01.969 1F10:360 TRACE :: Async work item posted for TLS negotiation: this 00C70768

16:56:01.969 1F10:360 ERROR :: ASYNC_SOCKET::SendHelperFn - send failed 0x2746 m_Buffer:
0x000AEA08, m_BytesSent: 0, m_BufLen: 70

16:56:01.969 1F10:360 ERROR :: ASYNC_SOCKET::CreateSendBufferAndSend
SendOrQueueIfSendIsBlocking failed 80072746
```

```
16:56:01.969 1F10:360 ERROR :: SECURE_SOCKET: failed to send security negotiation token

16:56:01.969 1F10:360 ERROR :: OUTGOING_TRANSACTION::OnRequestSocketConnectComplete
- connection failed error 80072746
```

Another case of sign-in failure occurs when the user is not SIP-enabled or the user is in the middle of being moved from one Office Communications Server pool to another. In this case, the log file is displayed as follows.

```
20:44:25.858 3E0:BEC INFO :: SIP/2.0 480 Target is either disabled or moving away

ms-diagnostics: errorId="2";reason="Unknown Failure";source="ocs.contoso.com";AppUri=http://
www.contoso.com/LCS/UserServices

20:44:25.858 3E0:BEC TRACE :: CUccServerEndpoint::UpdateEndpointState - Update state from
1 to 0. Status 80EF01E0. Status text Target is either disabled or moving away.
```

Common presence failures A client may be unable to publish presence information on behalf of a user because the to-be-published category instance is out of date, a custom category name is not registered with Office Communications Server, or the user's home server is being switched from one pool to another.

A to-be-published category instance is out of date if its version number is smaller than the one known to the server. In this case, the client receives a SIP 409 Conflict response, as shown in the following log entry, in which the ms-diagnositics header describes the reason for the failure as "Publication version out of date".

```
SIP/2.0 409 Conflict
FROM: "Adam Barr"<sip:adam@contoso.com>;epid=31042B029D;tag=d96d11467
TO: <sip:adam@contoso.com>;epid=31042B029D;tag=85D3331BB881E0D65FB5E2D463430003
CSEQ: 25 SERVICE
CALL-ID: b546a613be504ab69bdc4c624da1f4e7
VIA: SIP/2.0/TLS 192.168.0.112:58318;branch=z9hG4bK1830c3d1;received=66.235.34.227;
ms-received-port=58318;ms-received-cid=D34A900
CONTENT-LENGTH: 639
CONTENT-TYPE: application/msrtc-fault+xml
AUTHENTICATION-INFO: NTLM rspauth="0100000000000000009522756A0BAEF93B", srand="331B3862",
snum="71", opaque="CA63E6D4", qop="auth", targetname="ur.contoso.com", realm="SIP
Communications Service"ms-user-logon-data: RemoteUser
ms-diagnostics: 2044;reason="Publication version out of date";source="ocssrv.contoso.com"
```

When a client attempts to publish a custom presence that has not been registered with Office Communications Server, the publication fails. In this case, the client receives a 400 response that indicates an invalid request. This is illustrated in the following log entry, in which the ms-diagnostics header records the reason for the failure as "XML parse failure".

```
SIP/2.0 400 Bad request
FROM: "Adam Barr"<sip:adam@contoso.com>;epid=31042B029D;tag=4820e69f11
TO: <sip:adam@contoso.com>;epid=31042B029D;tag=85D3331BB881E0D65FB5E2D463430003
CSEQ: 24 SERVICE
```

```
CALL-ID: 954e54edbd3b4bc0b9f4bc7d6164d3b0
VIA: SIP/2.0/TLS 192.168.0.112:58318;branch=z9hG4bK71daaba6;received=66.235.34.227;
ms-received-port=58318;ms-received-cid=D34A900
CONTENT-LENGTH: 58
CONTENT-TYPE: application/msrtc-fault+xml
AUTHENTICATION-INFO: NTLM rspauth="0100000043453E3C0BFA9E530BAEF93B", srand="CA8F86D0",
snum="69", opaque="CA63E6D4", qop="auth", targetname="ur.contoso.com", realm="SIP
Communications Service"
ms-user-logon-data: RemoteUser
ms-diagnostics: 4001;reason="XML parse failure";source="ocssrv.contoso.com"
```

When the user's home server is being switched from one pool to another, the presence is treated as unknown and the client receives a 504 response from the server.

Common audio and phone calling failures Audio and phone calls may fail because clients have mismatched media capabilities. This can be corroborated by looking at the ms-client-diagnostics header that is generated by Office Communicator. The following log entry shows that a Voice over Internet Protocol (VoIP) call invitation is refused because the remote endpoint does not support audio.

```
SIP/2.0 488 Not Acceptable Here

From: <sip:john@clientee.contoso.com>;tag=dcf94;epid=a0894

To: "" <sip:jane@clientee.contoso.com>;epid=0a33;tag=0cd

Call-ID: 032acec238cd42eaa726b6077a3ce550

ms-client-diagnostics: 52036; reason="Audio is not configured"
```

When placing a call to a remote endpoint, the calling client may receive an exception that indicates that the remote user cannot be found, as shown in the following log file excerpt. This can occur because the called party is not SIP-enabled or the federation link is missing for the user from a federated network.

```
SIP/2.0 404 Not Acceptable Here

From: <sip:john@clientee.contoso.com>;tag=dcf94;epid=a0894

To: "" <sip:jane@clientee.contoso.com>;epid=0a33;tag=0cd

Call-ID: 032acec238cd42eaa726b6077a3ce550

ms-diagnostics: 1003;reason="User does not exist"
```

Debugging UCMA Core Applications

Debugging an application that is built on top of the UCMA Core involves enabling tracing. This results in the logging of application failures and error conditions. To handle run-time exceptions successfully, you should use *try/catch* blocks around your UCMA code. The following sections discuss how to enable tracing and handle exceptions for UCMA, before presenting some debugging examples to illustrate how to catch and handle run-time exceptions within the framework of the UCMA exception model.

Enabling Tracing

To debug protocol-level failures for a UCMA application, you must enable protocol tracing before you run the application. You can enable protocol tracing by using the Office Communications Server 2007 R2 Logging Tool. This tool traces a wide range of Office Communications Server components over various protocol stacks. The supported protocols include SIP, S4, UserServices, and other protocols.

When you install the UCMA SDK, it adds S4 and Collaboration components as the default logging options. You can select other supported components by using the advanced options in OCSLogger.

There are two versions of this tool available: OCSTracer.exe is the command-line version and OCSLogger.exe provides a graphical user interface. Both are installed as part of the UCMA 2.0 SDK.

By default, OCSTracer.exe is installed in the %Programfiles%\Microsoft Office Communications Server 2007 R2\Common\Tracing directory and OCSLogger.exe is installed in the %Programfiles%\Microsoft Office Communications Server 2007 R2\UCMA SDK v2.0 R2\UCMACore\Tracing directory. The remainder focuses on how to use OCSLogger. For more information about OCSTracer, see the "Unified Communications Managed API 2.0 Core SDK Documentation" at *http://go.microsoft.com/fwlink/?LinkID=133571*.

Enabling Tracing for UCMA Core Applications

1. Before you run a UCMA application, open Windows Explorer, right-click OCSLogger.exe, and then click Run As Administrator to start the Office Communications Server 2007 R2 Logging Tool. Note that the logging tool does not run unless it is started with elevated permissions.

2. In the Office Communications Server 2007 R2 Logging Tool, under Logging Options, in the Components section verify that Collaboration and S4 are selected, as shown here. In the Level section, select the level of tracing that you want. A tracing level specifies a type of tracing to be performed. The level is cumulative, which means that a selected level includes the selected message type plus all of the message types that are listed above

the selected message type. In the Flags section, select the trace flags that you want. You can also change the other options if the default settings do not meet your needs.

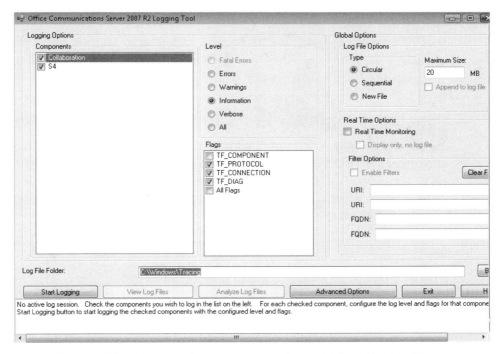

A trace flag specifies the type of messages to be logged. The following flags are supported:

- **TF_COMPONENT**   Logs miscellaneous messages that are not covered by the other trace flags.

- **TF_PROTOCOL**   Logs protocol messages or data, including SIP, Centralized Conferencing Control Protocol (CCCP), Session Description Protocol (SDP), or other text-based messages.

- **TF_CONNECTION**   Logs connection-related errors or warnings, including significant network-level information about components that do not support the concept of connection.

- **TF_DIAG**   Logs diagnostic events. For SIP, this includes certificate failures or errors and DNS warnings or errors.

Under Log File Options, in the Type section, Circular specifies that new records will be put at the beginning of the log file when the file size reaches the predefined maximum size. Sequential specifies that new records will be put at the end of the file. New File specifies that a new log file will be started when the file size reaches the maximum size. Filter Options lets you trace the messages to or from the specified Uniform Resource Identifier (URI) or the specified fully qualified domain name (FQDN).

3. Click Start Logging before you run your UCMA application.

4. Start your UCMA application.

5. After the application finishes running or being debugged, click Stop Logging, and then click Analyze Log Files to view the trace logs in Snooper, a log file parser.

   Alternatively, you can click View Log Files to view the raw data of the log file in Notepad.

Handling Exceptions Using the UCMA Core Exception Model

The UCMA Core defines a set of exception classes that can be used to debug a UCMA application. With the UCMA, all exceptions are derived from *RealTimeException*, which is the base exception class. More specific exception classes cover operation-specific error and failure conditions. Table 10-2 lists API-specific exception classes that cover the following types of failure or error conditions that may occur in a UCMA application. For the complete exception model, see "Unified Communications Managed API 2.0 Core SDK Documentation" at *http://go.microsoft.com/fwlink/?LinkID=126312*.

TABLE 10-2 Exception Classes Defined in the UCMA Core

UCMA Exception Class	Description
RealTimeException	The base class for all of the exceptions that are specific to UCMA. It is thrown when the error condition cannot be mapped to any other exception types.
ConferenceFailureException	Thrown when an error that is associated with conference scheduling or management occurs. Possible reasons for this exception include the following: ■ The conference could not be created, scheduled, retrieved, or removed. ■ The conference could not have its settings modified. ■ The check for whether a passcode is optional for a conference failed. ■ The conference passcode could not be verified. ■ A failure occurred in the attempt to get the available multipoint control unit (MCU) types for a conference.
OfferAnswerException	Thrown when one of the following errors occurs in a media provider: ■ The SDP offer could not be created. ■ The answer received from the remote side could not be accepted. ■ A collision in the offer/answer interchange occurred because a previous negotiation is still pending. ■ An answer for a received offer could not be sent.

TABLE 10-2 Exception Classes Defined in the UCMA Core

UCMA Exception Class	Description
ConnectionFailureException	Thrown when a network connection cannot be made. Applications might consider prompting for a new URI (host and name) or a server name.
FailureResponseException	Thrown when a 4xx, 5xx, or 6xx response is received for a request. This exception contains the *ResponseData* property. The *ResponsData* property contains the complete response. This includes response code, reason text, headers, and message body. In some rare cases, this exception may also be thrown when an error other than a 4xx or 5xx response occurs. In these cases, the *ResponseData* property is *NULL*.
AuthenticationException	Thrown when a 401 response is received in authentication. It exposes the failed response data (*ResponseData*) and parsed values of authentication-specific headers.
PublishSubscribeException	Thrown when a *FailureResponseException* occurs during requests for the *SERVICE* and *SUBSCRIBE* types. This class inherits from *FailureResponseException* and adds diagnostic information about the message body and the fault code that is parsed from the body.
RegisterException	Thrown by the *Register* or *Unregister* method when an error occurs in a registration-related operation.
ServerPolicyException	Thrown when an operation is rejected because of a server policy.
MessageParsingException	Thrown when an incoming response or message cannot be parsed. For example, an invalid custom signaling header was used.
MultipartContentException	Thrown when an error occurs during multipart body parsing.
OperationFailureException	Thrown when an operation fails. The *FailureReason* property indicates the cause of failure.
OperationTimeoutException	Thrown when an operation cannot be completed in a given reasonable time. For example, a response from another endpoint or server was not received after a request was sent.
TlsFailureException	Thrown when an error occurs during a Transport Layer Security (TLS) handshake. This error can occur when the remote server certificate is rejected. This can also occur when the Mutual TLS (MTLS) local certificate is rejected by the remote server.

At run time, instances of these exception classes contain valuable information that you can use for debugging an application's call stacks and the network protocol stacks. Detailed specification of these exception classes can be found in the "Unified Communications Managed API 2.0 Core SDK Documentation" at *http://go.microsoft.com/fwlink/?LinkID=133571*.

As with any .NET Framework application, these exceptions can be caught and handled by the *try/catch* blocks. The code fragment of interest is enclosed in a *try* block, while appropriate exceptions are handled in one or more *catch* blocks. Each *catch* block handles an exception of a specific type. The exception types handled in the *catch* blocks depend on the operations that are performed in the *try* block.

Errors may originate from a called member, or they may arise from lower-level APIs or other platform components. The API has little control over handling exceptions that originate from these other sources. For a UCMA application, this means that the developer cannot always anticipate what exception will be caught. For this reason, an error handler of the *RealTimeException* type can be used to catch any unhandled UCMA exceptions and to discover the exception type. Similarly, an error handler of the *System.Exception* type can be used to handle any .NET Framework exception and discover the exception type.

In Visual Studio, you can find documented exception types at design time by using the Visual Studio IntelliSense tooltip. In this situation, an exception is documented in the originating method only if the triple-slashed Extensible Markup Language (XML) comments preceding the member definition contain <exception> tags. Any exception that is not documented this way does not show up in the tooltip. Figure 10-4 provides an example of finding such documented exceptions. When the mouse pointer rests on the *EndStartup* member for the *CollaborationPlatform* instance, the documented exceptions, *InvalidOperationException* and *ConnectionFailureException*, thrown by this method are shown in the Intellisense tooltip that appears.

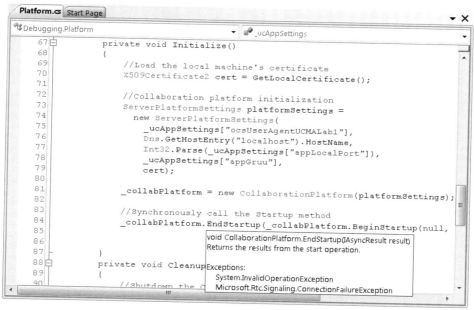

FIGURE 10-4 Finding documented exceptions for a given type member at design time.

Listing 10-3 is a C# code example in which a specific list of presence items is published. The *try* block encloses a synchronous call to publish presence information from a user's local endpoint. The failure or error conditions that are specific to such an operation are encapsulated by an exception of the *PublishSubscribeException* type. Therefore, the first *catch* block is used to handle this type of exception. To anticipate other miscellaneous exceptions that may occur, the error-handling code also includes a *catch* block for *RealTimeException*.

LISTING 10-3 C# Code Example That Illustrates the Types of Exceptions and Some Debugging Data That Can Be Caught in Presence Publication

```csharp
public void PublishPresence(List<PresenceCategoryiwthMetaData> categoryList)
{
    try
    {
        LocalOwnerPresence myPresence = _userEndpoint.LocalOwnerPresence;

        myPresence.EndPublishPresence(myPresence.BeginPublishPresence(categoryList, null, null));
    }
    catch (PublishSubscribeException psExcept)
    {
        // Examine user-defined information about the exception
        if (psExcept.Data != null)
        {
            foreach (object data in psExcept.Data)
            {

            }
        }

        if (psExcept.DiagnosticInformation != null)
        {
            // inspect ms-diagnostics or ms-diagnostics-public headers.
            string headerName = psExcept.DiagnosticInformation.HeaderName;
            string reason = psExcept.DiagnosticInformation.Reason;
            string source = psExcept.DiagnosticInformation.Source;
            int errrorCode = psExcept.DiagnosticInformation.ErrorCode;
            string value = psExcept.DiagnosticInformation.GetValue();
            string subErrorWarns = psExcept.DiagnosticInformation.SubErrorWarning;
        }

        if (psExcept.FaultCode != null || psExcept.FaultCode != string.Empty)
        {
            // Determine 4xx code returned from server.
            Console.WriteLine(psExcept.FaultCode);
        }

        if (psExcept.ResponseData != null)
        {
            SipResponseData responseData = psExcept.ResponseData;
            int cSeq = responseData.CSeq;
            string cType = responseData.ContentType.Name;
            StringDictionary cTypeParams = responseData.ContentType.Parameters;
            string userAgent = responseData.UserAgent;
            string requestUri = responseData.RequestUri;
```

```
                int responseCode = responseData.ResponseCode;
                string responseText = responseData.ResponseText;
            }
        }
        catch (RealTimeException rtExcept)
        {
            string msg = rtExcept.Message;
            string stackTrace = rtExcept.StackTrace;
        }
    }
}
```

When multiple *catch* blocks are used in a single *try/catch* sequence, the order in which the exceptions are handled should be such that the handling of a base exception type must follow that of its derived exception types. In the previous example, the *RealTimeException* type is the base class from which the *PublishSubscribeException* type is derived.

Valuable debugging information can be gleaned from a *PublishSubscribeException* instance. This information includes the following:

- User-defined data that is exposed by the *Data* property.
- Diagnostic information that is exposed by the *DiagnosticInformation* property. This property encapsulates the ms-diagnostics or ms-diagnostics-public header as reported by the server.
- SIP error code that is encapsulated in the *FaultCode* property.
- Data of a response, which is exposed by the *ResponseData* property, when a SIP request for presence publication or subscription is returned.

When an exception is thrown from a worker thread that is used by UCMA and the exception is not handled within that worker thread, the exception is not caught, even if the application has a *try/catch* block to handle the same type of exception. This applies even when the exception is thrown from the application's other threads, including the main thread or worker threads that are created by the application.

In UCMA, two likely sources of these exceptions are event handlers and callback functions that the API supports. An unhandled exception from such a worker thread causes the application to fail. In general, this result is expected because it ensures that the exceptions are investigated. An application should implement all of the required exception-handling routines in the event handlers and callback functions to catch these exceptions and to recover or report the errors. It is possible that some exceptions remain unhandled. However, to prevent applications from failing because of unhandled exceptions, you can use the *UnhandledExceptionManager* class that the UCMA API exposes.

To use the *UnhandledExceptionManager* class, you must create an instance of the *Unhandled-ThreadPoolExceptionHandler* delegate and register it with the *UnhandledExceptionManager* class. You can handle such exceptions in the delegate in two ways: ignore the exception to keep the application running, or allow the exception to make the application fail.

When an exception is raised in a worker thread and is not handled within the worker thread, the registered delegate is called. To ignore the exception, the delegate must return *True* so that the unhandled exception will not cause the application to fail. If the delegate returns *False*, the application fails.

The following C# code example shows how to register an *UnhandledThreadPoolException-Handler* delegate with the *UnhandledExceptionManager* class by setting the *VerifyAndIgnore UnhandledThreadPoolException* property.

```
// Configure UnhandledExceptionManager class to log uncaught exceptions
UnhandledExceptionManager.VerifyAndIgnoreUnhandledThreadPoolException =
        this.LogAndIgnoreUncaughtExceptions;
```

The delegate instance is the *LogAndIgnoreUncaughtExceptions* method of the calling class. The following C# code example shows one way to handle the exception: ignore the exception after you log it.

```
bool LogAndIgnoreUncaughtExceptions(Exception ex, WaitCallback method, Object state)
{
    System.Diagnostics.Trace.WriteLine(ex.ToString());
    return true;
}
```

Debugging UCMA Core Applications

The following sections look at some case studies in which some common bugs are detected and debugged.

Troubleshooting Application Initialization Failures

After a UCMA Core application is configured as a trusted application by Office Communications Server, you must start the UCMA Core application from an account that permits readable access to certificates and private keys or writable access to log files when using OCSLogger. Typically, an administrator account has these privileges and can be used to simplify testing the applications. However, in production mode, it is recommended that you create a separate account that has the required permissions to run the applications.

When applications are started from an account that does not have the required permissions, these applications fail during initialization, and the API throws an exception of the *TlsFailureException* type. Listing 10-4 shows a C# code example that attempts to start a UCMA application that uses the server platform and is configured as trusted by Office Communications Server.

LISTING 10-4 Example of C# Code That Creates a *CollaborationPlatform* Object Before Starting a UCMA Application

```
public void InitializeServerPlatform(X509Certificate2 cert,
                    string userAgent, string localhost, int port, string gruu)
{
        //Collaboration platform initialization
        ServerPlatformSettings platformSettings =
                new ServerPlatformSettings( userAgent, localhost, port, gruu, cert );

        _collabPlatform = new CollaborationPlatform(platformSettings);
}
```

When the code in Listing 10-4 is invoked without the proper permissions, an exception of the *TlsFailureException* type is thrown. Figure 10-5 shows this exception in Visual Studio Debugger.

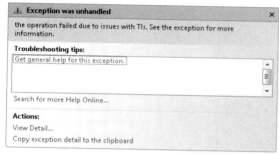

FIGURE 10-5 Visual Studio display of a *tls-failure* exception that is not handled by the application.

If you try to run or debug this code by using correct permissions, it executes without any exceptions. To open an application using the correct permissions in a test environment without creating a dedicated account, right-click the application executable file in Windows Explorer, and then select Run As Administrator.

If you use a console window to test the application, open the application by typing the command **runas/user:administrator myApp.exe** at a command prompt.

You are prompted to enter the correct administrator's password before the application starts. One way to remind the application user to run the application with the correct permissions is to catch and handle this exception by using a *try/catch* block. The following code example shows an example of this.

```
public void InitializeServerPlatform(X509Certificate2 cert,
                    string userAgent, string localhost, int port, string gruu)
{
        //Collaboration platform initialization
        ServerPlatformSettings platformSettings =
```

```
        new ServerPlatformSettings( userAgent, localhost, port, gruu, cert );
    try
    {
        _collabPlatform = new CollaborationPlatform(platformSettings);
    }
    catch (Microsoft.Rtc.Signaling.TlsFailureException tlsEx)
    {
        System.Windows.Forms.MessageBoxShow(
                "You must launch this application with elevated permissions.");
        System.Environement.Exit(-1);
    }
}
```

Troubleshooting Endpoint Connection Failures

A UCMA Core application must connect an endpoint to Office Communications Server before the endpoint can communicate with other endpoints. Such an operation is known as *signing in an endpoint* and must be performed after the endpoint is created and before any other operations are performed. Failures may occur if the server is not available or does not exist or if the endpoint cannot be authenticated or registered.

Programmatically, the operation of signing in an endpoint begins by calling the *BeginEstablish* method and ends after the *EndEstablish* method returns asynchronously. When failures are encountered, exceptions are thrown. Any error that occurs in the *BeginEstablish* method causes the *InvalidOperationException* to be thrown. In addition, unless a more specific exception is raised, any error that occurs in the *EndEstablish* method causes the *RealTimeException* to be thrown. When unhandled, an exception disrupts the execution of an application, which frequently leads to an unwanted user experience. In production code, you should always be prepared to catch exceptions at run time and handle them appropriately. You should also inform the user about the nature of the failures and provide information about how to recover from the problems.

When signing in an endpoint, a misspelled SIP URI will cause a *RegisterException* to be thrown at run time. In this case, it makes sense for the application to display the mistyped SIP URI and give the user an opportunity to correct the mistake. You can do this by using a *try/catch* block in any managed API application.

Generally, the call to start an asynchronous operation (*BeginXXX*) should have at least one *InvalidOperationException catch* block to handle otherwise unhandled UCMA Core exceptions. Similarly, a call to end an asynchronous operation (*EndXXX*) should have at least one *RealTimeException catch* block to handle otherwise unhandled UCMA Core exceptions.

More specific exceptions frequently provide additional information that may help you design a corrective measure programmatically. For example, if an *AuthenticationException* is caught and the data in the exception indicates that the credentials are missing, an application can retry by requiring the user to enter the correct credentials.

The C# code example in Listing 10-5 shows exceptions that may be caught when establishing an endpoint. In this example, the error-handling routines are rather trivial in that only error messages and other related information are displayed in a message box (*System.Windows.Forms.MessageBox*). The application proceeds only when the endpoint is established successfully and shows a SIP response of 200 OK.

LISTING 10-5 C# Code Example That Illustrates the Exception-Handling Pattern That Can Be Caught in the *LocalEndpoint.BeginEstablish* and the *LocalEndpoint.EndEstablish* Methods in a UCMA Application

```csharp
void BeginEndpointRegistration(LocalEndpoint endpoint)
{
    if (endpoint == null)
        return;
    try
    {
        AsyncCallback callback = new AsyncCallback(EndEndpointRegistration);
        endpoint.BeginEstablish(callback, endpoint);
    }
    catch (InvalidOperationException ioEx)
    {
        System.Windows.Forms.MessageBox.Show("InvalidOperationException: " + ioEx.Message);
    }
}

void EndEndpointRegistration(IAsyncResult result)
{
    try
    {
        LocalEndpoint endpoint = result.AsyncState as LocalEndpoint;
        SipResponseData response = endpoint.EndEstablish(result);

        if (response.ResponseCode == 200)
        {
            GetProvisioningData(endpoint);

            if (endpoint is UserEndpoint)
                SubscribeToContacts(endpoint as UserEndpoint);
        }
    }
    catch (AuthenticationException authEx)
    {
        string msg = "AuthenticationException while establishing endpoint" + Environment.NewLine;
        msg += "\tErrorMessage = " + authEx.Message + Environment.NewLine;
        msg += "\tFailureReason = " + authEx.FailureReason.ToString() + Environment.NewLine;
        msg += "\tErrorCode = " + authEx.ErrorCode + Environment.NewLine;
        msg += "\tSupported Protocols = " + authEx.SupportedAuthenticationProtocols.ToString();
        System.Windows.Forms.MessageBox.Show(msg);
        // Application can have logic to retry the endpoint establishing operation with
        // the correct credentials.
```

```csharp
            if (authEx.FailureReason == AuthenticationFailureReason.MissingCredentials)
            {
                // Retry endpoint prep with the correct credentials and start the endpoint
                // registration again.
                _userEndpoint.Credentials = CredentialCache.DefaultNetworkCredentials;
                this.BeginEndpointRegistration(_userEndpoint);
            }
        }
    }
    catch (OperationTimeoutException otEx)
    {
        string msg = "OperationTimeoutException while establishing endpoint" + Environment.NewLine;
        msg += "\tErrorMessage = " + otEx.Message;
        System.Windows.Forms.MessageBox.Show(msg);
        // This exception may indicate that the underlying OCS is down. In this case,
        // The application may want to poll the server to retry the endpoint connection.

    }
    catch (RegisterException regExcept)
    {
        string msg = "RegisterException while establishing endpoint" + Environment.NewLine;
        msg += "\tErrorMessage = " + regExcept.Message + Environment.NewLine;
        msg += "\tresponseCode = " + regExcept.ResponseData.ResponseCode + Environment.NewLine;
        msg += "\tresponseText = " + regExcept.ResponseData.ResponseText + Environment.NewLine;
        msg += "\tFrom: " + regExcept.ResponseData.FromHeader.ToString() + Environment.NewLine;
        msg += "\tTo: " + regExcept.ResponseData.ToHeader.ToString() + Environment.NewLine;
        System.Windows.Forms.MessageBox.Show(msg);
    }
    catch (RealTimeException exc)
    {
        string msg = "Caught RealTimeException while establishing endpoint" + Environment.NewLine;
        msg += "    Exception Name = " + exc.GetType().FullName + Environment.NewLine;
        msg += "    Exception Message = " + exc.Message + Environment.NewLine;
        if (exc.GetBaseException() != null)
            msg += "\tBaseException="+exc.GetBaseException().GetType().FullName+Environment.NewLine;
        if (exc.InnerException != null)
        {
            msg+="\tInner Exception Nmae="+exc.InnerException.GetType().FullName+ Environment.NewLine;
            msg+="\tInner Exception Message = " + exc.InnerException.Message + Environment.NewLine;
        }
        System.Windows.Forms.MessageBox.Show(msg);
    }
}
```

Figure 10-6 shows a *RegisterException* caught by the code example in Listing 10-5 when the owner URI ("sip:john@contoso.com") of the endpoint does not match any existing user. This incorrect URI string can be extracted from the *FromHeader* property of the *RegisterException*.

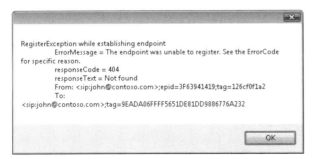

FIGURE 10-6 An application-generated display of an endpoint registration exception.

The *RegisterException* instance contains other information that can be useful in determining what went wrong. For example, the *Message* property gives the general description of the exception and the *StackTrace* or *DetectionStackTrace* property details the call stack tracing. The *DiagnosticInformation* property contains the ms-diagnostics or ms-diagnostics-public header of the SIP message returned from the server.

When a server name that does not exist is specified in an attempt to establish an endpoint, an exception of the *ConnectionFailureException* type is thrown. In the code example in Listing 10-5, this exception is caught by the *catch* block for the *RealTimeException* type, as shown in Figure 10-7.

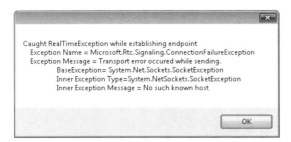

FIGURE 10-7 An application-generated display of an unspecified real-time exception.

If no *try/catch* block is implemented, you can catch unhandled exceptions and examine the failure conditions by using Visual Studio Debugger. Figure 10-8 shows an unhandled *ConnectionFailureException* instance that is caught by the Visual Studio Debugger when the execution breaks at a call to the *EndEstablish* method for an endpoint object. For more information about the exception, click View Detail.

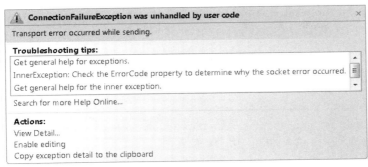

FIGURE 10-8 Exception caused by a connection failure that is unhandled by the application shown in Visual Studio.

The View Detail dialog box, seen in Figure 10-9, shows that the attempt to establish a connection failed because the specified server is unknown.

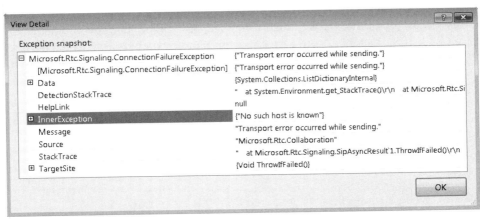

FIGURE 10-9 The View Detail dialog box of an unhandled exception caught by Visual Studio Debugger, showing that the connection-failing exception was caused by an unresolved server.

Connection attempts may fail in other situations. For example, if an incorrect port number (*serverPort*) is specified, the *ConnectionFailureException* instance is thrown when the execution breaks at the call to the *EndEstablish* method.

Troubleshooting Conversation Failures

After an endpoint is signed in to Office Communications Server, the UCMA Core application can initiate a conversation with another active endpoint. The conversation type or modalities can be any of the following: instant messaging, audio and video calling, data sharing, and conferencing. The process involves creating one or more conversations; inviting one or more participants to a conversation call; and exchanging text, media, or data through media-appropriate call flows.

Failures or errors can occur at each step. Some errors are of local origin caused by the application. Other errors are caused by network conditions or depend on a remote participant's state and capabilities. Programmatically, the errors can be caught by using various exception classes. For example, *ArgumentException* and *InvalidOperationException* encapsulate common errors of local origin. *ServerPolicyException* and *FailureResponseException* represent common failures of remote origin.

OperationTimeoutException reflects a failure of an unknown nature of remote origin, for example, when the network connection ends while the application is in the middle of an active conversation.

All of these exceptions can be handled by using appropriate *try/catch* blocks. You can log failures of remote origins by enabling tracing and examine them by using Snooper.

An *InvalidOperationException* is thrown when an application attempts to start a new conversation with existing conversation settings and then binds itself to the newly created *Conversation* instance. For example, calling the following C# code example two times throws an *InvalidOperationException* instance unless the existing conversation is terminated first or the *id* parameter of the conversation settings is updated.

```
string priority = ConversationPriority.Normal;
string subject = "Test";
string id = "MY_CONV";
ConversationSettings settings = new ConversationSettings(priority, subject, id);
_conversation = new Conversation(_userEndpoint, settings);
```

The following error message is produced from this exception.

```
"The conversation that is being bound already exists."
```

FailureResponseException may also be thrown during a conversation. This can occur when, for example, an application tries to invite a user to a conversation but the invited user is either offline or nonexistent. In this case, the exception is thrown when the application tries to make the call, as illustrated by the C# code example in Listing 10-6. This code example presents a simplified version of a functioning routine for starting an instant messaging conversation.

Note The following code example is for illustration only. A production code would most likely use an asynchronous programming pattern to establish the call and provide more meaningful exception-handling logic than displaying only the debugging information.

LISTING 10-6 C# Code Example That Illustrates *FailureResponseException* Thrown During a Conversation

```csharp
Conversation _conversation;
InstantMessagingCall _imCall;
UserEndpoint _userEndpoint;
public void StartImConversation(string sipUri)
{
    try
    {
        string priority = ConversationPriority.Normal;
        string subject = "Test";
        string id = "MY_CONV";
        ConversationSettings settings = new ConversationSettings(priority, subject, id);
        if (_conversation != null && _conversation.State != ConversationState.Terminated)
        {
            _conversation.EndTerminate(_conversation.BeginTerminate(null, null));
        }
        _conversation = new Conversation(_userEndpoint, settings);

        _imCall = new InstantMessagingCall(_conversation);
        _imCall.EndEstablish(_imCall.BeginEstablish(sipUri, null, null, null));
        _imCall.Flow.MessageReceived +=
                new EventHandler<InstantMessageReceivedEventArgs>(Flow_MessageReceived);
        // Bubble up the successful call-established event to the caller
        if (OnImCallEstablished != null)
            Utilities.RaiseEvent(this, OnImCallEstablished, null);
    }
    catch (FailureResponseException frEx)
    {
        string msg = "Caught while establishing IM Call." + Environment.NewLine;
        msg += "\tMessage=" + frEx.Message + Environment.NewLine;
        msg += "\tResponseCode=" + frEx.ResponseData.ResponseCode + Environment.NewLine;
        msg += "\tResponseText=" + frEx.ResponseData.ResponseText + Environment.NewLine;
        msg += "\tStackTrace=" + frEx.StackTrace + Environment.NewLine;
        System.Windows.Forms.MessageBox.Show(msg, frEx.GetType().FullName);
    }
}
```

In Listing 10-6, the invited party is specified by the *sipUri* parameter. If this user is from the same enterprise as the caller and is offline, a *FailureResponseException* instance is thrown at the *_imCall.EndEstablish(_imCall.BeginEstablish(sipUri, null, null, null))* statement. The explanation is provided in the *ResponseCode* and *ResponseText* properties of *FailureResponseException*, as shown in Figure 10-10. Their values are *480* and *Temporarily Unavailable*, respectively. Such failure conditions also show up in the log file that is generated by running the Office Communications Server Logging Tool (OCSLogger.exe).

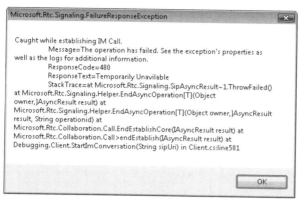

FIGURE 10-10 An application-generated display of a failure caused by inviting an unavailable user.

When the invited party is unknown, the *FailureResponseException* instance returns *404* and *Not Found* as the *ResponseCode* and *ResponseText* property values, respectively. A specified user is considered not found if the user is not assigned a SIP URI within the same enterprise as the caller or if the SIP URI refers to a federated user who is either offline or nonexistent.

Debugging UCMA Workflow Applications

Debugging a UCMA Workflow application involves the following three tasks:

- Debugging the workflow (.xoml) components of the application
- Debugging the code of the hosting application
- Debugging the code other than the workflow activities and events, including the UC-specific instant messaging, speech, and presence-querying activities and events

To debug the hosting application component and the code that is not part of the workflow activities and events, follow the process for debugging UCMA Core applications that was discussed earlier in this chapter. The following sections focus only on how to debug the workflow component.

Specifically, these sections discuss how workflow exceptions are handled by using the fault handler activities that are defined in the Windows Workflow Foundation. After that, the steps that are involved in debugging a workflow are explained.

Enabling Tracing

The UCMA Workflow relies on the UCMA Core to enable the local endpoint to communicate and collaborate with remote endpoints. Therefore, tracing the protocol stack in the workflow application is the same as in the UCMA Core application.

To enable tracing for the hosting application and the code besides the activities, follow the instructions given in the section titled "Debugging UCMA Core Applications" earlier in this chapter. To enable tracing for the workflow execution, add the following configuration settings shown in Listing 10-7 to your project's App.config file.

LISTING 10-7 Configuration Settings to Enable Tracing for the Workflow Application

```xml
<configuration>
    <system.diagnostics>
        <switches>
            <add name="System.Workflow.LogToTraceListeners" value="1"/>
            <add name="System.Workflow.RunTime.Hosting" value="All"/>
            <add name="System.Workflow.Runtime" value="All"/>
            <add name="System.Workflow.Activities" value="All"/>
        </switches>
        <trace autoflush="true" indentsize="4">
            <listeners>
                <add name="customListener" type="System.Diagnostics.TextWriterTraceListener"
                    initializeData="WFTrace.log" />
            </listeners>
        </trace>
    </system.diagnostics>
</configuration>
```

In Listing 10-7, the *<system.diagnostics>* element specifies trace levels and trace listeners. The specified trace levels are declared in the *<switches>* child element. The specified trace listeners are declared in the *<listeners>* child element and are used to collect, store, and route tracing messages. For more information, see "Trace and Debug Settings Schema" at *http://go.microsoft.com/fwlink/?linkid=143291*. In this example, if run-time exceptions are thrown from any workflow activity of your application, the error conditions are logged into the specified log file (WFTrace.log). The log file should be located in the application's current directory where the application executable is located. The following is an example entry in this log file.

```
System.InvalidOperationException: Accept Call activity 'acceptCallActivity1' cannot run.
The Call (AudioVideoCall or InstantMessagingCall) is not in the incoming state. The current
state is 'Terminated'.
   at Microsoft.Rtc.Workflow.Activities.AcceptCallActivity.AcceptCall()
   at Microsoft.Rtc.Workflow.Activities.AcceptCallActivity.Execute(ActivityExecutionContext
executionContext)
   at System.Workflow.ComponentModel.ActivityExecutor`1.Execute(T activity,
ActivityExecutionContext executionContext)
   at System.Workflow.ComponentModel.ActivityExecutor`1.Execute(Activity activity,
ActivityExecutionContext executionContext)
   at System.Workflow.ComponentModel.ActivityExecutorOperation.Run(IWorkflowCoreRuntime
workflowCoreRuntime)
   at System.Workflow.Runtime.Scheduler.Run()
```

The log file should provide a description of the message and the workflow activity call stack. This shows the location where an error is detected or an exception is thrown.

Handling Exceptions Using the Fault Handler Activity

In Windows Workflow Foundation, on which the UCMA Workflow is based, exceptions are handled by using fault handler activities. Each fault handler activity is an instance of the *FaultHandlerActivity* type. The *FaultHandlerActivity* type is supported in the Windows Workflow Foundation and corresponds to an error handler that has a specified .NET Framework exception type.

Adding a fault handler activity is similar to inserting a *try/catch* block within the activity. By default, the UC workflow application templates, which are made available in Visual Studio by the UCMA SDK, support a general fault handler for the *System.Exception* type. This general fault handler is added at the top level of a workflow and catches all unhandled exceptions that are encountered by any activities in the workflow.

Whenever an exception is thrown, this fault handler is called to invoke an event-handling code routine. This code routine is named *HandleGeneralFault* by the UCMA Workflow application templates. The event-handling code is bound to a *Code* activity that is part of the general fault handler. To better understand how this works, the implementation of the *HandleGeneralFault* method, as provided by the UC workflow application template, is shown in Listing 10-8.

LISTING 10-8 Implementation of the *HandleGeneralFault* Method as Provided by the UC Workflow Application Template

```
/// </summary>
/// <param name="sender"></param>
/// <param name="e"></param>
private void HandleGeneralFault(object sender, EventArgs e)
{
    // When an exception is thrown the actual exception is stored in the Fault property,
    // which is read-only.  Check this value for error information;
if it is an exception,
    // ToString() will include a full stack trace of all inner exceptions.
    string errorMessage = generalFaultHandler.Fault.ToString();
    Trace.Write(errorMessage);

    if (Debugger.IsAttached)
    {
        // If the debugger is attached, break here so that you can see the
error that occurred.
        // (Check the errorMessage variable above.)
        Debugger.Break();
    }
}
```

When an unhandled exception is thrown, the general fault handler is invoked and the code in Listing 10-8 is run. This fault handler method stops the debugger at the *Debugger.Break* statement so that the reported error message can be examined. The error message is also logged to a trace listener that is configured by the hosting application or the default trace listener. This is a simple approach to event handling, and it can be enhanced by adding application-specific behaviors.

Debugging UCMA Workflow Applications

The following sections demonstrate how to perform the two most basic tasks to debug the UC workflow application:

1. Step through the workflow in the Visual Studio Debugger.
2. Implement a custom fault handler activity to catch a specified exception.

Stepping Through the UC Workflow

Stepping through a workflow is similar to stepping through code in the Visual Studio Debugger, but with a somewhat different experience. A workflow defines the flow of activities or events as the execution proceeds. However, other information about the execution context is not available. This is the major difference between debugging a workflow and debugging the code. The following describes the steps for debugging a workflow in Visual Studio Debugger.

1. Set a breakpoint.

   Select a workflow activity in the workflow design (.xmol file), and then press the F9 key. Pressing F9 again toggles the breakpoint off. A red circle appears on the selected component when the breakpoint is enabled as shown here, where a breakpoint is set on the *speechQuestionAnswerActivity2* instance. This illustration is based on the lab exercise in Chapter 6, "Business Process Communication." The *speechQuestionAnswerActivity2* instance represents the question in the exercise that asks for the caller's student ID number.

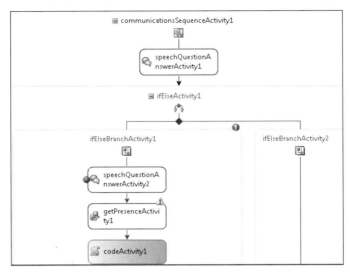

2. **Start debugging.**

   Pressing the F5 key starts the debugging process. In the previous illustration, the execution breaks before the question that is represented by the *speechQuestionAnswerActivity2* instance is asked. The execution continues until the first breakpoint is hit. The activity that has the shaded border around it (which appears yellow in Visual Studio) signifies this first breakpoint being hit in the Visual Studio Debugger for Windows Workflow.

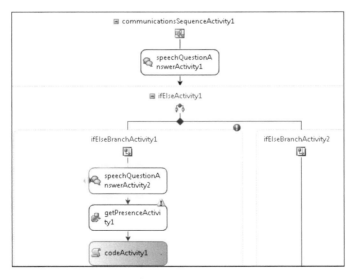

3. **Step through the workflow.**

To continue, you can step through the workflow by pressing the F5, F10, or F11 key. The F5 key causes the workflow to run until it hits the next breakpoint or until the workflow ends normally. The F10 or F11 key runs the activity and moves the workflow to the next activity. If a breakpoint has been set on any code routine besides the associated *Code* activity, that breakpoint will be hit.

The code routine besides the *codeActivity1* instance is shown here. When the workflow passes through *codeActivity1*, the breakpoint set on the first statement of the code snippet is hit and the execution is stopped at the boxed line.

```
private void codeActivity1_ExecuteCode(object sender, EventArgs e)
{
    Dictionary<RealTimeAddress, PresenceResult> results =
                            this.getPresenceActivity1.Results;

    foreach (RealTimeAddress address in results.Keys)
    {
        if (results[address].PresenceStatus == PresenceAvailability.Online)
        {
            this.blindTransferActivity1.CalledParty = address;
            this._sipOnlineFound = true;
            break;
        }
    }
}
```

Catching and Handling Custom Exceptions

Each fault handler has its scope of effectiveness. A scope defines a group of activities such that exceptions raised by any activity within the group are handled by the fault handler within the scope. In the Visual Studio Workflow Designer, activities that are enclosed in a box share the same scope. Activities of a certain scope can be grouped further into smaller scopes. This means that a scope can be embedded in other scopes.

The fault handlers of a given scope may be added to the first activity along an execution path fragment. In Workflow Designer, this activity is located at the top of a given box. A drop-down menu (indicated by an downward-pointing arrow) is displayed when the mouse pointer rests on the box. When you click the drop-down menu, it displays a list of menu items that lets you open the Fault Handlers View, where you can add, modify, or remove one or more fault handlers.

The following steps demonstrate how to add a fault handler for an exception of the *Microsoft.Rtc.Signaling.RealTimeException* type to an *IfElse* activity. The scope of the fault handler covers all of the activities along the pathways that originate from the *IfElse* activity. The workflow example is based on Lab 6, Exercise 1 in Chapter 6 of this book.

1. Click *ifEsleActivity2* in the Workflow Designer to select the activity. As shown here, the bounding outline indicates the scope of the fault handler to be added.

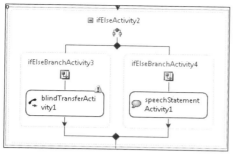

2. Click the arrow under *ifElseActivity2,* and then select View Fault Handlers to change the Designer view to display all fault handlers defined in your applications, as shown here.

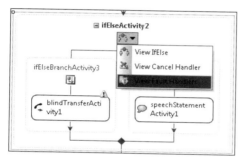

3. Drag a *FaultHandler* activity from the Windows Workflow v3.0 Tools panel in Visual Studio, as shown here. Drop this fault handler activity into the fault handler container named *faultHandlersActivity3,* which is defined for the IfElse activity named *ifElseActivity2*. When specifying a fault handler activity, it must be introduced in a fault handler container.

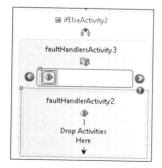

The added fault handler has a default name of *faultHandlerActivity2*. You can change this name by resetting the *Name* property value in the Properties window in Visual Studio.

4. In the Properties window, set the *FaultType* property for the fault handler activity that you just added (*faultHandlerActivity2*) to *Microsoft.Rtc.Signaling.RealTimeException,*

as shown here. This makes the fault handler catch all exceptions of type *RealTimeException*.

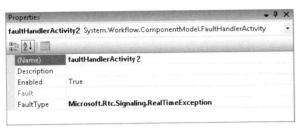

5. Drag a *Code* activity from the Tools panel (under Windows Workflow v3.0) in Visual Studio and drop it below *faultHandlerActivity2*. The newly added *Code* activity instance is named *codeActivity4*, as shown here.

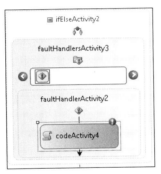

This *Code* activity is used to provide exception handling. You need to supply the exception-handling routine as the code-besides to the *Code* activity. You can also use another activity to handle the exception in other ways. For example, you can use *Compensate* to revert a failed transaction.

6. In the Properties window of *codeActivity4* in Visual Studio, enter the *HandleRtcFault* string as the value of the *ExecuteCode* property, as shown here, and then press Enter.

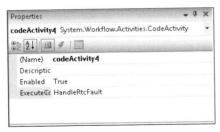

Setting the *ExecuteCode* property creates an event handler stub to which you can add application-specific exception-handling logic. The event handler stub is listed as the private *HandleRtcFault* method in the Windows1.xoml.cs file. The following is a sample implementation.

```
private void HandleRtcFault(object sender, EventArgs e)
{
    string errorMessage = this.faultHandlerActivity2.Fault.Message;
    string stackTrace = this.faultHandlerActivity2.Fault.StackTrace;
    Trace.WriteLine("Error message: {0}", errorMessage);
    Trace.WriteLine("Stack trace: {0}", stackTrace);
}
```

Repeat steps 1 through 6 to add another fault handler. However, the order in which fault handlers are added must be such that the more specific type of exception is added before the less specific type of exception. For example, if you add a *FaultHandler* activity for *System.Exception* before you add a *FaultHandler* activity for *Microsoft.Rtc.RealTimeException* to the same workflow activity, you get a compile-time error with the following error message.

```
Activity 'faultHandlersActivity3' validation failed: A FaultHandlerActivity for
exception type 'RealTimeException' must be added before the handler for exception type
'Exception'.
```

The reason this occurs is that *System.Exception* is more general than *Microsoft.Rtc.RealTimeException* because the latter is derived from the former. The fault handler for *Microsoft.Rtc.RealTimeException* never triggers because the fault handler for *System.Exception* catches all of the exceptions thrown. After the second *FaultHandler* activity is dragged to the beginning of the first one, the project compiles without error.

Summary

This chapter provided a general overview of debugging in the UC platform. Before covering the best practice for debugging UC applications, the source of errors and failures that may be encountered in your applications was examined. COM error codes and common run-time exceptions that may be raised by the UC APIs and the SIP error codes that will be reported by protocol tracing were also reviewed.

This chapter also examined how tracing is enabled and discussed how to examine the log files produced. The information presented should help you understand how to debug three types of UC applications: Office Communicator Automation API applications, UCMA Core applications, and UCMA Workflow applications.

Additional Resources

- "Microsoft Office Communicator Automation API Error Codes" documentation (*http://go.microsoft.com/fwlink/?linkid=143321*)

- ".NET Framework Class Library COMException Class" (*http://go.microsoft.com/fwlink/?linkid=143309*)

- "Unified Communications Client API Error Codes" (*http://go.microsoft.com/fwlink/?linkid=143308*)

- "Communicator Error Messages" (*http://go.microsoft.com/fwlink/?linkid=143307*)

- "Unified Communications Managed API 1.0 SDK" documentation (*http://go.microsoft.com/fwlink/?LinkID=133767*)

- "Unified Communications Management API 2.0 Core SDK Documentation" (*http://go.microsoft.com/fwlink/?LinkID=133571*)

- "Network Working Group: SIP: Session Initiation Protocol" (*http://go.microsoft.com/fwlink/?LinkID=143940*)

- "[MS-SIP]: Session Initiation Protocol Extensions" (*http://go.microsoft.com/fwlink/?linkid=143292*)

- "Trace and Debug Settings Schema" (*http://go.microsoft.com/fwlink/?linkid=143291*)

Glossary

2007 R2 release of Microsoft Office Communicator Mobile for Windows Mobile See Office Communicator Mobile for Windows Mobile.

2007 R2 release of Microsoft Office Communicator Web Access See Office Communicator Web Access.

Access Edge Server A server role that is located in the perimeter network and validates external users. In the Office Communications Server 2007 R2 consolidated edge topology, this server role is collocated on the same computer as the Office Communications Server 2007 R2 Edge Server.

Access levels A setting in Office Communicator 2007 R2 that enables you to control the level of your presence information that other people see. Access levels (from least restrictive to most restrictive) include Personal, Team, Company, Public, and Blocked.

ACD Automatic Call Distributor.

ACK A SIP request that is used in a three-way handshake, similar to Transmission Control Protocol (TCP).

ACP Audio Conferencing Provider.

Active Directory Domain Services (AD DS) The Windows-based directory service. AD DS stores information about objects on a network and makes this information available to users and network administrators. AD DS gives network users access to permitted resources anywhere on the network by using a single logon process. It provides network administrators with an intuitive, hierarchical view of the network and a single point of administration for all network objects.

Address Book service A service that provides Global Address List (GAL) information from Active Directory Domain Services (AD DS) to Office Communicator 2007 R2. This service can also perform phone number normalization for Office Communicator 2007 R2 telephony integration. The Address Book service runs on an Office Communications Server 2007 R2 Front End Server and synchronizes SIP user data on the client with updates from AD DS.

Advanced media gateway A category of media gateway vendors that does not require deploying Mediation Servers to interface with Office Communications Server 2007 R2.

Agent A user who is designated as a member of a Response Group. Response Group settings determine which calls will be routed to a particular agent or group of agents. There are two kinds of agents: formal and informal. A formal agent must sign in and sign out of the Response Group. An informal agent is not required to sign in or sign out and would typically be someone who covers calls on a part-time basis.

Agent Communications Panel for Microsoft Dynamics CRM 4.0 An application add-in that provides presence information in addition to instant messaging (IM) and call control capabilities within Microsoft Dynamics CRM 4.0. This application can also be used as a Response Group agent that enables users to sign in and sign out of Response Groups.

AJAX Asynchronous JavaScript and XML.

Allow/Block lists In the context of an individual user, a Block list refers to contacts to whom the user has assigned a permission level of Blocked, which means that the contact cannot view the user's presence information or contact the user. The Allow and Block list terminology is also used in the context of Office Communications Server 2007 R2 Edge Server configuration. Server administrators can configure the Access Edge Server properties to explicitly allow or block communication with other domains.

Anonymous user An external user who does not have credentials in Active Directory Domain Services (AD DS). Unlike a federated user, an anonymous user is not authenticated.

Answering agent Used with the Deployment Validation tool to simulate an answering machine. When users call an answering agent, those users are prompted to record a message. After the message has been recorded, it is replayed immediately. This gives the user an opportunity to verify the audio quality of the call.

Application Server A component of Office Communications Server 2007 R2 that provides a platform to deploy, host, and manage Unified Communications applications.

Application Sharing Server A component that resides on the Front End Server and is responsible for managing and streaming data for conferences that require desktop sharing.

Application Sharing Server A server role introduced in Office Communications Server 2007 R2 Front End Server that manages and streams data for conferences that share applications or an entire desktop.

ArchivingCdrReporter An Office Communications Server 2007 R2 Resource Kit Tool that enables you to quickly create reports drawn from either the Archiving Server database or the Monitoring Server database.

Archiving database A SQL Server database that, with the Archiving Service, comprises the Archiving Server role. This database stores instant messaging (IM) conversations and IM group conferences.

Archiving Server A server role in the internal network that captures all IM conversations and IM group conferences and stores them in a SQL Server database. With Office Communications Server 2007 R2, the Archiving Server role is separated from the Call Detail Record (CDR) collection, which is now included in the new Monitoring Server role.

Array A group of interconnected, identical processors operating synchronously, often under the control of the central processor. A group of servers that are clustered behind a load balancer and that are running the same server roles.

Attendant *See* Office Communicator 2007 R2 Attendant.

Audio conference A call that involves at least three people.

Audio Conferencing Provider (ACP) A third-party provider that enables PSTN conferencing.

Authenticated caller   A participant who joins a VoIP conference and is authenticated through Active Directory Domain Services (AD DS).

Authentication   A method of associating an identity with an entity. In a multiuser-server operating system, the process by which the system validates a user's logon information.

Authorization   The right granted to an individual to use the system and the data that is stored on it. Typically set up by a system administrator and verified by the computer based on some form of user identification, such as a code number or a password.

Auto attendant   A feature in Exchange Unified Messaging (UM) that supplies a caller with information and performs an action without the intervention of a human operator. It automatically routes calls based on selections made by the caller.

Autodiscover service   A service that returns server configuration information for a mobile device or other client by using only the user's Simple Mail Transfer Protocol (SMTP) e-mail address and password.

Automatic Call Distributor (ACD)   Functionality that classifies, queues, and distributes incoming calls to agents or outgoing calls to lines.

A/V   Audio/video.

A/V Conferencing Server   An Office Communications Server 2007 R2 server in the internal network that mixes and manages audio/video inputs from multiple audio/video conferences. In Office Communications Server, an A/V Conferencing Server must be deployed if you want users to be able to start a conference call by using Office Communicator Web Access.

A/V Edge Server   An Office Communications Server 2007 R2 server role in the perimeter network that provides a single trusted point through which media traffic can traverse NATs and firewalls. It enables audio and video conferencing and A/V peer-to-peer communications with external users equipped with the Office Communicator 2007 R2 or Office Live Meeting 2007 client. This server role can be collocated with the Access Edge Server and Web Conferencing Edge Server, or it can reside on a separate, dedicated server.

Back-End Database   A server role that hosts the SQL back-end database, which provides user information and conference state, including persistent user data, transient user data, and persistent Office Communications Server 2007 R2 settings to the Front End Server. The Back-End Database is collocated with a Standard Edition server. In an Enterprise pool, the Back-End Database is deployed on a separate, dedicated computer.

Basic Media Gateway   A category of media gateway vendors that requires the deployment of Office Communications Server, Mediation Server to work with the Office Communications Server 2007 R2.

Best Practices Analyzer (BPA)   The Office Communications Server 2007 Best Practices Analyzer Tool is a diagnostic tool that gathers configuration information from an Office Communications Server 2007 environment and determines whether the configuration is set according to Microsoft best practices.

BPA   Best Practices Analyzer.

BYE   A SIP request that is used to end a session.

C3P   Centralized Conferencing Control Protocol.

CA   Certificate authority.

Call Back Control   A feature that provides Enterprise Voice functionality and call control to mobile phones and other devices that are not otherwise enabled for Enterprise Voice.

Call deflection   The ability of a called party to redirect the calling party to a different phone number before picking up the call.

Call delegation   A feature that was introduced in Office Communications Server 2007 R2 that enables managers to delegate phone-call handling to one or more administrative assistants or other delegates. The receiver is notified when a delegate answers a call, together with which delegate answered.

Call Detail Record (CDR)   A record that contains information about a call. In Office Communications Server 2007 R2, a CDR is part of the Monitoring Server role, and it captures and reports information such as user logons and logoffs, instant messaging conversations, usage details about voice and video, and conference starts and joins.

Call forwarding   The process of automatically forwarding a missed call to a user-designated number, which can be a registered SIP device, a PSTN number, or voice mail. In Office Communications Server 2007 R2, one way to configure call forwarding is through the Response Group Service.

Call Me   Functionality that enables a Communicator Web Access user to specify a phone number for joining an audio conference, which is then used by Office Communications Server to conference in the user.

CANCEL   A SIP request that is used to cancel a session establishment process.

Category   An Enhanced Presence concept that is used by a Session Initiation Protocol (SIP) client to publish or subscribe to presence information. A category enables basic identification of the data that is being published. It implies an agreed-upon schema for interpreting the data. A category name identifies a contract between a publisher and a subscriber.

Category SUBSCRIBE   A SUBSCRIBE request that specifies the presentities and the categories for which information is requested.

Category subscriber   A Session Initiation Protocol (SIP) client that sent a category SUBSCRIBE request.

CDP   Certificate distribution point.

CDR   Call Detail Record.

Centralized Conferencing Control Protocol (C3P)   A new conference manipulation protocol that is used by the Office Communications Server conferencing servers to modify the conference state. C3P has *request/pending* and *response/ final response* semantics that are similar to SIP.

Certificate distribution point (CDP)   The location where you can download the latest certificate revocation list. A CDP is typically listed in the CRL Distribution Points field of the Details tab of the certificate.

Certificate revocation list (CRL)   A file that contains a list of revoked certificates, their serial numbers, and their revocation dates. Additionally, the CRL file contains the name of the issuer of the CRL, the effective date, and the next update date.

Certificate authority (CA)   An issuer of digital certificates, the cyberspace equivalent of identity cards. A certificate authority may be an external issuing company or an internal company authority that has installed its own server for issuing and verifying certificates.

Channel Server   A server role for Office Communications Server 2007 R2 Group Chat that provides core functionality for chat rooms, except for file posting, which is managed through the Web Service.

Chat room   A persistently available space for exchanging instant messages where authorized individuals can enter and leave at will. Unlike a group IM conference, the complete contents of the chat room remain available, even when there are no participants, as long as the chat room is open.

Chat Room History   A Group Chat feature that consists of saved messages that are no longer displayed in a chat room, but are searchable and readable.

CheckSPN   A tool that is part of the Office Communications Server 2007 R2 Resource Kit Tools, that validates service principal names (SPNs) to avoid authentication and topology errors.

Class   In Active Directory Domain Services (AD DS), the characteristics of an object and the kind of information an object can hold. For each object class, the schema defines the attributes that an instance of the class must have and the additional attributes that it might have.

Click to Call   A feature of Office Communicator 2007 R2 and the 2007 R2 release of Office Communicator Mobile for Windows Mobile that enables a cellular phone user to place a voice call by clicking a contact's single phone number or SIP URI.

ClientLogReader   A script that is part of the Office Communications Server 2007 R2 Resource Kit Tools that scans client trace log files to highlight errors, provide protocol summaries, or filter out specific protocol messages.

Codec   An algorithm that is used to convert media between digital formats, especially between raw media data and a format that is more suitable for a particular purpose. Encoding converts the raw data to a digital format. Decoding reverses the process.

COM interop   Component Object Model (COM) objects that have Visual Studio .NET. interoperability with COM. COM interop enables you to use existing COM objects while transitioning to the .NET Platform at your own pace.

CommunicationsWorkflowRuntimeService   Windows Runtime Service that provides a way to pass UCMA call objects, endpoint objects, and culture information from the hosting environment to a workflow instance.

Communicator call   A VoIP call that is routed to all of a contact's devices that are running Office Communicator 2007 R2.

Communicator Web Access Server   A service that is run by Microsoft Internet Information Services 6.0 and provides the Web access to the client functionality supported by Office Communications Server inside an enterprise network.

Compliance   Adherence to federal, state, and local statutory requirements with regard to the logging and archiving of corporate communications.

Computer-level settings   Settings that are applied to a specific server in an Enterprise Edition pool or to a Standard Edition server.

Computer Supported Telephony Applications (CSTA)   An international standard established by the Ecma International (formerly ECMA) that specifies an application's interface and protocols for monitoring and controlling calls and devices in a communications network.

Computer Telephony Integration (CTI)   A technology that allows interactions on a telephone and a computer to be integrated or coordinated. As contact channels have expanded from voice to include e-mail messages, Web, and fax, the definition of CTI has expanded to include the integration of all customer contact channels (voice, e-mail, Web, fax, and so on) with computer systems. Computer Telephony (CT) is the new term for this technology.

Conference   An instant messaging (IM), audio, or video session that is mediated by Office Communications Server.

Conference directory   A feature that is used to generate and to resolve personal identification numbers (PINs) used for PSTN conferencing. When a new pool is set up, one Conference directory is associated with the pool.

Conferencing Announcement Service   An application that uses a tone or a voice recording to announce when a phone user joins or leaves a conference.

Conferencing Attendant   An application that enables dial-in conferencing, whereby an enterprise user who does not have access to a Unified Communications client can join an audio/video conference by dialing in using a telephone on the Public Switched Telephone Network (PSTN).

Conferencing server   A server role that mixes and matches inputs from multiple clients in a group session. A conferencing server typically supports one or more media types such as audio, video, and data. Also known as a multipoint control unit (MCU).

Conferencing Server Factory   Provisions a conference for a particular media type on a conferencing server.

Consolidated Edge   A server that validates traffic from the edge network and connects to the internal servers in the organization. The internal servers could be a pool of servers or a single Office Communications Server 2007 R2 Standard Edition server.

Consolidated topology   An Enterprise pool configuration in which all server components, including Internet Information Services (IIS), the Web Conferencing Server, the Web Components Server, and the A/V Conferencing Server, are collocated on the pool's front-end servers. The Application Host, IM Conferencing Server, and Telephony Conferencing Server are also collocated with the Front End Server. The Monitoring, Archiving, and Mediation Servers are typically located on a separate computer (or separate computers).

Contact card   A feature that provides details about a contact's availability and activity. A contact card can be viewed by clicking the Presence button next to a contact in Office Communicator 2007 R2.

Contact group   A logical grouping of people in a contact list. Contact groups can be used to communicate with an entire group of people with one call.

Contact list   A list of coworkers, family, friends, and associates with whom you communicate most frequently.

Contact objects   Active Directory Directory Service objects that are used to identify and route to response groups.

Container   A data object that is used to store published presence information and a list of subscribers who are allowed to view the information. A container enables a publisher to publish different data values of the same category and instance, which enables different subscribers to *see* different values.

Credential Manager   One of several authentication services that can be used to authenticate clients accessing remote resources. Specifically, Credential Manager deals with managing credential information, such as user names and passwords. Credential Manager provides storage for cached credentials and enables the sharing of common credentials.

CRL   Certificate revocation list.

CSTA   Computer Supported Telephony Applications.

CTI   Computer Telephony Integration.

Custom authentication   An authentication method that allows administrators to use a third-party authentication system to enable single sign on (SSO) or two-factor authentication.

DbAnalyze   A tool that is part of the Office Communications Server 2007 R2 Resource Kit Tools that collects analysis reports from the Office Communications Server 2007 R2 database.

DCOM   Distributed Component Object Model.

DDR   Double data rate.

Delegate   A person designated through the Call Delegation feature of Office Communications Server 2007 R2 to answer the phone for someone else.

Denial of Service (DoS)   A category of threat in which a malicious user launches an attack against your servers that consumes server processing time and causes services to become unavailable.

Deployment Validation Tool (DVT)   Included in the Office Communications Server 2007 R2 Resource Kit Tools, this tool enables users or administrators to test the quality of audio connections in an Office Communications Server 2007 R2 infrastructure.

Desktop sharing   A feature of Office Communicator 2007 R2 that enables a user in an Office Communicator conference to share a view of his or her desktop with the conference attendees.

Destination Network Address Translation (DNAT) Not supported in Office Communications Server 2007 R2. A technique in which the destination IP address of an en route packet is transparently rewritten and then changed back on the reply to the packet.

Dial plan Basic unit of configuration in Exchange Unified Messaging that can be of the following types: telephone extensions, SIP URI, or E.164. The dial plan is an Active Directory container object that logically represents sets or groupings of PBXs that share common user extension numbers. An example of a dial plan is the 10-digit North American Numbering Plan (NANP) that includes a 3-digit area code and a 7-digit telephone number.

DID Direct Inward Dialing.

Digest authentication A protocol for use with HTTP and Simple Authentication Security Layer (SASL) exchanges, as documented in RFCs 2617 and 2831. An authentication method that prompts the user for a user name and a password, also called credentials, which are hashed with other data before being transmitted over the network. Digest authentication is available only on domains that have domain controllers that are running Microsoft Windows Server operating systems.

Direct Inward Dialing (DID) A service in which a local phone company provides a corporation with a block of phone numbers for calling into the corporation's Private Branch eXchange (PBX).

Director A server role in the internal network that authenticates internal and external users, and routes traffic between Edge Servers and the internal Office Communications Server deployment.

Distributed Component Object Model (DCOM) Technology that extends the Component Object Model (COM) to support communication among objects on different computers—on a local area network (LAN), a wide area network (WAN), or even the Internet.

Distribution group A group of users, stored in Active Directory Domain Services (AD DS), whose members can be contacted as a group.

DNAT Destination Network Address Translation.

Domain In Active Directory Domain Services, (AD DS) a collection of computer, user, and group objects that are defined by the administrator. These objects share a common directory database, security policies, and security relationships with other domains.

Domain controller In an Active Directory Domain Services (AD DS) forest, a server that contains a writable copy of the AD DS database, participates in AD DS replication, and controls access to network resources. Administrators can manage user accounts, network access, shared resources, site topology, and other directory objects from any domain controller in the forest.

Double data rate (DDR) A type of synchronous dynamic RAM (SDRAM) that supports data transfers on both edges of each clock cycle (the rising and falling edges), effectively doubling the memory chip's data throughput.

DTMF Dual-tone multifrequency.

Dual forking A configuration by which Office Communications Server 2007 R2 operates in co-existence with the PBX.

Dual-tone multifrequency (DTMF) In telephony systems, a signaling system in which each digit is associated with two specific frequencies. This system is typically associated with telephone touch-tone keypads.

DVT Deployment Validation Tool.

Dynamic Host Configuration Protocol (DHCP) A standards-based network protocol (IETF RFC 2131) used by computers to obtain an IP address and other network configuration information when they first connect to the network.

E.164 A standard industry format for number normalization. The E.164 format consists of a country code (1 to 3 digits) and a National Significant Number (12 to 14 digits) for a total of 15 digits. The National Significant Number consists in turn of a National Destination Number and a Subscriber Number (extension). For example, (425) 555-1212 ext. 3011 is represented by E.164 format as 42555512123011.

Ecma International (Ecma) Formerly European Computer Manufacturers Association.

Edge Server An Office Communications Server 2007 R2 server that is located in the perimeter network and provides connectivity for external users, federated users, and public IM connections. Each Edge Server has one or more of the following server roles: Access Edge Server, a Web Conferencing Edge Server, or an A/V Edge Server. An Edge Server is managed by using the Office Communications Server 2007 snap-in extension for the Computer Management snap-in, not the Office Communications Server 2007 Administrative snap-in.

EKU Enhanced Key Usage.

Endpoint The receiving client software of a communication, such as Office Communicator 2007 R2 or Office Live Meeting 2007.

Enhanced Key Usage (EKU) Both a certificate extension and a certificate extended property value. An EKU field specifies the uses for which a certificate is valid.

Enhanced Presence The publication of customized presence information to presence subscribers.

Enterprise cellular telephony Functionality that enables users of the 2007 R2 release of Office Communicator Mobile for Windows Mobile to set up and control Enterprise Voice calls over their cell phone providers' circuit-switched networks.

Enterprise pool Servers in the internal network that are running Office Communications Server 2007 R2, Enterprise Edition and host the necessary services, including IM, presence, and conferencing services. Depending on the pool configuration (consolidated configuration or expanded configuration), services can all be collocated on the Front End Server, or specific services can run on separate, dedicated computers. The back-end database must be run on a separate, dedicated computer. Typically, Monitoring, Archiving, and Mediation Servers are also on separate computers.

Enterprise user A user who has an identity in Active Directory Domain Services (AD DS).

Enterprise Voice A software solution from Microsoft that enables an enterprise to manage Voice over Internet Protocol (VoIP) and provide full unified communication capabilities. This includes the ability to make single-party and multiparty VoIP calls, configure robust call forwarding features, and receive voice mail in the Exchange Server mailbox.

Enterprise Voice Route Helper   *See* Route Helper tool.

ETW   Event Tracing for Windows.

Event Tracing for Windows (ETW)   A general-purpose, high-speed tracing facility provided by the operating system. By using a buffering and logging mechanism implemented in the kernel, ETW provides a tracing mechanism for events that are raised by user-mode applications and kernel-mode device drivers.

EWS   Exchange Web Services.

Exchange Management Console   The Exchange 2007 graphical user interface (GUI) from which administrators can perform tasks to configure and manage Exchange Servers. The Exchange Management Console is based on Windows Microsoft Management Console (MMC) 3.0.

Exchange Management Shell   A command-line interface and associated command-line plug-ins for Exchange Server that enable automation of administrative tasks. The Exchange Management Shell is built on Windows PowerShell technology.

Exchange Web Services (EWS)   An extensibility point for clients who connect to the computer that is running Exchange 2007 and consume information about user availability and the manipulation of items that are located in the Exchange data store.

Exchange Web Services Managed API   An application programming interface (API) that enables developers to create custom client, server, and middleware applications for Exchange without using Office Outlook or any server-side code. It provides a unified, cohesive, open standards–based API that replaces other Exchange APIs, while retaining much of their functionality, and extends them with an Outlook-compatible business logic layer.

Expanded topology   An Enterprise pool configuration in which the Front End Server, Application Sharing Server, Web Conferencing Server, Web Components Server, and the A/V Conferencing Server are installed on separate, dedicated computers. The Application Host, IM Conferencing Server, and Telephony Conferencing Server are collocated on the Front End Server.

Extensible Application Markup Language (XAML)   A markup language for declarative application programming.

External caller   A participant who joins a conference over the Public Switched Telephone Network (PSTN) and is not authenticated.

External user   A user who connects from outside the organization's firewall. External users include anonymous users, federated users, and remote users.

Federated Group Chat   A Group Chat feature that enables members of different organizations to post and access Group Chat content in Group Chat rooms.

Federated user   An external user who has valid credentials with a federated partner and who is authenticated on that basis by Office Communications Server 2007 R2.

Federation A trust relationship between two or more SIP domains that allows users in separate organizations to communicate in real time across network boundaries as federated partners. Internal users can communicate with external users of a federated partner by using IM, audio/video, or conferencing.

Filters In chat rooms, a feature that allows a user to selectively monitor chat room messages and execute a specific action if specified criteria are met.

Focus A conference state server that acts as the coordinator for all aspects of a conference. It is implemented as a SIP user agent that is addressable by using a conference URI. The Focus server runs in the User Services module of all front-end servers.

Focus Factory A conferencing component that provides the appropriate conferencing servers requested by the Focus for a conference and manages their state for the duration of the conference. The Focus Factory handles conference creation and deletion.

Forest Prep Action that prepares an Active Directory forest through the creation of objects, containers, and extended property rights. For Office Communications Server, Forest Prep creates universal groups for user and server administration.

Formal agent *See* Agent.

Forms-based authentication An authentication method by which a user enters a user name and password into a Web page. The Web server compares this information to a database or XML configuration file to determine whether to authenticate the user.

Front End Server An Office Communications Server 2007 R2 server in the internal network that hosts the Application Host, IM Conferencing Service, Address Book Service, and Telephony Conferencing Service to support registration, presence, IM, and conferencing. In an Enterprise pool, it can be collocated with the Web Conferencing Server and A/V Conferencing Server, or it can be deployed on a separate server.

GAL Global Address List.

Global Address List (GAL) A directory that contains entries for every group, user, and contact in an organization's implementation of Exchange Server.

Global group A security or distribution group that can contain users, groups, and computers from its own domain as members. Global security groups can be granted rights and permissions on resources in any domain in its forest.

Global-level settings Settings that apply to the entire Active Directory forest and affect all servers and users in the forest.

Globally Routable User Agent URI (GRUU) An Internet Engineering Task Force (IETF) standard that extends the Session Initiation Protocol (SIP) so that it is possible to reliably route to a specific device that belongs to a user anywhere on the IP network.

GPMC Group Policy Management Console.

Grammar files Files that provide rules that define all possible combinations of the words or phrases that a user can speak to an application. These rules enable the speech recognition engine to convert speech to text and add semantic information to the recognized text.

Group Chat room A topic-specific chat room, similar to discussion forums, for Group Chat sessions.

Group Policy A Windows Server feature that provides an infrastructure for centralized configuration management of the operating system and applications that run on the operating system.

Group Policy Management Console (GPMC) A downloadable console that simplifies the management of Group Policy by making it easier to understand, deploy, manage, and troubleshoot Group Policy implementations. GPMC also enables automation of Group Policy operations via scripting.

GRUU Globally Routable User Agent URI.

Hardware load-balancing device (HLD) A single-purpose piece of hardware that is used in a scaled single-site server topology on the external and internal side of the edge network. It manages connections only across a series of two or more computers to make it appear as one and handles the failure scenario when one computer fails, redirecting traffic to the rest of the devices.

Hidden Markov Model (HMM) A statistical model that is often applied to temporal pattern recognition, such as speech recognition.

HMM Hidden Markov Model.

Home, homed The user's home server is the Standard Edition server or Enterprise pool that is specified in the user's Office Communications Server 2007 R2 properties. The user is said to be "homed" on the specified server or pool.

HTTP reverse proxy A server in the perimeter network that is required to enable external users to download meeting content, to expand distribution groups, or to download files from the Address Book service. The reverse proxy does not run Office Communications Server and therefore is not an Office Communications Server role.

Hunt group A group of PBX or IP PBX resources or extension numbers that are shared by users. A hunt group is used to direct calls to identity-capable endpoints or to an application, such as voice mail.

Hybrid media gateway A media gateway vendor category that consists of a Basic Media Gateway with the functionality of the Mediation Server coexisting on the same physical server.

ICE Interactive Connectivity Establishment.

IM Instant messaging.

IM Conferencing Server An Office Communications Server 2007 R2 conferencing server that provides server-managed group IM. It runs as a separate process on the Standard Edition server or Enterprise pool Front End Server.

IM Conferencing Service A service that runs on an Office Communications Server 2007 R2 Front End Server that mixes and manages inputs from multiple clients in a group instant messaging (IM) session.

IM service provider A public or private organization that provides instant messaging services for multiple domains.

Informal agent *See* Agent.

Instant messaging (IM) A way to communicate interactively with one or more people by using a live text session. Unified Communications uses Office Communicator 2007 R2 as the client for IM.

Integrated Windows authentication (IWA) An authentication method in which a user name and password credentials are hashed before they are sent over the network. Also known as Windows NT Challenge/Response authentication.

Interactive Connectivity Establishment (ICE) A network protocol, developed by the Internet Engineering Task Force's (IETF) MMUSIC working group, that provides a mechanism for Network Address Translation (NAT) traversal, using various techniques. In particular, it is used to allow SIP-based VoIP clients to successfully traverse the many firewalls that might exist between a remote user and a network.

Interactive Voice Response (IVR) A feature of the Response Group Service that detects and recognizes both speech and dual-tone multifrequency (DTMF) keypad input. Response Group Service IVR also supports text-to-speech and the .WAV file format.

Internal IP address An IP address that can be accessed from the internal network of an organization. Also known as a private IP address.

Internet Protocol Private Branch eXchange (IP PBX) Device that acts as both a Session Initiation Protocol (SIP) server and a voice gateway.

INVITE A SIP request that helps establish sessions for client-to-client communication and to establish sessions with servers. Servers include the A/V Conferencing Server, the IM Conferencing Server, and the ACP Conferencing Server.

IP PBX Internet Protocol Private Branch eXchange.

IP/PSTN gate A media gateway that supports interoperation between IP-based and PSTN-based systems.

ISA Internet Security and Acceleration.

IVR Interactive Voice Response.

Join Conference An option in an Office Outlook 2007 meeting invite and reminder for joining a live meeting.

Kerberos An authentication protocol that builds on symmetric key cryptography and requires a trusted third party.

LCSCmd.exe An Office Communications Server tool that is used to configure Office Communications Server from the command line. Configuration options include setting up AD DS, creating Enterprise pools, activating and deactivating servers, requesting and assigning certificates, and exporting and importing server settings.

LCSDiscover A tool that is part of the Office Communications Server 2007 R2 Resource Kit Tools that discovers settings for previous and current versions of Live Communications Server 2005 and Office Communications Server R2.

Least-Cost Routing A version of the 2007 R2 release of Office Communicator Mobile for Windows Mobile in which Office Communications Server 2007 R2 performs reverse number lookup on one-number calls and routes the call over an IP connection to the recipient, routing the call over an IP connection to the Public Switched Telephone Network (PSTN) gateway that is nearest to the location of the destination number.

Line URI attribute An attribute in a user object that identifies the unique phone number assigned to a user that can be used for routing purposes. The Line URI contains the phone number that is assigned to an Enterprise Voice user. When a user is configured for dual forking, Office Communications Server 2007 R2 uses the user's Line URI attribute defined in Active Directory to route calls to the PBX.

Location normalization A rule that specifies how to convert numbers dialed in many formats to standard E.164 format. Normalization rules are required for call routing and authorization because users can, and do, use many formats when they enter phone numbers in their contact lists.

Location profile A container that holds a name, a description, and a set of normalization rules that are used to translate a phone number into E.164 format.

Lookup Server A server role for Group Chat that provides the chat room address, distributes sessions to Channel Servers, and manages load balancing in multiple-server topologies.

Match Making component The component of the Response Group Service that is responsible to match an incoming call (also known as a match request) with an available agent. It is also responsible to keep track of all of the agents in the system and their presence state. The matching is configured by the administrator—the administrator configures agent groups and assigns these groups to queues. When a call comes in, it goes into a queue, and then the Match Making component finds an available agent that serves this queue.

MCU Multipoint control unit.

MCU Factory Container that stores all instances of MCU Factories. An MCU Factory is created when the first instance of a specific vendor and type of MCU (such as Conferencing Server) is activated. An MCU Factory manages the set of MCUs of a specific type that belongs to a Standard Edition Server or Enterprise pool.

Media gateway A device that translates signaling and media between the PSTN or PBX and Office Communications Server Directors and front-end servers. Office Communications Server supports three types of media gateway:

- **Basic** Media gateway and Mediation Server deployed on separate computers.
- **Basic-Hybrid** Media gateway and Mediation Server deployed on the same computer.
- **Advanced** Mediation Server logic combined with the media gateway software.

Media Relay Authentication Server (MRAS) An internal component name for the A/V Edge Server Authentication Service. It provides users and servers that are authenticated with Office Communications Server with the credentials required for setting up sessions involving the A/V Edge Server.

Mediation Server An Office Communications Server 2007 R2 server role in the internal network that mediates signaling and media between the Enterprise Voice infrastructure (such as a Director or home server) and another gateway (such as a Basic Media Gateway). A Mediation Server is also used to link

Office Communications Server and a PBX in both departmental deployment and PBX integration topologies.

MESSAGE A SIP request that is used to exchange instant messaging (IM) messages within established sessions.

Message Queuing A message queuing and routing system for Windows that enables distributed applications running at different times to communicate across heterogeneous networks and with computers that may be offline. Message Queuing provides guaranteed message delivery, efficient routing, security, and priority-based messaging. Formerly known as MSMQ.

Microsoft Clustering Service (MSCS) Software that provides a clustering technology that keeps server-based applications highly available, regardless of individual component failures.

Microsoft Exchange Unified Messaging (UM) A server that can tie an office Private Branch eXchange (PBX) to the Exchange e-mail server to allow storing voice mail in Exchange, accessing calendar and e-mail from a phone, and generally enabling messaging across many interactive interfaces.

Microsoft Forefront Family of business security products that help provide greater protection and control over the security of a network infrastructure, including client, server, and edge.

Microsoft Internet Security and Acceleration (ISA) Server A family of Microsoft multilevel firewall and high-performance caching server software.

Microsoft Management Console (MMC) Management user interface (UI) and framework that is part of the Windows operating system. It enables snap-ins to be loaded for a more consistent management experience across several applications and services.

Microsoft Office Communications Server 2007 R2 *See* Office Communications Server 2007 R2.

Microsoft Office Communications Server 2007 R2 Best Practices Analyzer *See* Office Communications Server 2007 R2 Best Practices Analyzer.

Microsoft Office Communications Server 2007 R2 Group Chat *See* Office Communications Server 2007 R2 Group Chat.

Microsoft Office Communications Server 2007 R2 Group Chat Administration Tool *See* Office Communications Server 2007 R2 Group Chat Administration Tool.

Microsoft Office Communications Server 2007 R2 Group Chat Server *See* Office Communications Server 2007 R2 Group Chat Server.

Microsoft Office Communications Server 2007 R2 Group Chat Server Configuration *See* Office Communications Server 2007 R2 Group Chat Server Configuration.

Microsoft Office Communications Server 2007 R2 Logging Tool *See* Office Communications Server 2007 R2 Logging Tool.

Microsoft Office Communications Server 2007 R2 protocol analysis tool (Snooper.exe) Resource Kit Tool that can help you analyze SIP and C3P protocol logs, including those generated by OCSLogger.exe.

Microsoft Office Communications Server 2007 R2 Response Group administrative snap-in *See* Office Communications Server 2007 R2 Response Group administrative snap-in.

Microsoft Office Communications Server 2007 R2 Validation Wizard *See* Office Communications Server 2007 R2 Validation Wizard.

Microsoft Office Communicator 2007 R2 *See* Office Communicator 2007 R2.

Microsoft Office Communicator 2007 R2 Attendant *See* Office Communicator 2007 R2 Attendant.

Microsoft Office Communicator Automation API The Office Communicator Automation application programming interface (API) contains a set of Component Object Model (COM) interfaces, objects, events, enumerated types, and other related programming entities.

Microsoft Office Communicator 2007 R2 chat rooms *See* Office Communicator 2007 R2 chat rooms.

Microsoft Office Communicator 2007 R2 Phone Edition *See* Office Communicator 2007 R2 Phone Edition.

Microsoft Office Communicator Mobile for Windows Mobile *See* Office Communicator Mobile for Windows Mobile.

Microsoft Office Communicator Web Access plug-in A plug-in that allows desktop sharing in Communicator Web Access.

Microsoft Office Live Meeting 2007 *See* Office Live Meeting 2007.

Microsoft RoundTable conferencing device The 360-degree A/V conferencing unit that works as a Universal Serial Bus (USB) camera and microphone device for Live Meeting and that shows a panoramic video of a conference room to remote participants.

Microsoft SIP Processing Language (MSPL) A scripting language used specifically for filtering and routing SIP messages. Known as "message filters," such scripts are embedded in the application manifests of Office Communications Server applications.

Microsoft SQL Server Desktop Engine (MSDE) A redistributable database engine compatible with SQL Server that is designed primarily to provide a low-cost option for developers who need a database server that can be easily distributed and installed with a business solution. MSDE is not supported on the Microsoft Windows Vista operating system and was replaced by SQL Server 2005 Express Edition.

Microsoft SQL Server 2008 Reporting Services (SSRS) A server-based reporting platform that provides comprehensive reporting functionality for a variety of data sources. In Office Communications Server, it is used in conjunction with the Monitoring Server Report Pack.

Monitoring Server A server role in the internal network that collects Call Detail Record (CDR) information. Also, it can collect quality of experience (QoE) metrics data that is sent by participant endpoints at the end of each A/V session.

MPOP   Multiple Points of Presence.

MRAS   Media Relay Authentication Server.

MSCS   Microsoft Clustering Service.

MSDE   Microsoft SQL Server Desktop Engine.

MSPL   Microsoft SIP Processing Language.

MSTURN   Microsoft extensions to Traversal Using Relay NAT.

MTLS   Mutual Transport Layer Security.

MUI   Multilingual User Interface.

Multilingual User Interface (MUI)   A set of language-specific resource files that are applied as a language pack in Windows Vista.

Multiple Points of Presence (MPOP)   Enhanced presence model that aggregates a user's presence status from multiple endpoints, which can include IP phones, Office Communicator, Office Communicator Web Access, or Office Communicator Mobile for Windows Mobile.

Multipoint control unit (MCU)   A pluggable component that is responsible for managing one or more media types. Also known as a conferencing server.

Mutual Transport Layer Security (MTLS)   The TLS (Transport Layer Security) and MTLS protocols provide encrypted communications and endpoint authentication on the Internet. Office Communications Server 2007 R2 uses these two protocols to create its network of trusted servers and to ensure that all communications over that network are encrypted. All SIP communications between servers occur over MTLS. SIP communications from client to server occur over TLS.

My Chat   A feature of Group Chat that displays all of the chat rooms that you have joined and the filters that are active.

NAT   Network Address Translation.

Network Address Translation (NAT)   The process of converting network addresses. A NAT enables computers in organizations with private networks to access resources on the Internet or other public networks.

NOTIFY   *See* SUBSCRIBE.

NTBackup   A file system backup solution available in Microsoft Windows for backing up meeting content and meeting compliance logs.

NT LAN Manager (NTLM)   A Microsoft authentication protocol that uses a challenge-response sequence that requires the transmission of three messages between the user and the server.

NTLM   NT LAN Manager.

Office Communications Server 2007 R2   Part of the Unified Communications software solution from Microsoft that enables integration of VoIP, instant messaging (IM), chat, conferencing, presence, and other communication solutions. Office Communications Server 2007 R2 includes new features for Call Delegation, Team Call, server roles, and more.

Office Communications Server 2007 R2 Best Practices Analyzer   A tool for administrators who want to determine the overall health of their Office Communications Server 2007 R2 servers and topology.

Office Communications Server 2007 R2 Group Chat   The client application that is used to post and access channel content, including files, links, and text, as well as to exchange instant messages between two users.

Office Communications Server 2007 R2 Group Chat Administration Tool A component that enables a Group Chat administrator or delegated user to manage Group Chat categories and groups, as well as user and group accounts.

Office Communications Server 2007 R2 Group Chat Server A new server role introduced in Office Communications Server 2007 R2 that enables users to create and maintain conversations that persist over time and can be archived. This functionality is especially useful for compliance purposes.

Office Communications Server 2007 R2 Group Chat Server Configuration A component that allows an Office Communications Server 2007 R2 administrator or delegate to configure the Group Chat Server following installation.

Office Communications Server 2007 R2 Logging Tool A tool that starts and stops server logs as well as filters and displays logs.

Office Communications Server 2007 R2 Response Group administrative snap-in The Microsoft Management Console (MMC) interface that is used to manage how the system handles call routing. This snap-in is used only by the Office Communications Server administrator.

Office Communications Server 2007 R2 Validation Wizard A tool that analyzes and validates the current configuration and connectivity to detect errors, validate basic end-to-end scenarios, and provide recommendations.

Office Communicator 2007 R2 A Windows-based desktop client that enables users to access the communications and collaboration capabilities of a computer.

Office Communicator 2007 R2 Attendant An integrated client application for managing Response Group features.

Office Communicator 2007 R2 chat rooms A desktop client application that makes chat room conversations and features available to users. Chat rooms enable users to send and receive instant messages, either in a chat room or singly with another user outside the chat room.

Office Communicator 2007 R2 Phone Edition An intelligent IP phone designed to maximize the Unified Communications platform. It combines network voice, user-driven design, up-time reliability, quality audio, and the enhanced communication and collaboration of Office Communications Server 2007 R2.

Office Communicator Mobile for Windows Mobile The client for instant messaging and presence for Windows Mobile-powered devices.

Office Communicator Web Access Browser-based client software that enables users to access instant messaging, presence, and desktop sharing capabilities on a Windows, Mac, or Linux computer.

Office Live Meeting 2007 Client software that gives access to the conferencing and application sharing capabilities on a computer.

One-number calling   A feature of the 2007 R2 release of Office Communicator Mobile for Windows Mobile that allows users to have a unique single number through which all of the users' registered endpoints, including their mobile phone, can be reached.

Open authenticated   A meeting type in which all enterprise users can join the meeting. Users join as attendees unless they have been designated as presenters by the meeting organizer. Federated users can join the meeting as attendees if they are invited by the organizer. Federated users cannot join the meeting as a presenter, but they can be promoted to presenter during the meeting.

Organizer   The owner or creator of a conference.

Outside Voice Control   A service that connects a mobile device to the enterprise network. This service enables a mobile device user to send and receive calls that come through the enterprise network, in addition to calls that come through the cellular carrier network. The mobile device must be running the 2007 R2 release of Office Communicator Mobile for Windows Mobile.

Paired Mode   A mode of operation for Office Communicator and Communicator Phone Edition in which calls on the Office Communicator Phone Edition can be managed on Office Communicator. Office Communicator and Office Communicator Phone Edition automatically enter Paired Mode when a Universal Serial Bus (USB) cable is attached between them.

Participant   A user who is participating in a conference or peer-to-peer call or the object that is used to represent that user.

Participant List   A feature of Group Chat that shows the names and presence status of all members currently in the chat room.

PBX   Private Branch eXchange.

PCA   Personal Call Assistant.

Persistence   A Group Chat feature that enables Group Chat content to be retained after the end of a Group Chat session and accessed on an ongoing basis.

Persistent Shared Object Model (PSOM) protocol   A custom protocol for transporting conferencing content.

Personal Call Assistant (PCA)   A presence-enabled speech service that queues and routes calls based on an individual's availability and communications preferences.

Phone-only contact   A PSTN phone number that is stored as a contact in the contact list in Office Communicator 2007 R2.

Phone route   A rule that defines which Mediation Servers should be routed to by phone calls that match a specific number pattern.

PIA   Primary interop assembly.

PIDF   Presence Information Data Format.

Policy   A collection of user-specific settings abstracted by the name of the policy.

Pool   A Standard Edition server or an Enterprise server.

Pool-level settings   Settings for a specific server role that are applied to all computers in an Enterprise pool or to a specific server role on a Standard Edition server.

Pre-Call Diagnostics (PCD)   A tool that tests the last-hop wireless network conditions and provides guidance about possible quality issues before calls are placed. This tool is especially useful for mobile or remote users for whom the quality of the last hop network connection can vary widely.

Presence   The ability to *see* the status, such as Available, Busy, Away, and so on, of a person. Presence is a fundamental requirement for effective real-time communications.

Presence-Based Routing   A feature of the Response Group Service that allows call routing to be configured to take agent presence into account.

Presence Information Data Format (PIDF)   A data format for exchanging presence information.

Presence integration   An action that occurs in a dual forking scenario in which, when a call is answered from the PBX telephone, Office Communicator 2007 R2 automatically sets the user's presence to the "In a Call" state.

Presentity   An entity that provides presence information to a presence service.

Primary interop assembly (PIA)   An assembly that contains a signed set of wrapper classes that enables you to call unmanaged code from managed code.

Private Branch eXchange (PBX)   A switching system for voice communication that routes internal calls directly without access to the PSTN. A PBX can be a conventional phone network, a VoIP network, or a combination of the two.

PSTN   Public Switched Telephone Network.

PSTN agent   Used with the Deployment Validation tool to simulate a Public Switched Telephone Network (PSTN) telephone.

Public Switched Telephone Network (PSTN)   The telephone system standard that is used throughout the world.

Publicly routable IP address   An IP address that can be directly routed from outside an organization's firewall to the perimeter network or the internal network.

Quality of Experience (QoE)   A measurement of the overall user experience of a particular communication. For example, in a voice communication, QoE monitors items such as echoes and background noises (eg, hissing in the line). In Office Communications Server 2007 R2, QoE is monitored as part of the Monitoring Server role.

Quality of Service (QoS)   A metric that reflects or predicts the subjectively experienced quality. QoS is the cumulative effect on user satisfaction of all imperfections that affect the service and is determined by statistics that are collected on a media connection and information such as bytes sent, packets sent, lost packets, jitter, feedback, and round-trip delay.

Queue   A logical list of calls that are managed by the Response Group Service until an action is taken on these calls. For example, transfer the call to an agent or a different destination. Typically, the Response Group plays music-on-hold for the calls in the queue.

RCC   Remote Call Control.

Real-Time Audio (RTAudio)   An advanced speech codec that is designed for real-time two-way Voice over IP (VoIP) applications and is used by the A/V Conferencing Server.

Real-Time Control Protocol (RTCP)   A network transport protocol, specified in RFC 3550, that enables monitoring of Real-Time Transport Protocol (RTP) data delivery and provides minimal control and Identification functionality. The primary function of RTCP is to provide quality-of-service information for RTP.

Real-Time Streaming Protocol (RTSP)   A protocol for use in streaming media systems that allows a client to remotely control a streaming media server, specified by RFC 2326.

Real-Time Transport Protocol (RTP)   A network transport protocol that provides end-to-end transport functions that are suitable for applications that transmit real-time data, such as audio and video, as specified in RFC 3550.

Real-Time Video (RTVideo)   An advanced video codec that is designed for real-time video applications and is used by the A/V Conferencing Server.

REGISTER   A SIP request that is used by a SIP client to register the client address with a SIP server.

Regular expressions   Strings that are used to describe or match sets of strings according to certain syntax rules.

Remote Call Control (RCC)   The ability to send and receive calls on a desktop phone by using Computer Supported Telephony Applications (CSTA), such as Office Communicator. With Remote Call Control in Office Communicator 2007 R2, your phone system is integrated with a Private Branch eXchange (PBX) system and offers call forwarding features, but does not offer features such as ringing an additional number or redirecting unanswered calls.

Remote user   An external user whose account has a corresponding User object in the Active Directory Domain Services (AD DS).

Reporting Services   *See* Microsoft SQL Server 2008 Reporting Services (SSRS).

Response Group   Functionality that allows for incoming calls to be queued and routed to designated agents based on a set of defined routing rules.

Response Group Configuration tool   The Web interface that is used to manage the predefined templates. This is where an administrator can set up a team routing workflow, including announcement messages, working hours, questions and options given to the caller, music-on-hold, and to which queues to route calls. This interface is also used by users who are enabled as Response Group managers so as to manage specific Response Group settings, such as working hours and music-on-hold.

Response Group Deployment tool   A tool that is offered through the Web component of Response Group Service that enables administrators and Response Group Managers to administer Response Groups in a Web interface.

Response Group Manager   A user who is given the rights by the Response Group administrator to manage specific Response Group settings, such as business hours and music-on-hold, by using the Response Group Configuration tool.

Response Group Service   A service that is installed by default with Office

Communications Server 2007 R2 that enables administrators to create and configure one or more small Response Groups for routing and queuing incoming phone calls to one or more designated agents.

Response Group Service Contact Object Tool (RGSCOT.exe) Command-line utility that you can use to create and manage Response Group Service Contact objects.

Response Group tab The Office Communicator 2007 R2 tab that is used by the normal agent to sign in and sign out of Response Group groups.

Response Group templates Predefined templates that simplify creating a new Response Group. Templates define functionality such as questions asked to the caller, options given to the caller, music-on-hold options, and configuration of business hours and holidays.

Reverse number lookup (RNL) The functionality of matching an incoming E.164 number to an entry in the Global Address List (GAL) or local Office Outlook contacts.

Reverse proxy server A server in the perimeter network that is required if either of the following tasks are required: enabling external users to download meeting content or expand distribution groups, or enabling remote users to download files from the Address Book service. This can be a server that is running Internet Security and Acceleration (ISA) Server or another reverse proxy server.

RGSCOT.exe Response Group Service Contact Object Tool.

Rich presence Enhanced presence features that provide additional availability data, including next-available-meeting timing and out-of-office information.

RNL Reverse number lookup.

Route Helper tool Also known as Enterprise Voice Route Helper, this is a graphical user interface (GUI) tool that is part of the Office Communications Server 2007 R2 Resource Kit Tools. It provides everything that is required to create, modify, and analyze an Enterprise Voice routing configuration to deploy and maintain an Office Communications Server Enterprise Voice solution. The tool also simulates phone number normalization done by the client when the user's location profile is specified.

RTCP Real-Time Control Protocol.

RTP Real-Time Transport Protocol.

RTSP Real-Time Streaming Protocol.

SAN Subject Alternative Name.

Schema The set of definitions for the universe of objects that can be stored in a directory. For each object class, the schema defines which attributes an instance of a class must have, which additional attributes it can have, and which other classes can be its parent object class.

SCP Service connection point.

SDP Session Description Protocol.

Security association (SA) An establishment of shared security information between two user agents to enable them to communicate securely.

Security Support Provider Interface (SSPI) A common interface between transport-level applications that allows an application to use various security models available on a computer or network without changing the interface to the security system.

Server pool   A group that consists of multiple Lookup Servers and Channel Servers that supports communication and the sharing of data between servers, as well as the implementation of optional load balancing and failover.

Server role   A logical grouping of features and components in a software application. Server roles that are new to Office Communications Server 2007 R2 are Application Sharing Conferencing Server, Monitoring Server, and Group Chat Server.

SERVICE   A SIP request that is defined by the SIP extensions and is used by a SIP client to request a service from a server, such as changing a user's presence.

Service connection point (SCP)   In Active Directory, a marker that registers the kind of service installed on the computer that is joined to the Active Directory forest. Used to determine which services are running on every computer.

Service Level Agreement (SLA)   A written agreement that documents the required levels of service as agreed on by the IT service provider and the business or the IT service provider and a third-party provider.

Session   The time period during which the Group Chat client is connected to the Group Chat server.

Session Description Protocol (SDP)   A protocol that is used to announce sessions, manage session invitations, and perform other kinds of initiation tasks for multimedia sessions, as specified in RFC 3264.

Session Initiation Protocol (SIP)   A signaling protocol for Internet telephony.

Simple Object Access Protocol (SOAP)   Lightweight XML-based protocol for exchanging information in a decentralized, distributed environment.

Single sign on   A feature that enables users to sign in to Office Communications Server 2007 R2 by using their Windows credentials so that they do not have to manage separate credentials.

SIP   Session Initiation Protocol.

SIP address   A URI that identifies an end node in a Session Initiation Protocol (SIP) network. The format of a SIP address is identical to that of an e-mail address.

SIP/CSTA   A gateway that connects to the existing PBX or IP PBX that hosts a user's PBX or IP PBX phone. CSTA is an international standard that is set by Ecma International to combine computers that share resources with PBX or IP PBX environments.

SIP domain   A domain that is configured to accept SIP traffic.

SIP element   An entity that understands the Session Initiation Protocol.

SIP INFO message   A method for sending call-related information to and from the SIP/CSTA gateway.

SIPParser   A protocol parser for Session Initiation Protocol (SIP) that can be plugged into Network Monitor for viewing nonencrypted SIP over Transmission Control Protocol (TCP).

SIP proxy   In a Session Initiation Protocol network, a server that makes requests on behalf of other clients and routes SIP requests to another entity that is closer to the targeted user. SIP proxy is defined in RFC 2161.

SIP registrar   A server that accepts REGISTER requests and puts the information that it receives from those requests into the location service for the domain that it handles. SIP registrar is defined in RFC 3261.

SIP trunking   A mechanism that is used by an enterprise to connect its voice network to a service provider offering Public Switched Telephone Network (PSTN) origination, terminations, and emergency services without deploying IP-PSTN gateways, with or without Mediation Servers.

Smartphone   A Windows Mobile device that has telephony capability. Smartphone includes both telephones and Pocket PC devices that can function as a telephone.

SNAT   Source Network Address Translation.

Snooper   A graphical user interface (GUI) Resource Kit Tool that is used for summarizing, searching, and viewing client and server protocols and trace logs. This tool also works for Office Communications Server 2007 server logs.

SOAP   Simple Object Access Protocol.

Softphone   A multimedia application that works with VoIP technology to enable you to make calls directly from a computer. A softphone is typically used with a headset that is connected to the sound card of the computer or with a Universal Serial Bus (USB) telephone.

Source Network Address Translation (SNAT)   A process that involves rewriting the source or destination addresses of IP packets as they pass through a router or a firewall.

Spim   Spam over instant messaging. Unsolicited bulk commercial instant messages.

SRVLookup   A tool that queries relevant Domain Name System (DNS) Service Record Locator records for the specified domain. The tool is useful for federation and login diagnostics.

SSO   Single sign on.

SSPI   Security Support Provider Interface.

SSRS   SQL Server 2008 Reporting Services.

Standard Edition server   A server in the internal network running Office Communications Server 2007 R2, Standard Edition that hosts all of the necessary services, including IM, presence, and conferencing services as well as the database, on a single server.

Subject Alternative Name (SAN)   The field on a digital certificate that provides for a list of host names to be protected by a single Secure Sockets Layer (SSL) certificate.

Subject name (SN)   The text-based field in a certificate that identifies the name of the user or server that it refers to.

SUBSCRIBE   A SIP request that is used to set up event notifications from the server. SUBSCRIBE is used by clients to subscribe to information that can change because of updates. NOTIFY is used by the server to notify clients about information that has changed.

Subscriber access number   A number that is configured in a Private Branch eXchange (PBX) that allows a subscriber to access their mailbox over the telephone.

Subscription   In the context of the data that is stored for each user in the Office Communications Server 2007 R2 database, a subscription is a set of contacts for which a user wants to receive presence updates. Presence updates occur when one of the user's contacts changes state, such as when the contact signs in to Office Communications Server or joins a phone call.

System Center Operations Manager   Software that provides end-to-end service management that is easy to customize and extend for improved service levels across your IT environment.

Tagged contact   A contact in Office Communicator 2007 R2 whose presence status is displayed in a message when the contact goes offline or online.

TDM   Time Division Multiplexing.

Team Call   A feature that was introduced in Office Communications Server 2007 R2 that allows calls to be forwarded from a team leader to an entire team. Depending on the options that are specified when the feature is configured, phones of all of the team members will ring until someone answers.

Telephony Conferencing Server   An Office Communications Server 2007 R2 conferencing server that enables audio conference integration with Audio Conferencing Providers (ACPs). It runs as a separate process on the Standard Edition server or Enterprise pool Front End Server.

Telephony Conferencing Service   A service that runs on an Office Communications Server 2007 R2 Front End Server that enables multiparty conferencing with PSTN callers connecting through an Audio Conference Provider (ACP).

Third-Party Control Protocol (TPCP)   A protocol developed by Microsoft to remotely manage calls on a server endpoint from a client endpoint. The protocol enables a TPCP client to issue commands to a TPCP server to create a call between two endpoints or to answer or deflect calls.

Third-party request   A conference control request that modifies the state of participants other than the participant who sent the request.

Time Division Multiplexing (TDM)   A circuit-switched technology that converts one or more voice streams into a single stream for transmission.

Time-to-live (TTL)   An interval that is determined based on the registration refresh interval or any other session timers in the dialogs that traverse the server.

TLS   Transport Layer Security.

TPCP   Third-Party Control Protocol.

TrackingDataWorkflowRuntimeService   Windows Runtime Service that provides workflow instance-related storage in memory for storing *ActivityTrackingData* objects.

Transport Layer Security   Provides a mechanism for applications to communicate securely over IP networks. *See* Mutual Transport Layer Security.

TTL   Time-to-live.

UCMA Speech API   A server-grade speech API that allows developers to build multichannel speech recognition– and speech synthesis–enabled applications using Microsoft state-of-the-art speech technology.

UCMA Windows Workflow Activities Activities that can be used to build workflow-enabled speech and instant messaging applications on Office Communications Server.

UCMA Workflow API A higher API abstraction layer of the UCMA Core and Speech APIs.

UDP User Datagram Protocol.

Unauthenticated user A user who has not received a Session Initiation Protocol (SIP) 200 OK response from the server during registration. Except for federated users, all users start as unauthenticated and are authenticated only after providing the appropriate credentials to the server.

Unified Communications Agent Used with the Deployment Validation tool to simulate an Office Communicator 2007 R2 client.

Unified Communications AJAX API API to the Office Communicator Web Access Server. The AJAX API is based on the AJAX programming model. This AJAX API consists of a set of methods and events. The methods are client requests to have something done, such as querying the presence of a user, setting the presence of the caller, inviting a user to an instant messaging conversation, and so on. The events are the results of the method invocations and are returned to the client by the server. The methods and events use XML for their data format.

Unified Communications Managed API (UCMA) An endpoint API that allows advanced developers to build and integrate server applications into an existing Office Communications Server infrastructure.

Unified Communications Managed API Core A managed-code platform that provides access to and control over instant messaging, telephony, audio/video conferencing, and presence. It is intended to support the development of middle-tier applications targeting Office Communicator and Office Communications Server 2007 R2.

Unified Messaging An application that consolidates a user's voice mail, fax, and e-mail into one mailbox so that the user needs to check only a single location for messages, regardless of type. The e-mail server is used as the platform for all kinds of messages, making it unnecessary to maintain separate voice and e-mail infrastructures.

Uniform Resource Identifier (URI) A unique address for a resource on the Internet. For example, a user might use someone@domain.com as a SIP URI to log on to a SIP server.

Universal group A security or distribution group that can contain users, groups, and computers from any domain in its forest as members. A universal security group can be granted rights and permissions on resources in any domain in the forest.

URI Uniform Resource Identifier.

Usage name A friendly name that is associated with a telephone route to indicate its intent or usage. For example, an administrator may determine that the following will be some of the usages in the organization: "local," "domestic long distance," or "international long distance."

USB Audio Streaming audio over the Universal Serial Bus (USB) link between the computer and the phone in Paired Mode. The call is homed on the computer, but there is a minimal control path from the phone to computer.

USB Human Interface Device (USB HID) A protocol for sending and receiving control messages from a Universal Serial Bus (USB) device.

User agent client (UAC) A logical entity that creates a new request and then uses the client transaction state machinery to send it. The role of UAC lasts only for that transaction. If a process initiates a request, it acts as a UAC for that transaction. If a process receives a request later, it assumes the role of a user agent server for that transaction.

User Agent Server (UAS) A logical entity that generates a response to a Session Initiation Protocol (SIP) request. The response either accepts, rejects, or redirects the request. The role of the UAS lasts only for that transaction. If a process responds to a request, it acts as a UAS for that transaction. If it initiates a request later, it assumes the role of a user agent client for that transaction.

User Datagram Protocol (UDP) A connectionless TCP/IP protocol that corresponds to the transport layer in the International Standards Organization/Open Systems Interconnect (ISO/OSI) reference model and does not offer reliable delivery of data.

User replicator (UR) A component of the Office Communications Server service that synchronizes the database with user information and global Office Communications Server settings that are stored in Active Directory.

User Services module A module that provides closely integrated instant messaging, presence, and conferencing features that are built on top of the SIP proxy.

Video negotiation The process in which an endpoint that is proposing to send video can determine the video capabilities of the receiving endpoint before sending the video stream.

Voice mail An application that automatically answers calls and stores messages for retrieval in the future.

Voice over IP (VoIP) Office Communications Server 2007 R2 provides the ability to do Voice over IP, which is the basis of voice communication over computer networks. With VoIP, the audio traffic is carried over the IP network in contrast to Remote Call Control (RCC), which is related to controlling the PBX phone from a computer that is running Office Communicator 2007 R2.

VoIP Voice over Internet Protocol.

Web Components Server A server in the internal network that provides software Web components that require Internet Information Services (IIS) to support Office Communications Server 2007 R2. These Web components include IIS Virtual Directory setup to support Address Book Server, the Web Conferencing Server (downloading of meeting content), and the IM Conferencing group expansion Web service. It runs on each Standard Edition server and, for Enterprise pools, either on the Front End Server (in a consolidated configuration) or a dedicated server that is running IIS (in an expanded configuration).

Web Conferencing Edge Server   A server role that enables data collaboration with external users. In the Office Communications Server 2007 R2 consolidated edge topology, this server role is collocated on the same computer as the Office Communications Server 2007 R2 Edge Server.

Web Conferencing Server   A server role that manages data collaboration for online conferences. This server role is available on a Standard Edition server. In an Enterprise pool, it can be collocated with the Front End Server and A/V Conferencing Server, or it can be deployed on a separate server.

Web Farm Services   A collection of Internet Information Services (IIS) servers or an IIS server hosting content.

Web Service   A server role for Group Chat that is used to post files to group channels.

WFP   Windows Filtering Platform.

WFP Presence Controls   Controls that provide applications with presence information for contacts. The information is presented visually in a manner similar to Office Communicator 2007. The controls also have context menu items that launch Office Communicator dialog boxes for IM and audio conversations. The controls are implemented by using the Office Communicator 2007 Software Development Kit (SDK).

Windows Filtering Platform (WFP)   Platform that provides APIs for extending the TCP/IP filtering architecture so that it can implement packet filtering at all levels of the TCP/IP protocol stack to help protect services and support additional services that inspect and filter TCP/IP packets, such as Windows Firewall.

Windows Installer package   An .msi file that contains explicit instructions about installing and removing specific applications. The company or developer who produces the application provides the Windows Installer package .msi file and includes it with the application.

Windows Management Instrumentation (WMI)   Part of Windows that provides fully integrated operating system support for uniform system and applications management. WMI is the Microsoft implementation of Web-based Enterprise Management (WBEM), which is an industry initiative to develop a standard technology for accessing management information in an enterprise environment.

Windows Presentation Foundation (WPF)   Part of the WinFX platform that introduces a new application type: Web Browser Applications (WBAs). Web Browser Applications are online-only applications that run in the browser and are not installed. These applications execute in a security sandbox and harness the power of the WPF platform on the Web.

Windows Software Trace Pre-processor (WPP)   Part of the Windows operating system that enables applications to easily have a configurable high performance logging infrastructure.

Windows Workflow Foundation (WF)   A technology platform for building workflow-enabled applications. The platform includes a set of tools for designing and implementing workflows, a programming model for controlling and communicating with workflows, a rules engine, a workflow execution engine, and a set of workflow runtime services for persistence, tracking, transaction management, and more.

Work Call Functionality that enables a user to dial a number from the 2007 R2 release of Office Communicator Mobile for Windows Mobile, but have the server back end actually make the call.

Workflow Runtime services Part of the Windows Workflow Foundation that provides services for persistence, tracking, transaction management, and more.

Working Hours Office Outlook Calendar options that specify the hours that a person is typically at work. Working hours are used in advanced call handling to determine call forwarding rules and are also displayed as part of presence information.

WPF Windows Presentation Foundation.

XAML Extensible Application Markup Language.

Index

A

A record (DNS), 246
A/V (audio/video) communications
   API support, 13
   conference support, 195
   configuring conferencing, 269–272
   contextual collaboration, 6
   creating calls, 191
   MCU support, 184
   Office Communicator Automation API, 23, 87
   troubleshooting failures, 313
   UC support, 3
   UCC API support, 27
   UCMA Core API, 225
   UCMA support, 16
A/V Conferencing Server
   application development support, 219, 223
   configuring, 269–272
AcceptCallActivity
   Dequeue method, 117
   functionality, 18, 126–127
   properties, 127
access control entries (ACEs), 197, 205, 223
ACD (automatic call distributor)
   UC challenges, 4
   UCMA support, 11, 15
ACEs (access control entries), 197, 205, 223
acknowledgeSubscriber element (XML), 35
Active Directory Certificate Services (AD CS), 221
Active Directory Domain Services. *See* AD DS
Active Directory Domain Services Installation Wizard, 229–233
Active Directory Users and Computers snap-in, 224, 289
ActiveX control, 9

activities
   call control, 126–128
   call control communications events, 141–144
   command, 137–141
   custom, 18, 20, 113
   dialog, 128–137
   dialog communications events, 142, 144–147
   general, 120–125
   presence-related, 147–148
Activity class, 20
AD CS (Active Directory Certificate Services), 221
AD DS (Active Directory Domain Services)
   application development support, 219
   Contact object, 187, 221, 225, 285
   creating user accounts, 273
   managing networks, 221–223
   preparing for UC, 237–245
   TSE support, 285
   User object, 221
AD DS domains
   AD DS preparation, 223
   joining servers, 249–251
   preparing, 221, 244–245
   verifying settings, 245
AD DS forests
   assigning static IP addresses, 227–228
   building, 226–237
   installing domain CA, 234–237
   preparing, 221–223, 242–243
   promoting computers, 228–233
   verifying DNS server role, 233
   verifying settings, 244
AD DS preparation, 221–223
AD DS schemas
   extending, 221, 238–239
   verifying extended, 240–242
Add Roles Wizard, 253–254

addContact element (XML), 35
addGroup element (XML), 35
Address Book Server, 262
administrative tools
    installing, 224, 239–240
    server role requirements, 224
alerts
    Office Communicator Automation API, 23
    outbound, 14
    UCMA support, 11, 14
    UCMA Workflow support, 19
AnswerCallActivity class, 44
anywhere access information, 8
APIs (application programming interfaces).
        See also specific APIs
    additional resources, 46
    types supported, 8
application development. See also
        programming applications
    AD DS support, 221–223
    application components, 219–221
    building AD DS forest, 226–237
    configuring components, 279–291
    configuring DNS for automatic sign-in,
        245–248
    configuring UC user accounts, 272–278
    deploying OCS Standard Edition, 226
    installing/configuring OCS, 251–272
    Office Communications Server roles,
        223–224
    Office Communicator Automation API,
        224–225
    preparing AD DS for UC, 237–245
    setting up host computer, 248–251
    UCMA Core API, 225
    UCMA Workflow API, 225
    validating server functionality, 278–279
application programming interfaces.
        See APIs
application sharing
    UC support, 7
    UCC API support, 29
Application Sharing Server, 219, 224
ApplicationEndpoint class
    BeginEstablish method, 188–189
    creating calls, 191

creating instances, 183
enabling trusted applications, 285
EndEstablish method, 188–189
endpoint support, 187–189
functionality, 15
LocalEndpoint class and, 187
publishing presence, 198
UCMA support, 15
ApplicationEndpointSettings class, 188
ApplicationProvisioner.exe tool, 286
Approve action, 165–167
ArgumentException class, 328
ASR (automatic speech recognition). See
    speech recognition
audio/video communications. See A/V
    communications
Audio/Video Conferencing Server
    application development support, 219, 223
    configuring, 269–272
AudioVideoCall class
    creating calls, 191
    handling call flows, 193
    handling incoming calls, 193
    joining conferences, 197
AudioVideoFlowConfigurationRequested
    event, 193
authentication
    AD DS support, 221
    Unified Communications AJAX API, 33–34
AuthenticationException class, 317
automated agents
    UC opportunities, 7
    UCMA support, 11, 14
    UCMA Workflow support, 19
automatic call distributor. See ACD
automatic sign-in, 245–248
automatic speech recognition. See speech
    recognition

B

B2BUAs (back-to-back user agents), 14
binding
    conversations to endpoints, 184
    outbound calls, 161–162
    outbound IM calls, 168–169

blind transfers
    BlindTransferActivity, 44, 126, 128
    UCMA Workflow support, 21
BlindTransferActivity, 44, 126, 128
bots, query/response. See automated agents
breakpoints
    running applications to, 298
    setting, 297–298, 333
    stepping through code execution, 299
business process communication
    adding commands to dialog, 174–175
    adding events to dialog, 175–177
    additional resources, 179
    approver's presence information, 154–157
    building the application, 151–179
    business value, 149–150
    choice of technology, 150
    code structure, 150
    connecting to Office Communications Server, 151–152
    contacting approver by IM, 167–173
    contacting approver by phone, 160–167
    creating branches for modalities, 159–160
    creating Communication Workflow Project, 151
    defined, 8
    implementing branch logic, 155–157
    running applications, 178–179
    scenario, 149
    test environment, 150
    UC support, 7
    UCMA Workflow support, 19
    updating canBeContactedBranch, 158–177
    updating cantBeContactedBranch, 157–158
    user input to workflow instance, 152–154

C

CA (certificate authority)
    application endpoints, 187
    building AD DS forests, 227
    installing domain CA, 234–237
calendaring, 7
Call class, 191

call control activities
    functionality, 126–128
    Speech Server Managed API, 19
    UCMA support, 16
    UCMA Workflow support, 19, 21
    Unified Communications AJAX API, 32
call control communications event activities, 141–144
call deflection, 9, 32
call events, 177
call flows, 193–194
call routing, 14, 38
call state, 194
CallDisconnectedEventActivity, 142–143
CallOnHoldEventActivity, 142–143
CallOnHoldTimeoutEventActivity, 142, 144
CallProvider activity, 122–125
CallRetrievedEventActivity, 142, 144
calls. See also voice calls
    conferences and, 195
    creating, 191–192
    defined, 191
    handling incoming, 192–193
canBeContactedBranch
    adding commands to dialog, 174–175
    adding events to dialog, 175–177
    contacting approver by IM, 167–173
    contacting approver by phone, 160–167
    creating branches for modalities, 159–160
    disconnecting calls, 175
    updating, 158–177
cantBeContactedBranch, 157–158
CategoryNotificationReceived event, 201
CCCP (Centralized Conference Center Protocol), 28
certificate authority. See CA
Certificate Enrollment Wizard, 285
certificates
    configuring IIS, 266–268
    host computers, 282
    installing computer, 285
    installing domain CA, 234–237
    installing/configuring MTLS, 263–266
    installing/configuring TLS, 263–266
    modifying Program.cs, 152
    verifying installation, 283–284

Certificates snap-in, 283
CFG files, 45
class ID (CLSID), 280
ClientPlatformSettings class, 183–184
CLSID (class ID), 280
Code Activity, 20, 165
CollaborationPlatform class
    creating applications, 183–186
    creating/configuring, 185
    EndStartUp method, 185–186, 318
    functionality, 17
    publishing presence, 198
    starting process, 185–186
COM (Component Object Model)
    additional resources, 25, 85
    API support, 7
    interface IDs definition file, 225
    Office Communicator Automation API, 24, 224
    UCC API support, 29
COM interop service
    COM errors, 295, 310
    IMessengerAdvanced interface, 82
    Messenger class, 50
    Office Communicator Automation API, 225
Comet mechanism, 33
COMException class, 293, 295, 306
command activities, 137–141
command channel
    defined, 33
    Unified Communications AJAX API, 34
CommandActivity class, 45
commands
    adding to dialog, 174–175
    defined, 138
communication events, 142
Communication Workflow Project, 151
CommunicationsSequenceActivity
    binding outbound calls, 161–162
    binding outbound IM calls, 168–169
    call control communications event activities, 142
    CallProvider property, 168
    command activities and, 138, 140
    functionality, 120–125

CommunicationsWorkflowRuntimeService, 19, 21, 116–118
Communicator registry key, 50
Communicator Web Access server role, 9
Communicator Web Access snap-in, 224
CommunicatorAPI.dll library, 24
CommunicatorPrivate.dll library, 24
compile-time errors, 293
Component Object Model. *See* COM
computer certificates, 285
Computer Management console, 224, 266–268
Computer Supported Telephony Applications (CSTA), 28
Computers container, 223
Conditional Group Activity, 20
conference bridging, 14
Conference class, 196
conference element (XML), 35
Conference element (XML), 37
ConferenceFailureException class, 316
conferences/conferencing
    ConferenceServices class, 17
    defined, 195
    joining, 184, 196–197
    managing, 14
    multiparty sessions, 16
    recording calls, 14
    scheduling, 14, 184, 195–196
    UCC API support, 29
ConferenceScheduleInformation class, 195
ConferenceServices class, 17
Conferencing Server, 195
Configuration element (XML), 37
ConfigurationManager class, 185
Configure Office Communications Servers Users Wizard, 276–278
Configure Pool/Server Wizard, 259–262
ConnectionFailureException class, 317–318, 326
ConsecutiveNoInputsInstant-MessagingEventActivity, 145–146
ConsecutiveNoInputsSpeechEventActivity, 45, 145

ConsecutiveNoRecognitionsInstant-
    MessagingEventActivity, 145, 147
ConsecutiveNoRecognitions-
    SpeechEventActivity, 45, 145–146
ConsecutiveSilencesInstant-
    MessagingEventActivity, 145–147
ConsecutiveSilencesSpeechEventActivity,
    145–146
contact lists
    building, 23
    displaying application-specific, 92–97
    managing programmatically, 77–81
    Office Communicator Automation API,
        23, 87
contact management
    ContactGroupServices class, 17
    IMessengerContacts interface, 26
    retrieving information, 63–68
    UC support, 7
    UCC API support, 27
    UCMA support, 16–17
    working with information, 58–62
Contact object
    AppplicationEndpoint class and, 15, 187
    functionality, 15, 221
    trusted applications, 225, 285
contactGroup element (XML), 36–37
ContactGroupServices class, 17
containers
    adding ACEs, 223
    Office Communications Server
        support, 197
Containers element (XML), 37
context menus, displaying, 93–94
contextual collaboration
    accepting application-specific
        conversations, 104–108
    additional resources, 110
    application-specific contact lists, 92–97
    business value, 90
    choice of technology, 91
    code structure, 92–108
    defined, 6, 8, 87
    Office Communicator Automation API, 24
    overview, 87–90
    scenario, 90

starting application-specific conversations,
    97–104
test environment, 91
Conversation class
    BeginEscalateToConference method, 196
    functionality, 17
    joining conferences, 196
    session support, 191
CONVERSATION_TYPE enumeration, 83
conversations
    accepting application-specific,
        104–108
    binding to endpoints, 184
    Conversation class, 17
    destroyed, 99–100
    receiving notification, 99–100
    receiving with IMessengerAdvanced,
        105–108
    retrieving IM text, 104
    sending IM text, 100–104
    starting, 81–84
    starting application-specific, 97–104
    testing applications, 108
    troubleshooting failures, 327–330
    UCMA support, 190–191
CSTA (Computer Supported Telephony
    Applications), 28
custom activities
    defined, 20
    UCMA Workflow support, 18, 22, 113
custom presence
    additional resources, 216
    choice of technology, 206
    code structure, 207
    common scenario, 206
    creating categories, 205–206
    detailed code, 208–215
    test environment, 207–208
Custom Tab, 9
CustomPresenceCategory class, 199
cwaEvents element (XML)
    functionality, 35
    subelements supported, 36–37
cwaRequests element (XML)
    functionality, 34–35
    subelements supported, 35–36

cwaResponses element (XML)
   functionality, 35
   subelements supported, 36

D

data channel
   defined, 33
   Unified Communications AJAX API, 34
database requirements, 224
DCPromo tool
   assigning static IP addresses, 227
   promoting computers, 228
debugging applications
   additional resources, 339
   best practices, 301
   in UC platform, 293–301, 321–330
   Office Communicator Automation API, 302–313
   PublishSubscribeException example, 320
   tools supported, 297–301
   UCMA Core API, 314–330
   UCMA Workflow, 330–338
   Windows Workflow Foundation support, 20
Decline action, 165–167
DeclineCallActivity class, 44
deleteContact element (XML), 35
deleteGroup element (XML), 35
DequeueCallActivity, 117
destroyed conversations, 99–100
DetectAnsweringMachineActivity class, 44
device management, 30–31
DHCP (Dynamic Host Configuration Protocol), 227
dialog activities
   adding commands, 174–175
   adding events, 175–177
   functionality, 128–137
   Speech Server support, 37, 45
   UCMA Workflow support, 19
dialog communications event activities, 142, 144–147
Dialog Workflow Run Time, 42
DisconnectCallActivity, 44, 126–127, 175

disconnecting calls
   DisconnectCallActivity, 44, 126–127, 175
   UCMA Workflow support, 21
DLLs (dynamic-link libraries), 27
DMessengerEvents_AppShutdownEventHandler event handler, 54
DMessengerEvents_OnContactFriendlynameChangeEventHandler event handler, 64
DMessengerEvents_OnContactListAddEventHandler event handler, 78
DMessengerEvents_OnContactListRemoveEventHandler event handler, 79
DMessengerEvents_OnContactPhoneChangeEventHandler event handler, 65
DMessengerEvents_OnContactStatusChangeEventHandler event handler, 71
DMessengerEvents_OnIMWindowCreatedEventHandler event handler, 99
DMessengerEvents_OnIMWindowDestroyedEventHandler event handler, 99–100
DMessengerEvents_OnMyFriendlyNameChangeEventHandler event handler, 59
DMessengerEvents_OnMyPhoneChangeEventHandler event handler, 60
DMessengerEvents_OnMyStatusChangeEventHandler event handler, 71
DMessengerEvents_OnSigninEventHandler event handler, 53
DMessengerEvents_OnSignoutEventHandler event handler, 54
DNS (Domain Name System)
   application development support, 219
   building AD DS forests, 227
   configuring for automatic sign-in, 245–248
   creating SRV record, 246–247
   verifying records creation, 248

DNS Manager, 246–247
DNS servers
   joining to domains, 249–251
   verifying server role, 233
Domain Admins group
   becoming a member, 252
   installing server components, 251
   managing networks, 222–223
   trusted applications, 286
domain CA, 234–237
domain controllers
   AD DS preparation, 221
   building AD DS forests, 227
   promoting computers, 228–233
   static IP addresses, 227–228
Domain Controllers container, 223
domain functional levels, 221
Domain Name System. *See* DNS
domains. *See* AD DS domains
DTMF (dual-tone multifrequency)
   prompting for input, 163–164
   Speech Server support, 19, 38–39
   UCMA Workflow support, 19
DtmfRecognizer class, 43
dual-tone multifrequency. *See* DTMF
Dynamic Host Configuration Protocol (DHCP), 227
dynamic-link libraries (DLLs), 27

E

Edge Servers, 282, 292
e-mail, 6–7
Enable Office Communications Server Users Wizard, 274–275
endpoints
   ApplicationEndpoint class, 187–189
   assigning user rights, 221
   binding conversations, 184
   defined, 17, 187
   enabling, 30
   handling incoming calls, 192
   LocalOwnerPresence class, 17
   Presence Subscription objects, 17
   publishing presence, 205
   registering, 205, 207
   subscribing to presence, 200
   troubleshooting connection failures, 323–327
   UCC API support, 29–30
   UCMA support, 187–189
   Unified Communications AJAX API, 33
   UserEndpoint class, 189
   workflows and, 118
Enhanced Presence feature (OCS), 14, 16
Enterprise Admins group, 222–223
Enterprise CA, 227
Enterprise IM, 6
enterprise telephony, 7
error codes
   functionality, 294–295
   SIP, 295–296
Error element (XML), 36
ETL files, 296
ETW (Event Tracing for Windows), 296
event handlers, registering, 192
event ID (eid), 35
Event Tracing for Windows (ETW), 296
events
   adding to dialog, 175–177
   adding to IM dialog, 176–177
   adding to speech dialog, 176
   call, 177
   change notifications, 201
   communication, 142
   handling incoming, 194
exception classes
   additional information, 317
   overview, 294–295
   UCMA Core API, 316–321
exception handling
   catching custom exceptions, 335–338
   COMException class, 293, 295
   FaultHandlerActivity, 332–333
   HRESULT error codes, 293
   Office Communicator Automation API, 305–307
   sources of errors/failures, 293–294
   UCMA Core API, 316–321
   UCMA Workflow, 332–333

Exchange Server
  presence information, 206
  UC support, 5, 7
Extensible Application Markup Language.
  See XAML
Extensible Markup Language. *See* XML

F

FailureResponseException class, 317, 328
FaultHandlerActivity, 332–333, 336–338
firewalls
  application development support,
    219, 249
  starting OCS services, 268
Flow class, 191
forest functional levels, 221
forests. *See* AD DS forests
FormFillingDialog instance, 38
FormFillingDialogActivity class, 45
forms-based authentication, 33–34
FQDN (fully qualified domain name)
  AD DS forests, 226
  ApplicationEndpoint class, 188
  modifying Program.cs, 152
Front End Server, 219, 223
fully qualified domain name. *See* FQDN

G

general activities, 120–125
GetAndConfirmActivity class, 45
GetPresenceActivity, 118, 147–148
global catalog servers, 221
Global Positioning System (GPS), 15,
  206–207
Globally Routable User Agent URI.
  See GRUU URI
globally unique identifier (GUID), 280
GoToActivity
  functionality, 120, 122, 125
  Speech Server support, 45
  TargetActivityName property, 125
  UCMA Workflow support, 21
GPS (Global Positioning System), 15, 206–207
grammar, 18, 134

Grammar class, 43
grammar files, 42
groups
  ContactGroupServices class, 17
  IMessengerGroup interface, 26
  IMessengerGroups interface, 26
  managing, 16–17
GRUU (Globally Routable User Agent) URI
  ApplicationEndpoint class, 188
  endpoint support, 187
  modifying Program.cs, 152
  trusted applications, 285
  UCMA considerations, 15
GUID (globally unique identifier), 280

H

hardware requirements, 224
HelpCommandActivity class, 45
Hidden Markov Model (HMM), 17
HMM (Hidden Markov Model), 17
HMM-based Speech Synthesis (HTS), 17
hold events, 21
host computers
  certificates, 282
  installing computer certificates, 285
  joining servers to domains, 249–251
  setting up, 248–251
  verifying certificate installation,
    283–284
host name resolution, 251
HRESULT error code
  functionality, 293–295
  Office Communicator Automation API,
    305–307
HTS (HMM-based Speech Synthesis), 17
HTTP SSL, 249
HTTPS (Hypertext Transfer Protocol Simple)
  TLS/MTLS certificates, 263
  Unified Communications AJAX API,
    33–34
HttpWebRequest API, 32
Hypertext Transfer Protocol Simple (HTTPS)
  TLS/MTLS certificates, 263
  Unified Communications AJAX API,
    33–34

I

IApplicationHost interface, 19
ICE (Interactive Connectivity Establishment), 9, 28
IDisposable interface, 103
IfElse Activity, 20, 159, 335
IHostedSpeechApplication interface, 19
IID (interface ID), 280
IIS (Internet Information Services), 34, 249
IIS certificates, 266–268
IM (instant messaging)
   adding commands to dialog, 175
   adding events, 176–177
   API support, 9, 13
   asking questions, 136
   binding outbound calls, 168–169
   conference support, 195
   contacting approver, 167–173
   contextual collaboration, 6
   creating calls, 191
   introductory messages, 169
   MCU support, 184
   Office Communicator Automation API, 23, 87
   placing outbound calls, 167–173
   recognizing input, 170
   retrieving text from conversations, 104
   sending text to conversations, 100–104
   server roles and, 219
   UC support, 3–4, 6–7
   UCC API support, 27, 29
   UCMA Core API, 225
   UCMA support, 14–15
   UCMA Workflow support, 19, 21, 116
   Unified Communications AJAX API, 32
IMessenger interface, 25
IMessenger2 interface, 25
IMessenger3 interface, 26
IMessengerAdvanced interface
   creating conversations, 98
   functionality, 26
   receiving application-specific conversations, 105–108
   StartConversation method, 82–84, 98, 101, 310
   starting conversations, 81–84
IMessengerContactAdvanced interface
   accessing presence information, 68–76
   functionality, 26
   PresenceProperties property, 69–70, 73, 75
   retrieving contact information, 63–68
IMessengerContacts interface
   Counts property, 77
   functionality, 26, 310
   Item method, 77
   Remove method, 78, 80
   working with contact lists, 77–81
IMessengerConversationWndAdvanced interface
   functionality, 26
   History property, 107–108
   HWND property, 102, 106
   notification of destroyed conversations, 100
   notification of new conversations, 99
   sending IM text to conversations, 100–104
   SendText method, 100
IMessengerGroup interface, 26
IMessengerGroups interface, 26
IMessengerPrivate interface, 26
incoming calls
   handling, 192–193
   Office Communicator Automation API, 23
   UCMA Workflow support, 19
incoming events, 194
initiateSession element (XML), 34–35
instant messaging. *See* IM
InstantMessagingCall class, 191
InstantMessagingCommandActivity
   functionality, 137, 141
   properties supported, 141
InstantMessagingFlowConfiguration-Requested event, 193–194
InstantMessagingHelpCommandActivity, 137, 141
InstantMessagingQuestionAnswerActivity
   functionality, 129, 135
   IsDataTrackingEnabled property, 119
   properties supported, 135–136
   RecognitionResult property, 119
   setting prompts, 132, 135–136
   TrackingDataWorkflowRuntimeService, 119

InstantMessagingStatementActivity
functionality, 129, 134
IsDataTrackingEnabled property, 119
properties supported, 134
sending introductory messages, 169
setting prompts, 135
TrackingDataWorkflowRuntimeService, 119
integrated Windows authentication, 33–34
Interactive Connectivity Establishment (ICE), 9, 28
Interactive Voice Response. *See* IVR
interface ID (IID), 280
Internet Information Services (IIS), 34, 249
Internet Protocol. *See* IP
introductory messages, 163, 169
InvalidOperationException class, 318, 323, 328
InvokeWorkflowActivity class, 45
IP (Internet Protocol)
building AD DS forests, 227
contextual collaboration, 6
Speech Server support, 40
IPSEC Driver, 249
IQueryForm registry entry, 248, 292
IServiceProvider interface, 119
IVR (Interactive Voice Response)
Speech Server support, 38, 40, 42
UC challenges, 4
UC opportunities, 7
UCMA support, 11, 14
UCMA Workflow support, 19

L

language packs, installing, 291
load balancing, 15, 219
local users
checking status, 51–52
destroyed conversations, 99–100
displaying information, 58–62
LocalEndpoint class, 187
LocalOwnerPresence class
creating applications, 184
functionality, 17
subscribing to presence, 200

locationProfiles element (XML), 37
log files
examining, 304–305
Office Communicator Automation API, 304–305
troubleshooting common operational failures, 311–313
Logon element (XML), 35

M

MakeCallActivity class, 44
Marshal class, 50, 57, 82, 103
mashup, 32
Mayo, Chris, 87
MCUs (multipoint control units), 184, 195
media handling
defined, 28
UCC API support, 28–31
MediaFlowState enumeration, 194
MenuActivity class, 45
Message Queuing (MQ), 249
MessageParsingException class, 317
MessageReceived event, 194
messages
asynchronous processing, 184
introductory, 163, 169
UCMA Workflow support, 21
Messenger class
accessing presence information, 68–76
AddContact method, 77, 80
AppShutdown event, 54
assembly initialization failures, 307
AutoSignin method, 53, 57
displaying local user information, 58–62
functionality, 25–26, 49–50
get_MyPhoneNumber property, 60
GetContact method, 63, 67, 73
MyContacts property, 77
MyFriendlyName property, 59
MyServiceID property, 63
MySigninName property, 59
MyStatus property, 51–52, 57
OnContactFriendlyNameChange event, 64
OnContactListAdd event, 78

OnContactListRemove event, 79
OnContactPhoneChange event, 65
OnContactStatusChange event, 71
OnIMWindowCreated event, 99
OnIMWindowDestroyed event, 99–100
OnMyFriendlyNameChange event, 59
OnMyPhoneChange event, 60
OnMyStatusChange event, 71
OnSignin event, 53
OnSignout event, 54
retrieving contact information, 63–68
Signin method, 52
SignOut method, 54, 57
working with contact lists, 77–81
MessengerPriv class, 26
Microsoft Dynamics, 6
Microsoft Exchange Server. *See* Exchange Server
Microsoft Management Console. *See* MMC
Microsoft Office, 6, 23
Microsoft Office Communications Server. *See* Office Communications Server
Microsoft Office Communications Server 2007 R2 Deployment Wizard
accessing, 238–239, 254, 259, 263
Application Configuration page, 256
Component Service Account For This Standard Edition Server page, 256
Deploy Standard Edition Server page, 255, 263, 268, 278
Directory Location Of Schema Files page, 239
Domain Preparation Information page, 244
Domain Preparation Wizard Has Completed Successfully page, 244
Forest Preparation Wizard Has Completed Successfully page, 243
Location For Database Files page, 257
Location For Server Files page, 256
Location Of Universal Groups page, 243
Main Service Account For Standard Edition Server page, 256
Prepare Active Directory For Office Communications Server page, 238, 242, 244
Ready To Deploy Server page, 258
Ready To Prepare Domain page, 244
Ready To Prepare Forest page, 243
Ready To Prepare Schema page, 239
Schema Preparation Wizard Has Completed Successfully page, 239
Select Location To Store Global Settings page, 243
SIP Domain User For Default Routing page, 243
starting page, 238
Validate Pool Or Server Functionality page, 279
Web Farm FQDNs page, 257
Welcome To The Certificate Wizard page, 264
Welcome To The Configure Pool/Server Wizard page, 259
Welcome To The Deploy Server Wizard page, 256
Welcome To The Domain Preparation Wizard page, 244
Welcome To The Forest Preparation Wizard page, 243
Welcome To The Schema Preparation Wizard page, 239
Welcome To The Start Services Wizard page, 269
Microsoft Office Communicator. *See* Office Communicator
Microsoft Office Outlook, 87–90, 206
Microsoft SharePoint, 6, 23
Microsoft SpeechServer.Dialog namespace, 44
Microsoft Visual Basic, 20
Microsoft Visual Studio. *See* Visual Studio
Microsoft Windows, 6
Microsoft.Rtc.Collection namespace, 16
Microsoft.Rtc.Workflow.dll library, 23
Microsoft.Rtc.Workflow.Toolbox.dll library, 23
Microsoft.Speech namespace, 18
Microsoft.SpeechServer namespace, 44
Microsoft.SpeechServer.Dialog namespace, 44
Microsoft.SpeechServer.Recognition namespace, 45

Microsoft.SpeechServer.Recognition.
    SrgsGrammar namespace, 18, 45
Microsoft.SpeechServer.Synthesis
    namespace, 45
MIDL command, 225, 280
MISTATUS enumeration, 51–52
MMC (Microsoft Management Console)
    Active Directory Users and Computers
        snap-in, 224
    Certificates snap-in, 283–284
    Communicator Web Access snap-in, 224
    Computer Management console, 224
    Office Communications Server snap-in,
        224
    opening, 283
    Response Group Server snap-in, 224
Monitoring Agent service, 269
Monitoring server role, 269
MPHONE_TYPE enumeration, 60
MQ (Message Queuing), 249
MTLS (Mutual Transport Layer Security)
    application endpoints, 187
    enabling connections, 281–282
    installing/configuring certificates,
        263–266
    UCMA support, 184
MultipartContentException class, 317
multipoint control units (MCUs), 184, 195
Mutual Transport Layer Security. *See* MTLS
MyPersona control, 92–93

N

Name Control, 9
namespaces, 43–45
NavigatbleListActivity class, 45
network management, 221–223
New Object User Wizard, 273
New Project dialog box, 114–115, 151
notifications
    receiving for destroyed conversations,
        99–100
    receiving for event changes, 201
    receiving for new conversations, 99
    UCMA support, 11, 14
    UCMA Workflow support, 19
NTLM (Windows authentication), 33–34
NTLMSSP (NTLM Security Support
    Provider), 249

O

OCSLogger tool
    enabling logging, 300
    functionality, 293, 296–297
    installing, 314
    starting, 314
OCSTracer tool, 314
OfferAnswerException class, 316
Office Communications Server
    AD DS preparation, 222–223
    AD DS support, 221
    additional resources, 292
    API support, 9
    application development support, 220
    configuring connection, 151–152
    configuring users, 276–278
    container support, 197
    enabling connections, 281
    enabling users, 273–275
    Enhanced Presence feature, 14, 16
    installing administrative tools, 224,
        239–240
    joining to domains, 249–251
    registering endpoints, 205
    server roles, 223–224
    setting up host computer, 248–251
    UC support, 5
    validating functionality, 278–279
    verifying host name resolution, 251
    verifying replication of users, 275
Office Communications Server 2007 Speech
    Server. *See* Speech Server 2007
Office Communications Server Certificate
    Wizard, 264–266
Office Communications Server Installation
    Wizard, 240–242
Office Communications Server Logging Tool.
    See OCSLogger tool
Office Communications Server Standard
    Edition
    building AD DS forest, 226–237

configuring, 251–272
configuring DNS for automatic sign-in, 245–248
configuring IIS certificates, 266–268
configuring UC user accounts, 272–278
deploying, 226
installing, 251–272
installing/configuring TLS/MTLS certificates, 263–266
preparing AD DS for UC, 237–245
setting up host computer, 248–251
starting services, 268–269
validating server functionality, 278–279
Office Communicator
additional resources, 85, 204
application development support, 220
determining if running, 50
displaying local user information, 58–62
enabling tracing, 296
functionality, 3
installing, 280
Office Outlook support, 87–90
publishing contact presence, 68–76
retrieving contact information, 63–68
signing in, 49–53
signing out, 54–58
starting conversations, 81–84
subscribing to contact presence, 68–76
working with contact list, 77–81
Office Communicator Automation API
accepting application-specific conversations, 104–108
additional resources, 85, 110, 292
application architecture, 24–25
application development support, 220, 224–225
application-specific contact lists, 92–97
business value, 90
choice of technology, 91
code structure, 92–108
configuring, 280–281
considerations, 24
debugging applications, 302–313
displaying local user information, 58–62
downloading, 224

enabling tracing, 302–305
error codes supported, 294
examining log files, 304–305
examining Windows Event Log, 304
functionality, 9, 23, 87
generating class/interface IDs, 280–281
handling exceptions, 305–307
installing SDK, 280
Messenger class support, 49–50
object model, 25–26
publishing contact presence, 68–76
retrieving contact information, 63–68
scenarios, 23, 90
signing in to Office Communicator, 49–53
signing out of Office Communicator, 54–58
starting application-specific conversations, 97–104
starting conversations, 81–84
subscribing to contact presence, 68–76
test environment, 91
troubleshooting applications, 307–313
working with contact list, 77–81
Office Communicator Web Access client, 32–33
operating system requirements, 224
OperationFailureException class, 317
OperationTimeoutException class, 317, 328
outbound alerts, 14
outbound phone calls
binding, 161–162
binding IM calls, 168–169
creating, 21
OutboundCallActivity, 118, 126–128, 161
placing IM call to approver, 167–173
placing to approver, 160–167
Outbound Sequential Workflow Console Application project template, 114, 151
OutboundCallActivity, 118, 126–128, 161

P

PBX (Private Branch eXchange), 3, 40
PersonaList control
displaying context menus, 93–94
functionality, 92–93

phone calls. *See* voice calls
PIA (primary interop assembly), 225, 309
PKI (public key infrastructure), 219
pollFailed element (XML), 36
presence information
   API support, 9
   approver's, 154–157
   custom presence, 205–215
   defined, 5
   Enhanced Presence feature (OCS), 14
   implementing branch logic, 155–157
   LocalOwnerPresence class, 17
   Office Communicator Automation API, 23–24
   publishing, 68–76, 197–200
   publishing with UCMA, 205–215
   querying for, 21, 205
   RemotePresence class, 17
   subscribing to, 68–76, 198, 200–202
   troubleshooting failures, 312
   UCMA Core API, 225
   UCMA Workflow support, 19
   Unified Communications AJAX API, 32
   WPF Presence Controls, 92–97
Presence Subscription object, 17
PRESENCE_PROPERTY enumeration, 69–70
PresenceNotificationReceived event, 202
presence-related activity, 118, 147–148
PresenceResult object, 148
presenceSubscriptionState element (XML), 37
primary interop assembly (PIA), 225, 309
Private Branch eXchange. *See* PBX
private CA, 227
Program.cs file, 151–152
programming applications
   Office Communicator Automation API, 91
   publishing contact presence, 68–76
   retrieving contact information, 63–68
   signing in to Office Communicator, 49–53
   signing out of Office Communicator, 54–58
   starting conversations, 81–84
   subscribing to contact presence, 68–76
   testing process, 108
   working with contact information, 58–62
   working with contact list, 77–81
project templates
   accessing, 114
   Inbound Sequential Workflow Console Application, 114
   installing, 114
   Outbound Sequential Workflow Console Application, 114, 151
prompt database, 42
prompting
   for actions, 170
   for DTMF input, 163–164
   InstantMessagingQuestionAnswerActivity, 132, 135–136
   InstantMessagingStatementActivity, 135
   SpeechQuestionAnswerActivity, 131–133
   SpeechStatementActivity, 129–130
protocol analysis tool, 296
PSTN (Public Switched Telephone Network), 3
public key infrastructure (PKI), 219
Public Switched Telephone Network (PSTN), 3
publishing
   contact presence, 68–76
   LocalOwnerPresence class, 17
   troubleshooting failures, 312
   UCC API support, 27, 30–31
   Unified Communications AJAX API, 32
   user presence, 197–200
publishRawCategories element (XML), 35
publishSelfPresence element (XML), 35
PublishSubscribeException class, 317–318, 320

Q

query/response bots. *See* automated agents
querying for presence information, 21, 205
queryPresence element (XML), 35
QuestionAnswerActivity class, 45

R

Real-time Transport Control Protocol (RTCP), 28
Real-time Transport Protocol. *See* RTP
RealTimeException class, 295, 316, 318
Recognition class, 18
Recognizer class, 43
RecordAudioActivity class, 44
RecordMessageActivity class, 44
RegisterException class, 295, 317, 323
RegisterForIncomingCall event, 192–193
registry keys, 50
Remote Procedure Call (RPC), 249
remote users, 17
RemotePresence class
 creating applications, 184
 functionality, 17
 subscribing to presence, 202
RepeatCommandActivity class, 45
request ID (rid), 35
requestAccepted element (XML), 36
requestCancelled element (XML), 36
requestFailed element (XML), 36
requestRejected element (XML), 36
requestSucceeded element (XML), 36
Response Group Server snap-in, 224
retrieve events, 21
reverse proxy, 219
RMCAST Protocol Driver, 249
root domain controller, 227
routing, 11, 16
RPC (Remote Procedure Call), 249
RTCP (Real-time Transport Control Protocol), 28
RTCUniversalServerAdmins security group
 becoming a member, 252
 installing server components, 220
 trusted applications, 286
RTP (Real-time Transport Protocol)
 Speech Server support, 40
 UCC API support, 28
rules engine, 20
run-time exceptions, 294

S

SALT (Speech Application Language Tags)
 creating IVR application, 40
 dialog flows, 37
 Speech Server support, 19, 37, 43
 UC support, 11
SALT interpreter, 42
SaltInterpreterActivity class, 45
SAM (Security Account Manager), 249
scheduling conferences, 14, 184, 195–196
Schema Admins security group, 222
Schema Preparation Wizard, 239
schemas
 AD DS, 221–222
 Unified Communications AJAX API, 34
Search element (XML), 35
searchResult element (XML), 37
Secure Real-time Transport Protocol (SRTP), 28
Secure Sockets Layer (SSL), 263
Security Account Manager (SAM), 249
selfPresence element (XML), 37
selfRawCategories element (XML), 37
SequentialWorkflowActivity, 119
Server Manager, 253–254
server roles
 AD DS, 221
 additional information, 219
 administrative tools, 224
 database requirements, 224
 hardware requirements, 224
 IM support, 219
 Office Communications Server, 219–220, 223–224
 operating system requirements, 224
ServerPlatformSettings class, 183–184
ServerPolicyException class, 317, 328
SES (Speech Engine Services), 42
Session Initiation Protocol. *See* SIP
session management
 call support, 191
 UCC API support, 30–31
 Unified Communications AJAX API, 34

SetTaskStatusActivity class, 45
SetupSE.exe. *See* Microsoft Office
    Communications Server 2007 R2
    Deployment Wizard
Short Message Service. *See* SMS
signaling
  defined, 28
  Speech Server support, 40
  UCC API support, 28
signing in
  an endpoint, 323
  configuring DNS, 245–248
  to Office Communicator, 49–53
  troubleshooting failures, 312
signing out, 54–58
Simple Object Access Protocol (SOAP), 32
SIP (Session Initiation Protocol)
  ApplicationEndpoint class, 188
  endpoint support, 187
  error codes, 295–296
  functionality, 13
  modifying Program.cs, 152
  Office Communicator Automation
      API, 24
  publishing presence information, 198
  Speech Server support, 40
  trusted applications, 285
  UCC API support, 27–29
  UCMA support, 11, 13, 15–16
SMS (Short Message Service), 4, 15
Snooper tool, 296–297, 300–301
SOAP (Simple Object Access Protocol), 32
softphones, 40
Speech Application Language Tags.
      See SALT
speech dialog
  adding commands, 174–175
  adding events, 176
Speech Dialog Workflow Activities, 37
Speech Engine Services (SES), 42
speech recognition
  handling grammars, 18
  Speech Server support, 38, 40
  UCMA support, 11, 17
  UCMA Workflow support, 19

Speech Server 2007
  application architecture, 40–43
  considerations, 40
  Conversational Grammar Builder, 39
  Developer Edition, 37–46
  functionality, 11
  lexicon tools, 39
  Prompt Recording and Editing tool, 39
  scenarios, 38–39
  tuning tools, 39
  UCMA Workflow support, 225
Speech Server Managed API
  dialog flows, 37
  functionality, 38, 42
  namespace support, 43–45
  object model, 43
  UC support, 11
speech synthesis
  HMM-based, 17
  Speech Server support, 38–39
  UCMA support, 11, 16
  UCMA Workflow support, 19
Speech Synthesis Markup Language
    (SSML), 45
SpeechCommandActivity
  functionality, 137–139
  properties supported, 140
SpeechCompositeActivity class, 45
SpeechEventActivity class, 45
SpeechEventsActivity class, 45
SpeechHelpCommandActivity, 137,
    140–141
SpeechQuestionAnswerActivity
  command activities and, 138, 140
  functionality, 129–130
  IsDataTrackingEnabled property, 119
  properties supported, 130–131
  RecognitionResult property, 119
  setting prompts, 131–133
  TrackingDataWorkflowRuntimeService, 119
SpeechRecognizer class, 19, 43
SpeechRepeatCommandActivity, 137, 141
SpeechSequenceActivity class, 45
SpeechSequentialWorkflowActivity class,
    43, 45

SpeechStatementActivity
  functionality, 129
  IsDataTrackingEnabled property, 118–119
  properties supported, 129
  setting prompts, 129–130
  TrackingDataWorkflowRuntimeService, 119
  TurnStarting event, 130
SpeechSynthesizer class, 38
SPI URIs
  ApplicationEndpoint class, 188
  endpoint support, 187
  joining conferences, 196
  modifying Program.cs, 152
  UCMA support, 15
SQL (Structured Query Language), 219
SQL Server database, 207
SQLCMD tool, 207
SRTP (Secure Real-time Transport Protocol), 28
SRV record (DNS)
  creating, 246–247
  functionality, 246
  verifying creation, 248
SSL (Secure Sockets Layer), 263
SSML (Speech Synthesis Markup Language), 45
Start Services Wizard, 269
state engine, 21
StateChanged event, 193–194
StatementActivity class, 45
static IP addresses, 227–228
Structured Query Language (SQL), 219
subscribePresence element (XML), 35
Subscribers element (XML), 36
subscribing
  to contact presence, 68–76
  to user presence, 198, 200–202
  UCC API support, 27, 30–31
  Unified Communications AJAX API, 32
surveys, 11
Synthesis class, 18
Synthesizer class, 43
System.Runtime.InteropServices namespace, 306
System.Speech namespace, 18

T

TCP (Transmission Control Protocol), 189
TCP/IP Protocol Driver, 249
technology
  business process communication, 150
  contextual collaboration, 91
  custom presence, 206
  UCMA, 206
telephony
  API support, 13
  enterprise, 7
  Speech Server support, 40
  UCC API support, 28
  UCMA Core API, 225
Telephony Interface Manager Connector (TIMC), 40
TelephonySession class, 43
terminateSession element (XML), 34–35
testing
  application development and, 220
  application-specific conversations, 108
  business process communication, 150
  contextual collaboration applications, 91
  custom presence, 207–208
  UCMA, 207–208
text-to-speech (TTS)
  Speech Server support, 40
  UCMA support, 11, 17
TIMC (Telephony Interface Manager Connector), 40
TLS (Transport Layer Security)
  enabling connections, 281–282
  endpoint support, 189, 214
  installing/configuring certificates, 263–266
TlsFailureException class, 317, 321
touch-tone input, 19
tracing, enabling
  Office Communicator, 296
  Office Communicator Automation API, 302–305
  overview, 296
  UCMA applications, 314–316
  UCMA Core API, 296, 314–316
  UCMA Workflow, 296, 330–332

TrackingDataWorkflowRuntimeService, 19, 21, 116, 119–120
Transmission Control Protocol (TCP), 189
Transport Layer Security. *See* TLS
troubleshooting applications
    additional resources, 339
    application initialization failures, 321–322
    assembly initialization failures, 307–309
    best practices, 301
    COM interop errors, 310
    common operational failures, 311–313
    conversation failures, 327–330
    endpoint connection failures, 323–327
    ETL file support, 296
    Office Communicator Automation API, 307–313
trusted applications
    enabling, 283, 285–290
    UCMA Core API, 225
trusted service entry (TSE), 285–286
try/catch blocks, 295, 307, 317
TSE (trusted service entry), 285–286
TTS (text-to-speech)
    Speech Server support, 40
    UCMA support, 11, 17
turn, 130

U

UC (Unified Communications)
    additional resources, 12, 292
    challenges, 4–5
    configuring user accounts, 272–278
    debugging applications, 293–301
    defined, 3
    goal, 3
    opportunities, 5–7
    platform overview, 7
    preparing AD DS, 237–245
    stepping through workflow, 333–335
UCC (Unified Communications Client) API
    additional resources, 46
    application architecture, 28–29
    application development support, 220
    considerations, 28
    error codes documentation, 295
    functionality, 9–10, 27
    object model, 29–31
    presence considerations, 206
    scenarios, 27
UCMA (Unified Communications Managed API)
    additional resources, 46, 110, 179, 204, 216, 292
    API architecture, 16–17
    availability, 15
    call flows, 190–191
    choice of technology, 206
    code structure, 207
    CollaborationPlatform class, 184–186
    common scenario, 206
    conferences, 195–197
    conversations, 190–191
    creating applications, 183–184
    creating calls, 191–194
    creating categories, 205–206
    custom activities, 18
    custom presence, 205–215
    debugging applications, 295, 321–330
    detailed code, 208–215
    enabling tracing, 314–316
    endpoints, 187–189
    extensibility, 15
    functionality, 13
    installing SDK, 114, 148, 281–282
    Media stack, 16
    object model, 17–18
    publishing presence, 197–200
    scalability, 9, 14
    scenarios, 14
    SIP support, 11, 13, 15–16
    subscribing to presence, 200–202
    test environment, 207–208
    workflow architecture, 22
UCMA Core API
    additional information, 225
    application development support, 220, 225
    configuring, 281–290
    debugging applications, 314–330
    enabling tracing, 296, 314–316
    exception classes, 316–321

functionality, 16
handling exceptions, 316–321
workflow architecture, 22
UCMA Speech API
  functionality, 16–17
  UCMA Workflow support, 19
  workflow architecture, 22
UCMA Workflow API and Workflow Activities
  additional resources, 148
  application development support, 220, 225
  business process communication, 19
  call control activities, 126–128
  call control communications event activities, 141–144
  command activities, 137–141
  configuring, 291
  considerations, 21
  custom activities, 20, 113
  debugging applications, 330–338
  dialog activities, 128–137
  dialog communications event activities, 144–147
  enabling tracing, 296, 330–332
  functionality, 11, 18, 21–22, 111
  general activities, 120–125
  handling exceptions, 332–333
  infrastructure, 21
  object model, 22–23
  persistence considerations, 21
  presence-related activity, 118, 147–148
  project templates, 114–115
  scenarios, 19
  selecting workflow language, 115
  support restrictions, 20
  Windows Workflow Activities, 11
  Windows Workflow Foundation, 20–21
  workflow architecture, 22
  Workflow Runtime Services, 113–114, 116–120
UCMA Workflow Runtime Services. *See* Workflow Runtime Services
UnhandledExceptionManager class, 320
Unified Communications. *See* UC

Unified Communications AJAX API
  additional resources, 46
  application architecture, 33–34
  considerations, 32–33
  functionality, 9, 32
  presence considerations, 206
  scenarios, 32
  XML elements, 35
Unified Communications Managed API. *See* UCMA
Uniform Resource Locators (URLs), 34
unsubscribePresence element (XML), 35
updateContact element (XML), 35
updateContainer element (XML), 35
updateGroup element (XML), 35
URLs (Uniform Resource Locators), 34
user accounts
  configuring, 272–278
  enabling, 273–275
  verifying replication, 275
User object, 221
user presence. *See* presence information
user rights, 221
UserEndpoint class
  creating calls, 191
  creating instances, 183
  endpoint support, 189, 207
  functionality, 15
  LocalEndpoint class and, 187
  publishing presence, 198
  UCMA support, 15
UserEndpointSettings class, 189
userPresence element (XML), 37
userRawCategories element (XML), 37
Users container, 223

V

ValidatorActivity class, 45
Visual Studio
  building contextual collaboration solution, 92–108
  documented exception types, 318
  Microsoft.Rtc.Workflow.Toolbox.dll library, 23

MIDL command support, 280
Speech Server support, 40
UC support, 6
UCMA Workflow support, 225
Visual Studio Debugger, 297–301, 308
Visual Studio Workflow Designer
  additional resources, 46
  functionality, 115, 335
voice calls. *See also* incoming calls;
    outbound phone calls
  accepting, 21
  multiple, 21
  music on hold functionality, 14
  Office Communicator Automation API, 24
  Speech Server support, 44
  troubleshooting failures, 313
  UCC API support, 29
voice mail, 7
Voice over Internet Protocol (VoIP), 40
voice recognition, 11
VoiceXML (VXML)
  additional resources, 46
  creating application, 40
  dialog flows, 37
  Speech Server support, 19, 37, 40, 43
  UC support, 11
VoiceXML interpreter, 42
VoiceXMLInterpreterActivity class, 45
VoIP (Voice over Internet Protocol), 40
VXML (VoiceXML)
  additional resources, 46
  creating application, 40
  dialog flows, 37
  Speech Server support, 19, 37, 40
  UC support, 11

W

W3C (World Wide Web Consortium), 37
Web Components Server
application development support, 219, 224
  configuring IIS certificates, 266–268
Web conferencing
  conference support, 195
  configuring, 269–272
  UC support, 7

Web Conferencing Server
  application development support,
    219, 223
  configuring, 269–272
Web Server (IIS) server role
  installing, 251, 253–254
  Speech Server support, 42
While activity, 20, 121
Windows Event Log
  examining, 269
  Office Communicator Automation API,
    304
  trace statements, 293
Windows Firewall, 249, 268
Windows Management Instrumentation
  (WMI), 249
Windows Presentation Foundation (WPF),
  92, 110
Windows Server
  AD DS preparation, 221
  building AD DS forest, 226
  joining servers to domains, 250–251
  promoting computers, 228–233
  server role support, 220
  UCMA Core API, 225
  UCMA support, 14
Windows Vista, 226
Windows Workflow Activities, 20, 111.
  See also UCMA Workflow API
Windows Workflow Foundation
  additional resources, 20–21, 46, 148
  functionality, 20–21
  handling exceptions, 332
  Speech Server support, 37
  UCMA Workflow support, 11, 18, 113,
    225, 291
  workflow architecture, 22
Windows XP, 226
WMI (Windows Management
  Instrumentation), 249
Workflow Runtime Services
  CommunicationsWorkflowRuntimeService,
    19, 21, 116–118
  functionality, 19, 22, 113–114, 116–120
  TrackingDataWorkflowRuntimeService, 19,
    21, 116, 119–120

Workflow1.xoml file, 151
Workflow1.xoml.cs file, 151
WorkflowPersistenceService class, 21
workflows. *See also* UCMA Workflow API and Workflow Activities
   activities and, 20
   allowing user input, 152–154
   creating, 20
   debugging, 20
   endpoints and, 118
   persistence considerations, 21
   rules engines, 20
   stepping through, 333–335
   Windows Workflow Foundation, 20
World Wide Web Consortium (W3C), 37
WPF (Windows Presentation Foundation), 92, 110

WPF Presence Controls
   functionality, 92
   MyPersona control, 92–93
   PersonaList control, 92–93

X

XAML (Extensible Application Markup Language)
   functionality, 20
   MyPersona control, 93, 95
XML (Extensible Markup Language)
   custom presence support, 207
   presence information, 5
   Unified Communications AJAX API, 32–37
XMLHTTPRequest API, 32–33

About the Authors

Rui Maximo is a senior technical writer in the Office Communications Group. He has worked on different aspects of the Microsoft Office Communications Server product suite (management, migration, topology, VoIP, Communicator Web Access) and shipped Microsoft Live Communications Server 2003, Live Communications Server 2005 (the original version and the SP1), and Office Communications Server 2007 as a lead program manager and program manager. With 13 years of experience at Microsoft, Rui has been fortunate to work in diverse roles (program management, software engineering, and technical writer) and various products (including Microsoft Windows, Windows Mobile, and Microsoft Office), primarily focusing on security. Prior to Microsoft, Rui worked at IBM as a software tester and at Brigham Young University as a UNIX administrator. Rui holds a master's degree in mathematics, specializing in abstract algebra and cryptography. You can reach him at ruim@ruimaximo.com. Please send your comments!

Kurt De Ding is a senior programming writer in the Office Communications Group. As the pioneering member of the SDK documentation team, he was instrumental in the initial design, authoring, and delivery of the SDK documentation for the Microsoft Unified Communications APIs, including Microsoft Office Communicator Automation API, Unified Communications Client API, Unified Communications Managed API v 1.0, and Unified Communications AJAX API, as well as Live Meeting Service API. Before joining the Office Communications Group, Kurt had worked on various Microsoft technologies, including Windows CE SDK, Windows Platform SDK, and Microsoft SQL Server SDK.

Vishwa Ranjan is a program manager in the Unified Communications Group. Most recently, Vishwa has worked on the Unified Communications Platform API Workflow Activities, which is available as part of Office Communications Server 2007 R2. Previously, he worked on Microsoft Speech Server 2004 and Office Communications Server 2007 Speech Server. He has more than 7 years of experience as a software design engineer in test, a technical lead, and a program manager.

Chris Mayo is a technical evangelist in the Developer and Platform Evangelism group. Chris focuses on the Unified Communications products (Office Communications Server 2007 R2, Office Communicator 2007 R2, and Microsoft Exchange Server 2007) and platform software development kits (SDKs), working with the Office Communications Group since the early betas of Office Communications Server 2007. Chris has been with Microsoft for 8 years as an evangelist working with the developer and independent software vendor communities. Chris has experience as both a writer for developer publications and a public speaker at professional events, such as the Professional Developers Conference and TechEd. Prior to joining Microsoft, Chris served as a developer and architect in the IT departments of Fortune 500 companies in the retail and finance industries. Keep up with Chris at his Unified Communications Development blog at *http://blogs.msdn.com/cmayo/*.

Oscar Newkerk is a Unified Communications Architect at Unify Square Inc. working in the area of Unified Communications, with an emphasis on integrating and enhancing business processes with collaboration technologies. With 14 years of experience at Microsoft, Oscar worked in various roles and groups within the company. Most recently, he was a technical evangelist in the Unified Communications Group, helping the developer community to plan, design, develop, and deploy solutions that integrate with Office Communications Server. Prior to Microsoft, Oscar worked for Digital Equipment Corporation as a software specialist and in software engineering in the areas of systems and network management. Oscar holds a bachelor of science degree in physics from Guilford College and holds patents in the areas of systems management and speech recognition.

Albert Kooiman is a member of the Unified Communications Marketing team and has been responsible for product management of the Unified Communications Developer Platform since the group was created in 2006. He works on both Exchange Server and Office Communications Server. With 14 years of experience in the telecommunications and speech technology industry, Albert has been involved in a wide range of projects encompassing the broad spectrum of Unified Communications solutions currently in the market. Albert holds a master's degree of the Medical Faculty of the University of Amsterdam, specializing in medical informatics.

Mark Parker is a programming writer in the Office Communications Group. Most recently, Mark was responsible for the Unified Communications Managed API 2.0 Core SDK documentation and part of the Unified Communications Managed API 1.0 SDK documentation. Before joining the Office Communications Group, Mark worked as a writer on the Speech Server 2007 documentation team and was a lead programming writer on the Windows Device Driver Kit documentation team. Prior to Microsoft, Mark taught mathematics and a number of programming languages at Shoreline Community College. Mark holds a master of science degree in mathematics.

What do you think of this book?

We want to hear from you!

Your feedback will help us continually improve our books and learning resources for you.
To participate in a brief online survey, please visit:

microsoft.com/learning/booksurvey

...and enter this book's ISBN-10 or ISBN-13 number (appears above barcode on back cover). As a thank-you to survey participants in the U.S. and Canada, each month we'll randomly select five respondents to win one of five $100 gift certificates from a leading online merchant. At the conclusion of the survey, you can enter the drawing by providing your e-mail address, which will be used for prize notification only.*

Thank you in advance for your input!

Where to find the ISBN on back cover

* No purchase necessary. Void where prohibited. Open only to residents of the 50 United States (includes District of Columbia) and Canada (void in Quebec). For official rules and entry dates see: **microsoft.com/learning/booksurvey**

Stay in touch!

To subscribe to the *Microsoft Press* Book Connection Newsletter—for news on upcoming books, events, and special offers—please visit: **microsoft.com/learning/books/newsletter**